Advances in Image Understanding

A Festschrift for Azriel Rosenfeld

Advances in Image Understanding

A Festschrift for Azriel Rosenfeld

Edited by
Kevin Bowyer and Narendra Ahuja

IEEE Computer Society Press
Los Alamitos, California

Washington • Brussels • Tokyo

IEEE Computer Society Press
10662 Los Vaqueros Circle
P.O. Box 3014
Los Alamitos, CA 90720-1314

IEEE Computer Society Press Order Number BP07644
Library of Congress Number 96-76772
ISBN 0-8186-7644-2

Additional copies may be ordered from:

IEEE Computer Society Press
Customer Service Center
10662 Los Vaqueros Circle
P.O. Box 3014
Los Alamitos, CA 90720-1264
Tel: +1-714-821-8380
Fax: +1-714-821-4641
Email: cs.books@computer.org

IEEE Service Center
445 Hoes Lane
P.O. Box 1331
Piscataway, NJ 08855-1331
Tel: +1-908-981-1393
Fax: +1-908-981-9667
mis.custserv@computer.org

IEEE Computer Society
13, Avenue de l'Aquilon
B-1200 Brussels
BELGIUM
Tel: +32-2-770-2198
Fax: +32-2-770-8505
euro.ofc@computer.org

IEEE Computer Society
Ooshima Building
2-19-1 Minami-Aoyama
Minato-ku, Tokyo 107
JAPAN
Tel: +81-3-3408-3118
Fax: +81-3-3408-3553
tokyo.ofc@computer.org

Assistant Publisher: Matt Loeb
Acquisitions Editor: Bill Sanders
Production Editor: Lisa O'Conner
Acquisitions Assistant: Cheryl Smith
Advertising/Promotions: Tom Fink
Cover Autostereogram by A.M. Brucstein and T.J. Richardson (see pages 158-176)

The Institute of Electrical and Electronics Engineers, Inc.

Contents

Section 4—Object Recognition

Section 5—Computer Vision Technology in Applications

Section 6—Image Understanding in Human and Machine Intelligence

Preface

This volume of papers has been assembled to honor Azriel Rosenfeld on his 65-th birthday. For about half of these 65 years, he has been a dominant figure in the field of computer vision and image processing. Azriel is a distinguished University Professor and the Director of the Center for Automation Research at the University of Maryland in College Park. Over a period of 30 years he has made many fundamental and pioneering contributions to nearly every area of computer vision and image processing. He wrote the first textbook in the field (1969), was founding editor of its first journal (1972), and was co-chairman of its first international conference (1987). He has published over 25 books and over 500 book chapters and journal articles, and has directed about 50 Ph.D. dissertations.

Azriel's 65 years symbolize coming of age of a discipline which he has grown with and which has mostly been viewed as a nascent field. The contributions to this volume illustrate to some degree the changes that have occurred in the research problems of interest and the methodologies employed. The papers are by a handful of the many people who have known and worked with Azriel over the years. All of the researchers we were able to contact were overwhelmingly supportive of the idea of this Festscrift honoring Azriel. Many others would have contributed if not for the tight publication schedule. The contributed papers address five major themes: image segmentation, feature extraction, three-dimensional (3-D) shape estimation from two-dimensional (2-D) images, object recognition, and applications technologies. These themes have been used to organize the papers into five main sections.

The first section on segmentation of images into regions consists of six papers. Jacob Beck presents results of an experiment that delves into the ability of humans to segregate texture patterns in color images. The experiment looks at the effects of hue, spatial scale, and background luminance. Rama Chellappa and S. Krishnamachari describe the use of Markov random field models for image segmentation in a multi-resolution framework. Mike Brady and Zhi-Yan Xie address the problem of selecting wavelet-based features for segmentation of texture images. Narendra Ahuja describes a new transform for extraction of low-level image structure at all of the unknown scales present in an image. Anil Jain and Pat Flynn give an overview of unsupervised clustering techniques used as a basis for image segmentation. Herb Freeman revisits one of the earliest topics in image analysis, that of encoding the boundary of a 2-D image region.

The four papers in the second section address different aspects of the problem of extracting features, such as edges, from 2D images. The paper by Richard Qian and Tom Huang presents a new algorithm for edge detection in 2D images using a scale-space framework. Jan Koenderink and Andrea van Doorn investigate what is needed in order to be able to make general statements about local image operators, such as edge detectors, in a scale space framework. Steve Zucker's paper is also concerned with multiscale structure and argues that the scale space concept is broadly applicable. Robert Haralick looks at the problem of propagating the effects of noise at one level of processing in a vision system to the next level.

The third section is concerned with the estimation of 3-D Shape from 2-D images. The two papers in this section present very different techniques for perceiving 3-D shape from

2-D images. Freddy Bruckstein, R. Onn and T.J. Richardson discuss the generation and perception of depth in "autostereograms." An autostereogram is a single 2-D image that contains information from which surface shape (depth) can be perceived. Liangyin Yu and Charles Dyer present a method that an "active observer" can use to recover 3-D surface shape from the projected deformation of stationary contours and markings on object surfaces.

Object recognition is the focus of the papers in the fourth section. Bob Hummel advances the view that matched filtering in the feature domain can be viewed as performing what we typically think of as pattern recognition or model-based vision. Michael Kelly and Martin Levine present an approach to extracting the location and approximate structural description of objects in the imaged scene. The paper by Kah-Kay Sung and Tommy Poggio describes some networks that can learn to recognize faces, or to find specific objects in cluttered scenes. Jake Aggarwal, J. Ghosh, D. Nair and I. Taha present a comparative study of three popular paradigms for object recognition: Bayesian statistics, artificial neural networks, and expert systems.

The fifth section consists of papers about some major applications areas of computer vision technologies. D. M. Gavrila and Larry Davis describe a system that performs 3-D model-based tracking of unconstrained human motion. Ruzena Bajcsy and Hans-Hellmut Nagel examine the role of computer vision technology in formulating mobility tasks for indoor and outdoor robots. Ramesh Jain and Amarnath Gupta discuss the role of computer vision in visual information management for multimedia systems. Hanan Samet presents work on interior-based representations of objects in images aimed at answering geometric queries.

The final paper by Sandy Pentland is as much about the role of perception in intelligence as it is about computer vision specifically. In this paper, Pentland argues that progress in building computer vision systems is best made by "... creatively assembling 2-D image processing modules of the sort developed by Azriel Rosenfeld and other pioneers, rather than by developing exotic new mathematical formulations, detailed understanding of photometry, or using sophisticated 3-D representations."

As an old Urdu poem goes, "... ask time the joys of being a traveler, for it has no destination but journeys day and night ..." – we are bound to witness more milestones of familiar research personalities reaching 65, 70 and beyond, only more frequently. We are happy about the ripening of the field that this means, and we look forward to many more opportunities to ruminate upon and take pride in the achievements of the likes of Azriel Rosenfeld.

Kevin Bowyer
University of South Florida

Narendra Ahuja
University of Illinois

Section 1

Segmentation of Images into Regions

Texture Segregation in Chromatic Element-Arrangement Patterns

Jacob Beck
Department of Cognitive and Neural Systems
Boston University
Boston, MA 02215

Abstract

An experiment compared the perceived segregation of element-arrangement patterns composed of the same two element types arranged in vertical stripes in the top and bottom regions and in a checkerboard pattern in the middle region. The elements in the patterns differed in hue. Patterns were equated for either luminance or for brightness. The experiment investigated the effects of the hues of the squares, pattern size, and background luminance on the segregation of the element-arrangement patterns. Perceived segregation was strong with a low luminance black background but was greatly decreased by a high luminance white background. Perceived segregation on a black background was stronger for blue and green patterns than for green and yellow patterns. Perceived segregation increased with a decrease in pattern size. Hue similarity, as rated by subjects in a separate procedure, was a relatively weak factor for predicting perceived segregation. The results are consistent with the hypothesis that perceived segregation is a function of cone contrasts.

1: Introduction

An element-arrangement pattern is composed of two types of elements that differ in the ways in which they are arranged in different regions of the pattern. Figure 1 illustrates an element-arrangement pattern in which the elements are red and blue squares arranged in a striped pattern in the top and bottom regions and in a checkerboard pattern in the middle region. The pattern is shown with black and white interspaces. Perceived segregation is strong with black interspaces but is diminished by white interspaces. A brief review of past research is presented to provide context for the experiment reported.

 0-8186-7644-2 $5.00

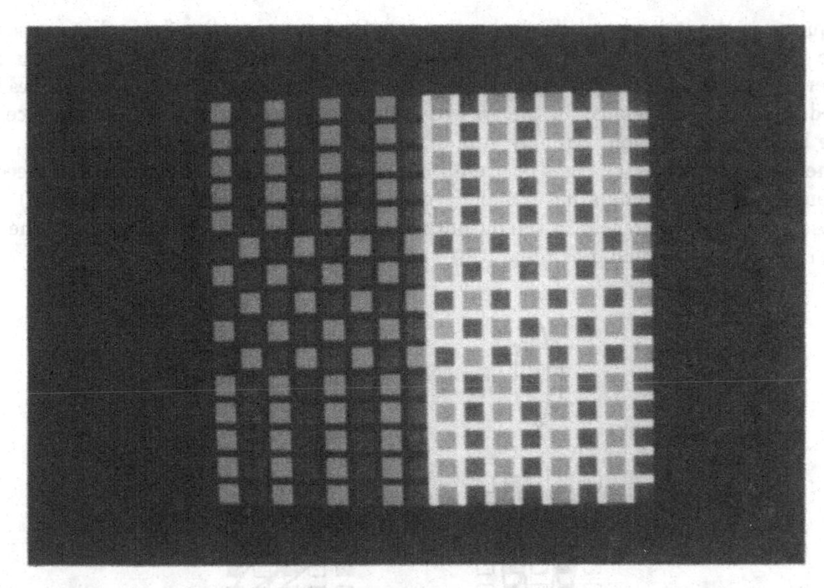

Figure 1. An element-arrangement pattern composed of red and blue squares with black and white interspaces. Perceived segregation with black interspaces is strong while perceived segregation with white interspaces is reduced. Unfortunately the photographic reproduction process distorts the image and the effect may be diminished.

1.1: Achromatic Element-Arrangement Patterns

Research with achromatic element-arrangement patterns indicates that the information for texture segregation takes place at a level of representation preceding the specification of the squares and their properties. First, texture segregation in element-arrangement patterns is not a direct function of the lightness differences of the squares. A striking finding reported in [1] was that in element-arrangement patterns of light and dark squares a large lightness difference could fail to yield strong texture segregation while a smaller lightness difference could yield strong segregation. Second, texture segregation is not impaired by contour misalignment. Judgments of perceived segregation were the same when the elements composing an element-arrangement pattern were aligned squares, misaligned squares, circles or blobs [2]. Third, texture segregation in an element-arrangement pattern fails to scale. Sutter, Beck and Graham [3] found that perceived segregation was strongest for patterns whose fundamental spatial frequency (the distance between the centers of two columns of the same type of square) was approximately 4 cycles per degree. Patterns with higher or lower fundamental spatial-frequencies segregated less strongly.

Beck, Sutter, and Ivry [4] showed that the perceived segregation of achromatic element-arrangement patterns was qualitatively consistent with the hypothesis that differences in the outputs of spatial frequency channels underlie the perceived segregation. They proposed that the differential responses of oriented simple cell-like mechanisms to the striped and checked regions of an element-arrangement pattern is the basis for the perceived segregation. For achromatic element-arrangement patterns, Sutter, Beck, and Graham [3] showed that the receptive fields that show strikingly different outputs to the different arrangements of the squares in the striped and checked regions are the large receptive fields that are sensitive to the fundamental spatial frequency of the texture pattern. These receptive fields match the period of the pattern and signal the differences in the overall pattern of squares in the striped and checked regions. In the striped region the changes of overall luminance occur in the horizontal direction, strongly stimulating vertically oriented receptive fields, and in the checkerboard region changes of overall luminance occur in a direction 45 degrees from horizontal, strongly stimulating obliquely oriented receptive fields (see Figure 2). They proposed that the differences in the outputs of these receptive fields are used by the visual system to establish boundaries separating the regions of the pattern.

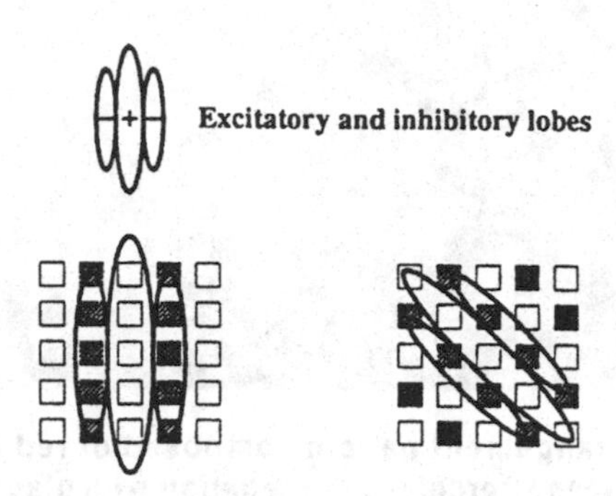

Figure 2. An illustration of how the responses of large oriented receptive fields sensitive to the fundamental frequency of a pattern can account for the segregation of achromatic element-arrangement patterns composed of light and dark squares. Top: Excitatory and inhibitory areas of an even symmetric receptive field. Bottom left: Large vertical receptive fields respond strongly to the vertical columns of squares in the striped region. Bottom right: Large oblique receptive fields respond strongly to the diagonal columns of squares in the checkerboard region.

The large receptive fields that match the period of a pattern and respond differentially to the striped and checkerboard arrangements of the squares do not have the right properties to signal the lightness of the squares because they average over several squares. Perceived segregation would thus not be expected to be a simple function of the lightness differences of the squares. Similarly, large receptive fields would also not be sensitive to edge alignment. Perceived segregation would therefore not be impaired by the misalignment of the squares. Perceived segregation would also not be expected to scale since perceived segregation is a function of the visual system's sensitivity to the fundamental spatial frequency. Proportionally reducing the overall size to a pattern would increase perceived segregation up to the point where the fundamental spatial frequency of the pattern has a spatial frequency at the peak of the contrast sensitivity function.

Sutter, Beck, and Graham [3] further found that the perceived segregation in an element-arrangement pattern is minimal when the area times contrast of large and small squares were equal. The area times contrast of the large and small squares is the same when the greater area of the large square is compensated for by the higher contrast of the small square. Squares that have the same area times contrast produce the same output at the fundamental frequency of the pattern, i.e., the frequency which when the excitatory region of a receptive field falls on one column of squares, the inhibitory region of the receptive field falls on the adjacent column of squares (see Figure 2). Although the contrast ratio—the ratio of the contrasts of the two square types with the background—at which the minimum perceived segregation occurred was correctly predicted by the outputs of simple cell-like mechanisms, the amount of segregation at this minimum was incorrectly predicted. The amount of perceived segregation depended also on the difference in the sizes of the squares. When the area times contrast of the large and small squares were equated, perceived segregation was greater as the size difference between the large and small squares increased. One way of accounting for this discrepancy is by a more complicated spatial-frequency model in which the initial linear filtering is followed by a rectification and a second filtering at a lower spatial frequency [3]. Graham, Beck, and Sutter ([5]; see also [6]) showed that texture segregation in element-arrangement patterns cannot be explained in terms of solely linear operations, and the application of spatial frequency analysis to texture segregation involves at least two nonlinearities. One nonlinearity is an intensity-dependent nonlinearity which can be accounted for by either sensory adaptation occurring before the channels or by a compressive intracortical interaction among neuronal responses which normalizes the responses [7]. The second nonlinearity is a rectification-like nonlinearity that is like that presumed to occur in complex cells [8].

1.2: Chromatic Element-Arrangement Patterns

Beck [9] and Pessoa, Beck, and Mingolla [10] investigated the perceived segregation of element-arrangement patterns composed of equiluminant squares differing in hue. They found that the luminance of the interspace region (i.e., the spaces between squares) strongly affected perceived segregation, whereas the luminance of the surround (i.e. the space surrounding a pattern) affected perceived segregation to a minor degree. For element-arrangement patterns composed of squares differing in hue, perceived segregation was strongly interfered with by high luminance interspaces but not by low luminance interspaces (see Figure 1). It is important to note that the squares composing an element-arrangement pattern do not have to be at precise equal luminance for perceived segregation to be diminished by a high luminance interspaces. Pessoa, Beck, and Mingolla [10] found that perceived segregation tended to decrease with increasing luminance of the interspaces when the squares composing an element-arrangement pattern differed in luminance. Perceived segregation varied approximately inversely with the ratio of the background luminance to the higher luminance square. Pessoa, Beck, and Mingolla [10] also found that perceived segregation was approximately constant for constant ratios of interspace luminance to square luminance. Stereoscopic cues that caused the squares composing the element-arrangement pattern to be seen in front of the interspaces did not greatly improve perceived segregation with high luminance interspaces. As in the case of achromatic element-arrangement patterns, these results suggest that the explanation of the perceived segregation of chromatic element-arrangement patterns is in terms of the early visual mechanisms that encode hue.

2: Blue and Green and Green and Yellow Element-Arrangement Patterns

Scott Oddo, Ennio Mingolla and I are investigating texture segregation in chromatic element-arrangement patterns. We are studying the degree to which brightness differences, hue similarity, and differences in cone contrasts account for differences in the perceived segregation of chromatic element-arrangement patterns. I present a preliminary report of our results.

Five subjects rated the perceived segregation of element-arrangement patterns composed of equal luminance blue ($x = .146$, $y = .061$) and green ($x = .294$, $y = .551$)[1] squares and of green ($x = .295$, $y = .529$) and yellow ($x = .457$, $y = .411$)[1] squares on a rating scale from 0 to 4. The luminances of the blue square in the element-arrangement pattern of blue and green squares and of the green square in the element-arrangement pattern of green and yellow squares were set at 2.5 ft.-L. Equiluminance of the blue and green hues and of the green and yellow hues was established using the criterion of minimally distinct borders [11]. The procedure used is described in Pessoa, Beck and Mingolla [10]. Hues of equal luminance are often not of equal brightness. The subjects also rated the perceived segregation of element-arrangement patterns composed of blue and green squares and of green and yellow squares judged equal in brightness. Equal brightness values were determined for each subject by having the subject make heterochromatic brightness judgments. Subjects adjusted the brightness of the green squares in an element-arrangement pattern composed of blue and green squares to be equal in brightness to that of the blue squares, and the brightness of the yellow squares in an element-arrangement pattern composed of green and yellow squares to be equal in brightness to that of the green squares. The subjects also rated the similarity of the two hues composing an element-arrangement pattern for similarity on a scale from 0 to 4. The overall size of the patterns was varied by proportionally decreasing the sizes and the separations of the squares making up the pattern. The fundamental spatial frequency of the patterns (the period between two columns of squares of the same hue) was .5, 1, and 2 cycles per degree. The stimuli were presented on a high luminance white background (16.3 ft.-L., $x = .312$, $y = .325$) and on a low luminance black background (.23 ft.-L., $x = .248$, $y = .254$).

2.1: Results and Discussion

Figure 3 shows the mean segregation ratings with a black background and Figure 4 with a white background as a function of pattern size. Perceived segregation was significantly greater with a black background than with a white background for both the blue and green and green and yellow element-arrangement patterns ($F(1,4) = 93.4$; $p < .05$). The interactions of Background $\times$ Hue was significant ($F(1,4) = 14.3$; $p < .05$). On a black background the element-arrangement patterns composed of blue and green squares segregated significantly more strongly than the element-arrangement patterns composed of green and yellow squares ($F(1,4) = 41.7$; $p < .05$). On a white background the blue and green element-arrangement pattern failed to significantly segregate more strongly than the green and yellow element-arrangement pattern ($F(1,4) = 1.02$; $p > .05$) On both black and white backgrounds, perceived segregation significantly increased with a decrease in pattern size (an increase in the number of cycles per degree) ($F(2,8) = 46.0$; $p < .05$). The interaction of Background Luminance $\times$ Cycles per Degree was significant ($F(2,8) = 30.1$; $p < .05$). This interaction reflects the greater increase in perceived segregation with decreased pattern size on a black background than on a white background.

Brightness: Chromatic hues of equal luminance need not be of equal brightness. A blue hue is commonly seen as brighter than an equiluminant green hue while equiluminant green and yellow hues are commonly seen to be more nearly equal in brightness [12]. One possibility, therefore, is that the greater perceived segregation of the blue and green element-arrangement patterns than of the green and yellow element-arrangement patterns is due

[1] These are the mean chromaticity coordinates of the hues judged by subjects to be of equal luminance. The chromaticity coordinates for the equal brightness hues were similar.

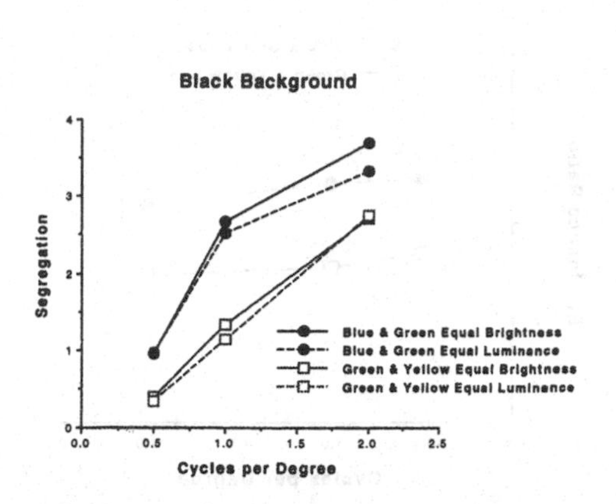

Figure 3. Mean segregation ratings on a black background plotted as a function of pattern size.

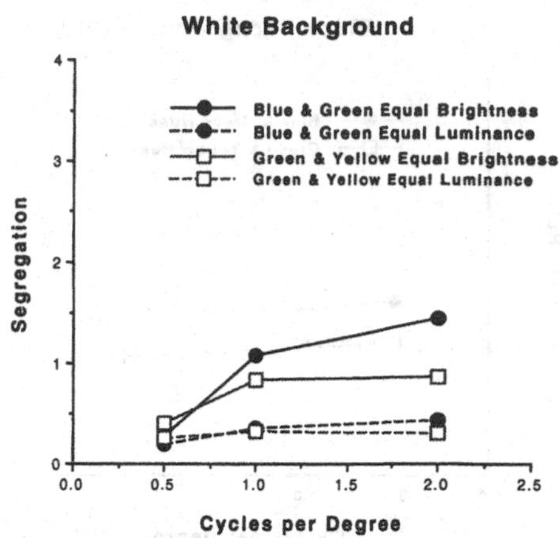

Figure 4. Mean segregation ratings on a white background plotted as a function of pattern size.

to the greater brightness difference between the equiluminant blue and green hues than between the equiluminant green and yellow hues. The ratio of the luminance of the green square to the luminance of the blue square and of the luminance of the yellow square to the luminance of the green square at equal brightness is an ordinal measure of the brightness difference between equiluminant hues. Figure 5 shows the mean ratios of the luminance of the green square to the luminance of the blue square and of the luminance of the yellow square to the luminance of the green square at equal brightness on a black background. Figure 6 shows the mean luminance ratios on a white background. The brightness difference between blue and green squares was significantly greater than between the green and yellow squares on both black and white backgrounds ($F(1,4) = 8.51; p < .05$). The brightness differences are in accord with the judgments of greater segregation for the blue and green equiluminant element-arrangement patterns than for the green and yellow equiluminant element-arrangement patterns. However, Figures 3 and 4 show that the equal brightness and equal luminance stimuli segregate alike. This clearly indicates that brightness differences are not the principal factor underlying the difference in the perceived segregation of the blue and green and the green and yellow element-arrangement patterns.

The greater luminance ratios yielding equal brightness of the hues with a white background than with a black background is the result of the inhibition of brightness by the higher luminance white background. Without knowing the degree of inhibition one can not determine whether the relative brightness differences of the hues on the black background were greater than the brightness differences on a white background. One is not able therefore to decide whether the greater segregation on a black background than on a white background reflects the greater brightness difference between the hues on a black background than on a white background. In a second experiment subjects rated the perceived segregation of element-arrangement patterns composed of equiluminant purple (x = .310, y = .162) and gray (x = .297, y = .301)[1] squares on black and white backgrounds. The purple square was set at 2.5 ft.-L. The luminances and chromaticity coordinates of the back

[1] These are the mean chromaticity coordinates of the hues judged by subjects to be of equal luminance. The chromaticity coordinates for the equal brightness hues were similar.

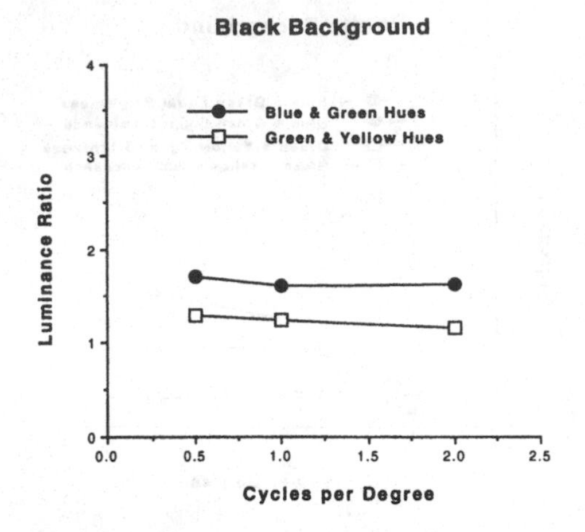

Figure 5. Mean ratios of the luminance of the green square to the luminance of the blue square and of the luminance of the yellow square to the luminance of the green square when the blue and green squares and the green and yellow squares were judged to be equally bright. The mean ratios are for a black background and are plotted as a function of pattern size.

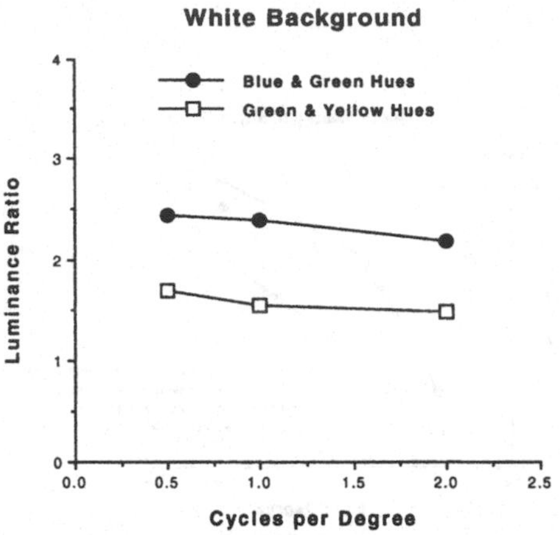

Figure 6. Mean ratios of the luminance of the green square to the luminance of the blue square and of the luminance of the yellow square to the luminance of the green square when the blue and green squares and the green and yellow squares were judged to be equally bright. The mean ratios are for a white background and are plotted as a function of pattern size.

and white backgrounds were as in the first experiment. Figure 7 shows the mean segregation ratings with black and white backgrounds as a function of pattern size. As with the blue and green and green and yellow element-arrangement patterns, perceived segregation was dramatically better on a black background than on a white background ($F(1,4) = 54.07$; $p < .05$). The gray square in an element-arrangement pattern was assigned a value of 100 and subjects also made magnitude estimations of the brightnesses of the purple squares. Figure 8 shows the magnitude estimations of brightness. The purple squares were judged to be slightly brighter than the gray squares on a black background than on a white background. However, there is not enough difference in the magnitude estimations of brightness to explain the effect of background luminance on perceived segregation.

Similarity: A second possibility is that perceived segregation is a function of hue similarity. Figure 9 shows the mean similarity ratings with a black background and Figure 10 shows the mean similarity ratings with a white background as a function of pattern size. The green and yellow hues overall were judged slightly more similar than the blue and green hues. An Anova of the similarity ratings showed that the differences in the similarity ratings of the blue and green and the green and yellow hues were not significant ($F(1,4) = 2.53$; $p > .05$). The correlations of the mean segregation judgments with the mean similarity judgments was not significant with a white background ($t(10) = 1.13$, $p > .05$). The correlation of the mean segregation judgments with the mean similarity judgments was $-.69$ with a black background ($t(10) = 3.01$, $p < .05$). However, the slope of the regression equation was -3.5 and the intercept of the regression equation was 6.6. If perceived segregation were

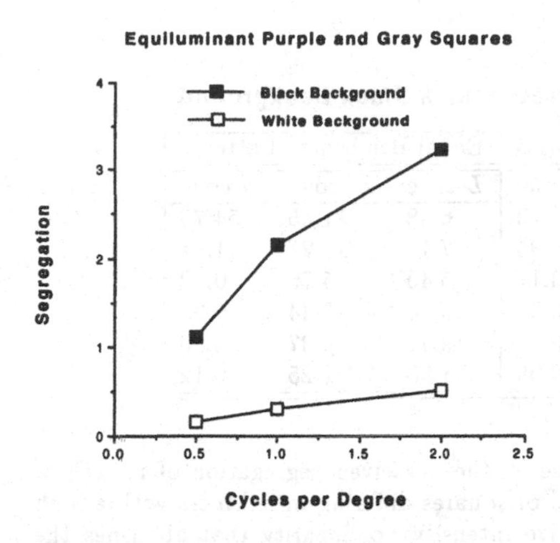

Figure 7. Mean segregation ratings on black and white backgrounds as a function of pattern size.

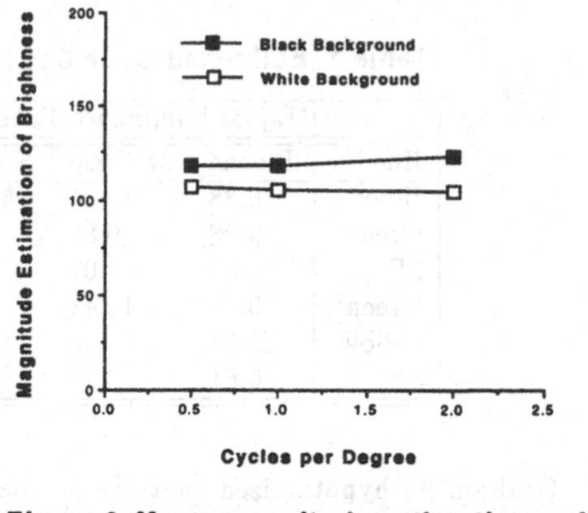

Figure 8. Mean magnitude estimations of the brightnesses of the purple squares plotted as a function of pattern size. The gray squares were assigned a value of 100.

a linear function of the similarity of the hues, the slope of the regression equation should be -1.0 and the intercept of the regression equation should be 4.0. This suggests that hue similarity is not the explanation for the greater perceived segregation of the blue and green element-arrangement patterns than of the green and yellow element-arrangement patterns.

Cone Contrasts: A third possibility is that what is important are the early visual mechanisms that encode the differences in hue of the squares composing an element-arrangement texture pattern. For the perceived segregation of an achromatic element-arrangement pattern, the luminance contrasts of the squares with the background is an important factor [1]. The analogous variable for chromatic patterns are the chromatic contrasts of the squares with the background. Cone contrasts are a measure of the hue differences encoded by the early visual mechanisms. The cone contrasts can be estimated from the long (L-cone), middle (M-cone), and short-wavelength (S-cone) cone responses. Macleod and Boynton [13] present a transformation for converting the Judd color matching functions (or tristimulus values $\bar{x}$, $\bar{y}$, and $\bar{z}$) into L-, M-, and S-cone responses. The cone contrasts are $\Delta L/L$, $\Delta M/M$ and $\Delta S/S$—the numerators are the differences between the long, middle, and short-wavelength cone responses to the squares and their responses to the background; the denominators are the cone responses to the background. Table 1 shows the mean cone contrasts for the equal luminance and equal brightness patterns on black and white backgrounds. The cone contrasts are consistent with the greater segregation of the blue and green element-arrangement patterns than of the green and yellow element-arrangement patterns. For both the equal luminance and equal brightness element-arrangement patterns, there is a large difference between the S-cone contrasts of the blue and green hues. For the green and yellow hues, which showed weaker texture segregation, there are no strikingly large differences in the contrasts of the three cones. The cone contrast responses of the equiluminant and equal brightness stimuli are also similar. The main difference is that the cone contrast differences for the L- and M-cones are greater for the equal brightness stimuli than for the equiluminant stimuli. This is to be expected since the equal brightness stimuli differ in luminance.

Table 1. Estimated Cone Contrasts with a Black Background

	Equal Luminance Patterns			Equal Brightness Patterns		
Hue	L-cone	M-Cone	S-cone	L-cone	M-cone	S-cone
Blue[2]	6.38	11.35	54.73	6.38	11.35	54.73
Green	9.28	9.80	0.96	17.14	18.27	1.61
SE	1.02	1.07	0.17	5.40	5.79	0.52
Green[2]	9.88	10.44	0.91	9.88	10.44	0.91
Yellow	10.54	7.35	0.67	13.11	9.37	0.86
SE	0.61	0.36	0.05	1.68	1.25	0.12

Graham [6] hypothesized that the decrease in the perceived segregation of an achromatic element-arrangement pattern composed of squares differing in lightness with a high background luminance is due to a compressive intensity nonlinearity that abolishes the differences in the neural responses to the high and low luminance squares composing the pattern. Analogously, a high luminance background which strongly stimulates the L-, M-, and S-cones would be expected to decrease the differences in the cone responses to the hues of the two squares. Table 2 shows the cone contrasts for the blue and green, and green and yellow element-arrangement patterns with white backgrounds. Note that the differences in the cone contrast on a white background are not nearly as great as those with a black background. As mentioned above, Pessoa, Beck, and Mingolla [10] found that equal ratios of background luminance to hue luminance yielded approximately the same perceived segregation. It should be noted that keeping the ratio of the background luminance to the hue luminance constant leaves the cone contrasts constant. The overall results are therefore consistent with the hypothesis that perceived segregation in element-arrangement patterns is determined by the encoding of hue by early visual processes.

Table 2. Estimated Cone Contrasts with a White Background

	Equal Luminance Patterns			Equal Brightness Patterns		
Hue	L-cone	M-Cone	S-cone	L-cone	M-cone	S-cone
Blue[2]	–0.90	–0.81	–0.37	–0.90	–0.81	–0.37
Green	–0.86	–0.84	–0.95	–0.66	–0.60	–0.92
SE	0.01	0.02	0.00	0.10	0.12	0.02
Green[2]	–0.85	–0.83	–0.95	–0.85	–0.83	–0.95
Yellow	–0.84	–0.87	–0.96	–0.75	–0.80	–0.95
SE	0.01	0.01	0.00	0.11	0.09	0.01

Pattern Size: Contrast sensitivity for chromatic gratings tend to decrease beyond 1 cycle/deg [14]. The increase in perceived segregation with decreasing pattern size is therefore not ascribable to the chromatic contrast sensitivity function. It is also not explainable by any of the three factors we have studied, i.e., differences in the brightness, similarity, or

[2]The hues were set by the experimenter. SE = the standard error of the mean cone contrasts calculated from subjects' equal luminance and equal brightness judgments.

cone contrasts of the hues in an element-arrangement pattern. The magnitude estimations of brightness presented in Figure 8 were similar across the three spatial scales and do not account for the effect of pattern size. The rated similarity of the blue and green hues and of the green and yellow hues shown in Figures 9 and 10 also did not vary with pattern size, and would not explain the effects of pattern size on perceived segregation of the element-arrangement patterns. Scaling of the patterns also leaves the cone contrasts constant. One hypothesis is that the high spatial frequency information in the element-arrangement patterns interferes with perceived segregation and that this interference is reduced when the high spatial frequencies are outside the range of visual sensitivity. We are examining this possibility.

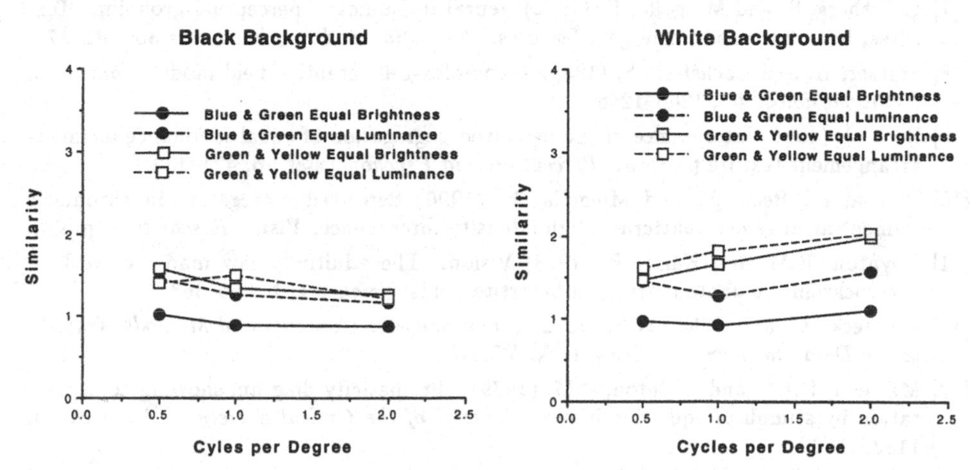

Figure 9. Mean similarity ratings on a black background plotted as a function of pattern size.

Figure 10. Mean similarity ratings on a white background plotted as a function of pattern size.

Acknowledgments

We thank Rhea Eskew for helpful discussions concerning the calculations of the cone contrasts, and Ken Nakayama for the use of the Minolta CS-100 chroma-meter. The research was supported in part by the Office of Naval Research (ONR N00014-J-4100, ONR N00014-J-4015) and the Air Force Office of Scientific Research (AFOSR F49620-92-J-0334).

References

[1] Beck, J., Graham, N., and Sutter, A. (1991) Lightness differences and the perceived segregation of regions and populations. *Perception and Psychophysics* **49**, 257–269.

[2] Beck, J. (1993) Visual processing in texture segregation. In: *Visual Search 2*, D. Brogan, A. Gale, and K. Carr (Eds.), London: Taylor and Francis (pp. 1–35).

[3] Sutter, A., Beck, J., and Graham, N. (1989) Contrast and spatial variables in texture segregation: Testing a simple spatial frequency channels model. *Perception and Psychophysics* **46**, 312–332.

[4] Beck, J., Sutter, A., and Ivry, R. (1987) Spatial frequency channels and perceptual grouping in texture segregation. *Computer Vision, Graphics, and Image Processing* **37**, 299–325.

[5] Graham, N., Beck, J., and Sutter, A. (1992) Nonlinear processes in spatial frequency channel models of perceived texture segregation: Effects of sign and amount of contrast. *Vision Research* **32**, 719–743.

[6] Graham, N. (1994) Nonlinearities in texture segregation. In: Ciba Foundation Symposium, 184, *Higher Order Processing in the Visual System*, Chichester: Wiley (pp. 309–329).

[7] Grossberg, S. and Mingolla, E. (1985) Neural dynamics of perceptual grouping: Textures, boundaries, and emergent features. *Perception and Psychophysics* **38**, 141–171.

[8] Spitzer, H. and Hochstein, S. (1985) A complex-cell receptive field model. *Journal of Neurophysiology* **53**, 1266–1286.

[9] Beck, J. (1994) Interference in the perceived segregation of equal luminance element-arrangement texture patterns. *Perception and Psychophysics* **56**, 424–430.

[10] Pessoa, L., Beck, J., and Mingolla, E. (1996) Perceived segregation in chromatic element-arrangement patterns: High intensity interference. *Vision Research* (in press).

[11] Boynton, R.M. and Kaiser P. (1968) Vision: The additivity law made to work in heterochromatic photometry with bipartite fields. *Science* **161**, 366–368.

[12] Wyszecki, G. and Stiles, W.S. (1982) *Color Science: Concepts and Methods, Quantitative Data and Formulae.* New York: Wiley.

[13] Macleod, D.I.A. and Boynton, R.M. (1979) Chromaticity diagram showing cone excitation by stimuli of equal luminance. *Journal of the Optical Society of America* **69**, 1183–1186.

[14] Mullen, K.T. (1985) The contrast sensitivity of human color vision to red-green and blue-yellow chromatic gratings. *Journal of Physiology* (London) **359**, 381–400.

Multiresolution GMRF Models for Image Segmentation*

Rama Chellappa† and Santhana Krishnamachari ‡

† Department of Electrical Engineering and
Center for Automation Research
University of Maryland
College Park, MD 20742

‡ Image Processing Department
Communication Technology Division
COMSAT Laboratories
Clarksburg, MD 20871

Abstract

A multiresolution model for Gauss Markov random fields (**GMRF**) *with application to texture segmentation is presented. Coarser resolution sample fields are obtained by subsampling the sample field at the fine resolution. Although the Markov property is lost under such resolution transformation, coarse resolution non-Markov random fields can be effectively approximated by Markov fields. We present a local conditional distribution invariance approximation to estimate the GMRF parameters at coarser resolutions from the fine resolution parameters. Our experiments with synthetic, Brodatz texture and real satellite images show that this multiresolution technique results in a better segmentation and requires lesser computation than the single resolution algorithm.*

1: Introduction

There has been an increasing emphasis on using statistical techniques for modeling and analyzing images. Typical image processing problems have the following aspects to be dealt with: the identification of an appropriate model that reflects the prior beliefs and knowledge about the family of images that are to be analyzed, the selection of a proper observation model that reflects the nature of the transformations these images undergo during observation, and the selection of a suitable error criterion to be optimized. For many image processing problems, such as image enhancement, image restoration, texture identification and segmentation, prior and observation models and error criteria can be very efficiently selected in a statistical framework. Using statistical models in a Bayesian framework enables posing many image processing problems as statistical inference problems.

Among several possible 2-D models for images most of the research has been restricted to Markov random field (**MRF**) models, because of the local statistical dependence of images.

* This work was supported in part by Grant #ASC 9318183 from National Science Foundation. A longer version of this chapter is due to appear in the IEEE Transactions on Image Processing.

0-8186-7644-2 $5.00 © 1996 IEEE

MRF models have been used to characterize prior beliefs about various image features such as textures, edges and region labels. Since MRF models express global statistics in terms of the local neighborhood potentials, all computations are restricted to a local window. This spawned a lot of interest in developing algorithms that utilize local computations to achieve global optimization [9]. But the main drawback is that the optimization schemes associated with MRF energy functions are iterative. Typical MRF algorithms visit all lattice sites in a specific order and perform a local computation at each site; this is repeated until some form of convergence is reached. Even though the individual iterations involve only simple local computations, the iterative nature of these algorithms contributes to the computational burden. Two different approaches have been used to reduce the computational requirement. The first is to use non-optimal, deterministic methods that converge to a local optimal point, but still provide reasonably good results. Geiger and Girosi [8] and Zhang [24] use mean field approximations that lead to deterministic relaxation algorithms. Wu and Doerschuk [23] use a tree approximation that replaces the lattice on which an MRF is defined by an acyclic tree which allows replacing the iterative MRF computations by recursive computations.

The second approach is to use multiresolution techniques. Two important aspects of multiresolution approaches are: (1) divide and conquer and (2) action at a distance [21]. Research efforts on multiresolution models and analysis can be found in [22],[2],[10],[4] and, [16]. We elaborate on the following, because of their relevance to our work.

Jeng, in [12], discusses the effect of subsampling resolution transformation on MRFs and presents two results: first, the Markov property is not preserved for a general subsampling scheme and, second, it is preserved under some specific subsampling schemes depending on the size and shape of the neighborhood. In [15] Lakshmanan and Derin present an excellent discussion on multiresolution GMRF models. It is shown that the GMRFs lose their Markov property under subsampling and expressions for the power spectral density functions at coarser resolution are obtained. It is also shown that for the special case of second order GMRFs with separable autocovariances, the Markov property is retained under subsampling. In addition, a *covariance invariance* approximation is presented to approximate the coarser resolution data by GMRFs. Many interesting properties of this estimator such as maximizing the entropy and minimizing the Kullback-Leibler **(KL)** distance can be found in [15].

We present a multiresolution model based on a KL distance measure. Given that the data at the fine resolution is a GMRF, the goal is to obtain suitable models at coarser resolutions. Data at coarser resolutions are obtained by subsampling the fine resolution data. Under these resolution transformations, coarser resolution data are non-Markov. We present an estimator to compute the parameters corresponding to the "best" GMRF approximation at lower resolutions from the parameters at the fine resolution based on minimizing the KL distance between the conditional densities (conditional relative entropy). We also show that the computations for this estimator turn out to be similar to the psuedo likelihood estimator [1], except that the sample covariances are replaced by covariances calculated with respect to the non-Markov measure that is being approximated. We present results on the existence of different sets of GMRF parameters at fine resolution that result in statistically identical coarser resolution random fields. As an application, we consider the texture segmentation problem and performe segmentation over multiple resolutions using our multiresolution GMRF model. We show that the multiresolution technique performs better than the single resolution approach.

The rest of the chapter is organized as follows. Section 2 introduces the GMRF and the basics of the resolution transformation. Section 3 presents the Markov approximation for non-Markov fields based on local conditional distribution invariance approximation. Section 4 presents the many-to-one nature of transformation of the GMRF parameters from the fine to coarse resolution. Section 5 presents various aspects of the multiresolution segmentation and Section 6 gives synthetic and real experiments. Section 7 concludes the paper.

2: GMRFs and Resolution Transformation

In this section we introduce basic notations used in the rest of the chapter and also present results on loss of the Markov property under resolution transformation [15].

2.1: The GMRF Model

We use the following notation :
$t = (t_1, t_2), s = (s_1, s_2)$: coordinates of grid points on a 2-D lattice
$\Omega = \{s : 0 \leq s_1 \leq M-1, 0 \leq s_2 \leq N-1\}$: a two dimensional lattice
$\underset{\sim}{X}$: a random field on Ω, represented as a vector by a row-wise scan ordering
X_s: the random variable at site s
η, ψ, ξ: neighborhood sets

The set of lattice points that are contained in the neighborhood of a site s is denoted by η_s. The elements that are included in the neighborhood of the site marked s for different neighborhood orders can be found in [6].

For the first order case, $\eta = \{(1,0),(0,1),(-1,0),(0,-1)\}$, and $\eta_s = \{s+r : r \in \eta\}$.

If $\underset{\sim}{X}$ is modeled by a GMRF with a symmetric neighborhood η, then $\underset{\sim}{X}$ can be written as [13]:

$$X_s = \sum_{r\in\eta} \theta_r X_{s+r} + e_s$$

where $\underset{\sim}{e}$ is a zero mean, Gaussian noise, with autocorrelation given by :

$$E[e_s e_{s+r}] = \begin{cases} \sigma^2 & \text{if } r = (0,0) \\ -\theta_r\sigma^2 & \text{if } r \in \eta \\ 0 & \text{otherwise.} \end{cases} \tag{1}$$

Hence the GMRF can be completely characterized by the set of parameters $\{\underset{\sim}{\theta}, \sigma^2\}$. The parameter set $\underset{\sim}{\theta}$ should satisfy the following conditions :

$$\begin{aligned} &1.\quad \theta_r = \theta_{-r} \quad \forall r \in \eta \\ &2.\quad 1 - \underset{\sim}{\theta}^T \underset{\sim}{\phi}_s > 0 \quad \forall s \in \Omega \end{aligned} \tag{2}$$

where $\underset{\sim}{\phi}_s$ is a vector whose length is equal to the number of elements in the neighbor set η. The individual elements of $\underset{\sim}{\phi}_s$ are given by:

$$\cos\left(\left(\frac{2\pi s_1}{M} \frac{2\pi s_2}{N}\right) \begin{pmatrix} r_1 \\ r_2 \end{pmatrix} \right) \quad r \in \eta.$$

The first condition is necessary to ensure stationarity and the second to ensure that the covariance matrix of $\underset{\sim}{X}$ is positive definite.

$\underline{X}$ exhibits the Markov property [13],

$$\begin{aligned} p(x_s|x_t, \forall t \neq s, t \in \Omega) &= p(x_s|x_{s+r}, r \in \eta) \\ &= \frac{1}{\sqrt{2\pi\sigma^2}} \exp\{-\frac{[x_s - \sum_{r\in\eta} \theta_r x_{s+r}]^2}{2\sigma^2}\}. \end{aligned} \quad (3)$$

The power spectrum $S_x(\omega)$ of $\underline{X}$ can be shown to be [13]:

$$S_x(\omega) = \frac{\sigma^2}{1 - \sum_{r\in\eta} \theta_r \cos[\frac{2\pi}{M} r_1\omega_1 + \frac{2\pi}{N} r_2\omega_2]} \quad (4)$$

where $\omega = \{\omega_1, \omega_2\}$, and $0 \leq \omega_1 \leq M-1, 0 \leq \omega_2 \leq N-1$.

2.2: GMRFs and Resolution Transformation

Let $\Omega^{(0)} = \Omega = \{(s_1, s_2) : 0 \leq s_1 \leq M-1, 0 \leq s_2 \leq N-1\}$ be a rectangular lattice and M and N are assumed to be powers of 2. The superscript stands for the level in the image pyramid, $\Omega^{(0)}$ being the lattice at the fine resolution, $\Omega^{(k)}$ represents the lattice which is obtained by subsampling $\Omega^{(0)}$, k times. Let $\underline{X}^{(k)}$ represent a random field, obtained by ordering the random variables on $\Omega^{(k)}$. The parameters of a GMRF defined on a lattice $\Omega^{(k)}$ are denoted by $\{\underline{\theta}^{(k)}, [\sigma^2]^{(k)}\}$ and the associated neighborhood is denoted by $\eta^{(k)}$. The covariance matrix and the power spectrum associated with $\underline{X}^{(k)}$ are denoted by $\Sigma^{(k)}$ and $S_x^{(k)}(\omega)$ respectively. The probability distributions defined on a lattice $\Omega^{(k)}$ are indexed by $p^{(k)}(.)$.

Let $\underline{X}^{(0)}$ be a GMRF defined on $\Omega^{(0)}$ with parameters $\{\underline{\theta}^{(0)}, [\sigma^2]^{(0)}\}$ and a neighborhood $\eta^{(0)}$. The power spectrum of $\underline{X}^{(0)}$ can be written as in Eq. (4) :

$$S_x^{(0)}(\omega) = \frac{[\sigma^2]^{(0)}}{1 - \sum_{r\in\eta^{(0)}} \theta_r^{(0)} \cos[\frac{2\pi}{M} r_1\omega_1 + \frac{2\pi}{N} r_2\omega_2]} \quad (5)$$

where $\omega = \{(\omega_1, \omega_2) : 0 \leq \omega_1 \leq M-1, 0 \leq \omega_2 \leq N-1\}$. The subsampling resolution transformation is defined as:

$$X_s^{(k)} = X_{2s}^{(k-1)}$$

defined for all $s \in \Omega^{(k)}$.

The power spectrum of $\underline{X}^{(k)}$ can be shown to be [15]:

$$S_x^{(k)}(\omega) = \frac{1}{2^{2k}} \sum_{r\in C_k} S_x^{(0)}(\omega + r') \quad (6)$$

where $r' = (\frac{M}{2^k} r_1, \frac{N}{2^k} r_2)$ and $C_k = \{r : 0 \leq r_1 \leq 2^k - 1, 0 \leq r_2 \leq 2^k - 1\}$.

It can be observed that $S_x^{(k)}(\omega)$ cannot be written in the form of Eq. (4) with a finite neighborhood. Therefore, the subsampled fields $\underline{X}^{(k)}$ are non-Markov, except for the special case of second order separable correlation GMRFs [15].

3: Local Conditional Distribution Invariance Approximation

As mentioned in the last section, GMRFs become non-Markov when subsampled. However, if the coarser resolution data are modeled by the exact non-Markov Gaussian measures, conventional optimization techniques based on Markov properties cannot be applied. In this section we show that it is possible to obtain good Markov approximations for coarser resolution fields.

In this section we present a technique to estimate the best GMRF parameters of a non-Markov random field, based on a KL distance measure between local conditional distributions (conditional relative entropy) [5]. In MRF applications all optimizations are performed based on the local conditional distribution, so, we believe an estimator based on it should be well suited for image analysis applications. We also exemplify the connection between this estimator and the pseudo likelihood estimator [1].

The Markov approximation presented in this section is based on linear estimation. Before presenting the details, we will provide a known result regarding the linear estimation of a GMRF. Let $\underline{Z}$ be a GMRF defined by $(\underline{\theta}, \sigma^2)$ with a neighborhood ψ. Then the best estimate of Z_s based on the elements of ψ is given by [3]:

$$\hat{z}_s = \sum_{r\in\psi} \theta_r z_{s+r}$$

and the mean square error

$$E(Z_s - \hat{Z}_s)^2 = \sigma^2.$$

The conditional density $p(z_s|z_r, r \in \psi)$ is Gaussian with conditional mean $\sum_{r\in\psi} \theta_r z_{s+r}$ and conditional variance σ^2.

Let $\underline{X}$ be a random field with a stationary non-Markov probability measure $p(\underline{x})$ *and let* $q^*(\underline{x})$ *be a GMRF approximation such that:*

$$q^*(x_s|x_{s+r}, r \in \eta) = \arg\min_q D[p(x_s|x_{s+r}, r \in \eta) \;||\; q(x_s|x_{s+r}, r \in \eta)], \tag{7}$$

where the minimization is performed over the entire family of GMRF pdfs with a chosen neighborhood η. In addition, under certain conditions (given at the end of the section), $q^*(x_s|x_{s+r}, r \in \eta)$ is exactly equal to $p(x_s|x_{s+r}, r \in \eta)$.

Since $q(\underline{x})$ belongs to the family of GMRF densities, $q(x_s|x_{s+r}, r \in \eta)$ will be of the form given in Eq. (3).

$$q(x_s|x_{s+r}, r \in \eta) = \frac{1}{\sqrt{2\pi\sigma^2}} \exp\{-\frac{[x_s - \sum_{r\in\eta} \theta_r x_{s+r}]^2}{2\sigma^2}\}.$$

Let $(\underline{\theta}^*, [\sigma^2]^*)$ be the parameters corresponding to $q^*(\underline{x})$. To simplify the notation, let $\underline{Y}$ be the vector containing the neighborhood random variables in a proper order. For a first order neighborhood,

$$\underline{Y}^T = \begin{pmatrix} X_{s+(1,0)} & X_{s+(0,1)} & X_{s+(-1,0)} & X_{s+(0,-1)} \end{pmatrix}.$$

Now performing the minimization in terms of the parameters,

$$(\underline{\theta}^*, [\sigma^2]^*) \;=\; \arg\min_{(\underline{\theta},\sigma^2)} D[p(x_s|\underline{y}) \;||\; q(x_s|\underline{y})]$$

$$
\begin{aligned}
&= \arg\min_{(\underline{\theta},\sigma^2)} E_p\left[\log \frac{p(X_s|\underline{Y})}{q(X_s|\underline{Y})}\right] \\
&= \arg\max_{(\underline{\theta},\sigma^2)} E_p\left[\log q(X_s|\underline{Y})\right] \\
&= \arg\max_{(\underline{\theta},\sigma^2)} E_p\left[-\frac{1}{2}\log\sigma^2 - \frac{1}{2\sigma^2}(X_s - \sum_{r\in\eta}\theta_r Y_r)^2\right] \\
&= \arg\min_{(\underline{\theta},\sigma^2)} \frac{1}{2}\log\sigma^2 + \frac{1}{2\sigma^2}(E_p[X_s - \sum_{r\in\eta}\theta_r Y_r]^2).
\end{aligned}
\tag{8}
$$

It can be seen that the $\underline{\theta}^*$ parameters corresponding to $q^*(\underline{x})$ are obtained by minimizing the second term in the Eq. (8)

$$
\underline{\theta}^* = \arg\min_{\underline{\theta}} E_p[X_s - \sum_{r\in\eta}\theta_r Y_r]^2 \tag{9}
$$

and using the $\underline{\theta}^*$ obtained, we can estimate the $[\sigma^2]^*$ that minimizes Eq. (8),

$$
[\sigma^2]^* = E_p[X_s - \sum_{r\in\eta}\theta_r^* Y_r]^2. \tag{10}
$$

then,

$$
\begin{aligned}
\underline{\theta}^* &= \arg\min_{\underline{\theta}} E_p[X_s - \underline{\theta}^T\underline{Y}]^2 \\
\underline{\theta}^* &= [E_p(\underline{Y}\underline{Y}^T)]^{-1}E_p(X_s\underline{Y})
\end{aligned}
\tag{11}
$$

and,

$$
\begin{aligned}
[\sigma^2]^* &= E_p(X_s^2) - E_p(X_s\underline{Y}^T)[E_p(\underline{Y}\underline{Y}^T)]^{-1}E_p(X_s\underline{Y}) \\
&= E_p(X_s^2) - [\underline{\theta}^*]^T E_p(X_s\underline{Y}).
\end{aligned}
\tag{12}
$$

In addition, the estimated $\underline{\theta}^*$ parameters should satisfy the positivity conditions in Eq. (2).

Now, returning back to multiresolution discussion, let $X^{(0)}$ be a GMRF defined by $(\underline{\theta}^{(0)}, [\sigma^2]^{(0)})$ and $X^{(k)}$ be the field obtained by subsampling $X^{(0)}$, k times. The non-Markov $X^{(k)}$ can be approximated by a GMRF by minimizing Eq. (7). The minimization requires the autocorrelation values $E_{p^{(k)}}[X_s^{(k)} X_{s+r}^{(k)}]$ which can be computed, given the GMRF parameters for $X^{(0)}$ as shown below.

$$
\begin{aligned}
X_s^{(k)} &= X_{2^k s}^{(0)} \\
E_{p^{(k)}}[X_s^{(k)} X_{s+r}^{(k)}] &= E_{p^{(0)}}[X_{2^k s}^{(0)} X_{2^k(s+r)}^{(0)}].
\end{aligned}
$$

For any two lattice sites u and v in $\Omega^{(0)}$ the correlation is given by [13]:

$$
E_{p^{(0)}}[X_u^{(0)} X_v^{(0)}] = \frac{[\sigma^2]^{(0)}}{MN}\sum_{s\in\Omega^{(0)}} \frac{(\lambda_M^{s_1u_1}\lambda_N^{s_2u_2})(\lambda_M^{-s_1v_1}\lambda_N^{-s_2v_2})}{1-[\underline{\theta}^{(0)}]^T\underline{\phi}_s} \tag{13}
$$

where $\lambda_n^i = \exp(\sqrt{-1}\frac{2\pi i}{n})$.

Under the assumption that the covariance matrix with respect to p - measure is positive definite, the function in Eq. (9) to be minimized is convex and is minimized over a convex set defined by $1 - [\underline{\theta}^{(k)}]^T \underline{\phi}_s > 0$, for $\forall s \in \Omega$. If the solution lies inside the convex set, it can obtained from Eq. (11). Otherwise, a gradient descent procedure can be used.

Remarks:

1. If the $\underline{\theta}^*$ obtained from Eq. (11) satisfies the positivity conditions and if p is Gaussian, then $p(x_s|x_{s+r}, r \in \eta) = q^*(x_s|x_{s+r}, r \in \eta)$. Since $p(\underline{x})$ is Gaussian, $p(x_s|x_{s+r}, r \in \eta)$ is also Gaussian with conditional mean $\sum_{r\in\eta} \theta^*_r x_{s+r}$ (which is the best linear estimate of X_s in terms of $X_{s+r}, r \in \eta$) and conditional variance $[\sigma^2]^*$ (which is the corresponding minimum mean square error of the estimator) [19]. $q^*(\underline{x})$ being a GMRF with parameters $(\underline{\theta}^*, [\sigma^2]^*)$, from the discussion at the beginning of this section, has the conditional distribution $q^*(x_s|x_{s+r}, r \in \eta)$ with the conditional mean $\sum_{r\in\eta} \theta^*_r x_{s+r}$ and conditional variance $[\sigma^2]^*$. However, the joint densities $p(\underline{x})$ and $q(\underline{x})$ on the whole lattice are not the same, $p(\underline{x})$ is a non-Markov density and $q(\underline{x})$ is a Markov density.
2. It is worth observing that Eq. (8) is similar to the pseudo likelihood estimate [3], [1] where the GMRF parameters are obtained by minimizing the products of local conditional densities over the entire lattice. The pseudo likelihood estimator uses the sample covariances obtained from the observed sample field, whereas our local conditional distribution invariance estimator uses the covariances calculated with respect to the p - measure.

4: Parameters Resulting in Identical PDFs at Coarser Resolutions

In the previous section, we presented methods to approximate subsampled random fields by GMRFs assuming that data at the fine resolution is modeled by a GMRF. It is also necessary to analyze if different GMRF parameters at the fine resolution can result in probabilistically identical coarser resolution random fields. Since we are dealing with Gaussian fields, it suffices to check the covariance matrices of the subsampled fields instead of the pdfs. However, the covariance elements are complicated functions of the parameters (see Eq. (13)). Therefore, we look at the power spectrum of the subsampled random fields which are simpler functions of the parameters. We show that there exists different sets of GMRF parameters, which on subsampling result in the same pdf at the lower resolution. Since the parameter $[\sigma^2]^{(0)}$ is a multiplicative factor in the power spectral function, we assume it be equal to one and investigate the existence of different sets of $\underline{\theta}$ parameters that result in the identical coarser resolution random fields.

Case 1: First order GMRF on $\Omega^{(0)}$

The first order GMRF model is defined by the parameters $(\theta_{(1,0)}, \theta_{(0,1)}, 1)$. For a first order GMRF at the fine resolution, the only set of parameters that results in the same power spectrum at $\Omega^{(1)}$ is $(\theta_{(1,0)}, \theta_{(0,1)})$, $(-\theta_{(1,0)}, \theta_{(0,1)})$, $(\theta_{(1,0)}, -\theta_{(0,1)})$, $(-\theta_{(1,0)}, -\theta_{(0,1)})$.

Case 2: Second order GMRF on $\Omega^{(0)}$

The second order GMRF model is defined by the parameters $(\theta_{(1,0)}, \theta_{(0,1)}, \theta_{(1,1)}, \theta_{(-1,1)}, 1)$. For a second order GMRF at the fine resolution, the only set of parameters that result in the same power spectrum at $\Omega^{(1)}$ is,
$(\theta_{(1,0)}, \theta_{(0,1)}, \theta_{(1,1)}, \theta_{(-1,1)})$, $(-\theta_{(1,0)}, \theta_{(0,1)}, -\theta_{(1,1)}, -\theta_{(-1,1)})$,
$(\theta_{(1,0)}, -\theta_{(0,1)}, -\theta_{(1,1)}, -\theta_{(-1,1)})$, $(-\theta_{(1,0)}, -\theta_{(0,1)}, \theta_{(1,1)}, \theta_{(-1,1)})$.

Proof: The proof can be found in [14].

Similar results can be obtained for higher order cases.

5: Texture Segmentation

Computer vision and image analysis algorithms use various visual cues to analyze and interpret an image of a complex scene. These visual cues include, among others, photometric and geometric cues. Photometric cues include shading, texture, etc., from which features such as edges and regions are obtained. Texture is one of the basic characteristics of a visible surface and provides useful information for scene segmentation and understanding. Texture is a very important property for the analysis of remote sensed satellite images, their segmentation into various vegetation classes. Texture classification and segmentation problems have been addressed by several authors with different approaches that can be broadly classified into two, namely, structural [20],[11] and statistical [6],[7],[18],[17] approaches.

Texture segmentation problem is the labeling of pixels in a lattice to one of V texture classes, based on a texture model and the observed intensity field. Each site in the lattice carries a class label (say $L_s = v, v \in \{1, 2, \ldots, V\}$) and this label field is modeled by an MRF. We do not directly observe the label field, but a function of the labels, the intensity field. The intensity field is modeled by a GMRF, whose parameters depend on the value of label field at that site. The goal is to estimate the unobserved label field from the observed intensities by optimizing a suitable error criterion.

We model the label field $\underline{L}$ by an MRF with a neighborhood ψ:

$$p(\underline{L} = \underline{l}) = \frac{1}{Z} \exp \left[\beta \sum_{s \in \Omega} U(l_s) \right] \tag{14}$$

where $U(l_s)$ is the number of neighbors in ψ that have the same label as l_s. This model is also called a pairwise interaction model.

The local conditional probability of the label field is given by:

$$p(l_s | l_{s+r}, r \in \psi) = \frac{\exp\left[\beta U(l_s)\right]}{\sum_{l'_s = (1,2,\ldots V)} \exp\left[\beta U(l'_s)\right]}. \tag{15}$$

The GMRF parameters corresponding to a label v are denoted by $(\underline{\theta}(v), \sigma^2(v))$. The conditional density of the intensity field can be written as follows, from Eq. (3):

$$\begin{aligned} p(X_s = x_s | L_s = v, X_{s+r}, r \in \eta) \\ = \frac{1}{\sqrt{2\pi\sigma^2(v)}} \exp\{-\frac{1}{2\sigma^2(v)}[x_s - \sum_{r \in \eta} \theta_r(v) x_{s+r})]^2\}. \end{aligned} \tag{16}$$

We restrict ourselves to the iterated conditional mode method (ICM) . The ICM solution is obtained by performing the following optimization at each lattice site [1]:

$$\max_{L_s} \quad P(X_s | L_s, X_{s+r}, r \in \eta) P(L_s | L_{s+r}, r \in \psi).$$

This is equivalent to,

$$\min_{L_s} \ \frac{1}{2} \log(\sigma^2(v)) + \frac{1}{2\sigma^2(v)} [x_s - \sum_{r \in \eta} \theta_r(v) x_{s+r}]^2 - \beta U(L_s = v) \tag{17}$$

the minimization is performed by visiting the pixels in raster scan order for all $s \in \Omega$ and stopped when no further changes in the labels occur.

5.1: Multiresolution Segmentation

The segmentation algorithm presented above is a single resolution algorithm. As we have discussed before, data at lower resolutions can be approximated by a GMRF. Thus the same algorithm can be applied at lower resolutions too. Our multiresolution algorithm includes the following steps. First, given the number of classes and the associated parameters at the fine resolution, the GMRF parameters at lower resolutions are obtained by the local conditional distribution invariance approximation. Then, segmentation is performed at the coarsest resolution using Eq. (17) with the corresponding parameters and the results of segmentation are passed on to the immediate higher resolution. This is repeated until the fine resolution is reached. At each resolution a confidence measure is attached to the segmentation result at each pixel and propagated to the finer resolution. We address issues regarding confidence measures in this section. After obtaining the segmentation result by ICM convergence at one resolution, the results have to be propagated to the immediate higher resolution. Since we obtain resolution transformation by subsampling, we have a quad tree type of graph. If $\underline{L}^{(k)}$ is the segmentation result at the kth resolution, the labels in the $(k-1)$th level are initialized as:

$$L_s^{(k-1)} = L_{\lfloor s/2 \rfloor}^{(k)}. \tag{18}$$

In addition, at level k, after the ICM converges, we attach a confidence measure $C_s^{(k)}$ to the segmentation result obtained at site s.

At level k, after the convergence of ICM iterations, let $\underline{\dot{v}}$ and $\underline{\ddot{v}}$ be such that,

$$\begin{aligned}
\dot{v}_s &= \arg \max_{v \in \{1,2,\ldots,V\}} P(X_s | L_s = v, X_{s+r}, r \in \eta) P(L_s = v | L_{s+r}, r \in \psi) \\
\ddot{v}_s &= \arg \max_{v \in \{1,2,\ldots,V\}, v \neq \dot{v}_s} P(X_s | L_s = v, X_{s+r}, r \in \eta) P(L_s = v | L_{s+r}, r \in \psi)
\end{aligned}$$

and the confidence measure is defined as,

$$C_s^{(k)} = \frac{P(X_s | \dot{v}_s, X_{s+r}) P(\dot{v}_s | L_{s+r})}{P(X_s | \ddot{v}_s, X_{s+r}) P(\ddot{v}_s | L_{s+r})}. \tag{19}$$

These confidence measures at level k are propagated upwards to level $k-1$ in the same manner as in Eq. (18). At level k, ICM is restricted to only those pixels with the confidence measure such that, $\frac{1}{C_s^{(k)}} \geq c^{(k)}$, where $c^{(k)}$ is a confidence threshold at level k. Also from the definition, $0 < \frac{1}{C_s^{(k)}} \leq 1.0$. For the coarsest resolution $c^{(\cdot)} = 0$, i.e., ICM is performed over all sites in the lattice.

6: Experiments

We present experimental results with simulated, Brodatz texture images and real satellite images to show that the multiresolution algorithms perform better than the single resolution both in terms of the classification accuracy and computational requirements. In all the experiments, the confidence threshold $c^{(k)} = \{0.6, 0.25, 0.0\}$, is used for the different levels with smaller values used at coarser resolutions. Multiresolution results presented in this section are obtained by performing the algorithm over three resolutions. In all cases, percentages of correct classification and computational requirements are given in parenthesis.

To compare the computational requirements between the single resolution and multiresolution approaches, we define a unit of computation to be the computation required to perform ICM at a single pixel site.

We generated texture images using the technique given in [3]. Three third order GMRF textures are generated with parameters { ($\theta_{(1,0)}$ = 0.0934154, $\theta_{(0,1)}$ = 0.520252, $\theta_{(1,1)}$ = 0.0303413, $\theta_{(-1,1)}$ = 0.0180476, $\theta_{(2,0)}$ = -0.0216434, $\theta_{(0,2)}$ = -0.148331), σ^2 = 0.9342 }, {$\underline{\theta}$= (0.308257, 0.468389, -0.0755398, -0.0755797, -0.0407557, -0.100678), σ^2 = 1.8472 }, {$\underline{\theta}$= (0.406875, 0.423393, -0.178478, -0.188702, -0.0649544, -0.121439), σ^2 = 1.264811 }. Figure 1(a) shows the composite image with these three textures. Figure 1(b) shows the single resolution segmentation result (classification accuracy = 89.84%, computational requirement = 2686976) and Figure 1(c) shows the result for multiresolution segmentation (96.75%, 431031).

We have tested our algorithm on textures from the Brodatz texture album. Figure 2(a) contains grass, calf leather, wool, and wood textures. The original GMRF parameters are estimated by maximum likelihood estimation. Figure 2(b) shows the single resolution segmentation (86.04%, 1114112) and Figure 2(c) shows the multiresolution segmentation (92.75%, 679444). We have another interesting plot of $\frac{1}{C_s^{(k)}}$ for the level $k = 1$ in Figure 2(d). The brighter points in this image correspond to points of low confidence measure. As expected, all the boundary regions between different textures have low confidence measures. In texture segmentation, classification near the texture boundaries is usually more ambiguous.

Figure 3(a) shows a section of a single channel of a multispectral sensor (MSS) image over Africa. We chose three classes corresponding to river, forest, and deforestation. The GMRF parameters obtained from small sections of a different part of the image are used to classify the image shown. Unfortunately exact class maps are not available. Figure 3(b) shows the single resolution result (unknown, 2160000) and Figure 3(c) shows the multiresolution result (unknown, 426105). Clearly, we can see that the multiresolution algorithm has performed better, with lesser computation, than the single resolution algorithm.

Finally, we present results of multiresolution segmentation on a thematic mapper (TM) image consisting of four classes corresponding to river, forest, deforestation and regrowth. Figure 4(a) shows a section of thematic mapper (TM) data and Figure 4(b) shows the 4-class multiresolution segmentation result.

7: Summary

Multiresolution models and algorithms play an important role in image analysis. These algorithms not only help to reduce the computational time, but also help to analyze the given information at different spatial scales. We have presented a technique based on minimizing the KL distance, to estimate the parameters of GMRFs at coarser resolutions and have used it for texture segmentation. GMRFs are widely used in many image processing applications including restoration, segmentation, compression, etc., and the proposed models can be used for these applications. Also, this can be extended to perform unsupervised texture segmentation. However, as mentioned in Section 4, GMRF parameters at a lower resolution can correspond to more than one set of parameters at fine resolution. Hence the problem of retrieving the GMRF parameters at fine resolution given the parameters at coarse resolution has to be addressed for unsupervised segmentation.

References

[1] J. Besag, "On the Statistical Analysis of Dirty Pictures," *Journal of the Royal Statistical Society*, Vol. 48, pp. 259–302, 1986.

[2] C. Bouman and B. Liu, "Multiple Resolution Segmentation of Textured Images," *IEEE Trans. Patt. Anal. Mach. Intell.*, Vol. 13, pp. 99–113, Feb. 1991.

[3] R. Chellappa, "Two-dimensional Discrete Gaussian Markov Random Field Models for Image Processing," in *Progress in Pattern Recognition* (L. N. Kanal and A. Rosenfeld, eds.), pp. 79–112, Elsavier, 1985.

[4] F. S. Cohen and D. B. Cooper, "Simple Parallel Hierarchical and Relaxation Algorithms for Segmenting Noncausal Markovian Random Fields," *IEEE Trans. Patt. Anal. Mach. Intell.*, Vol. 9, pp. 195–219, March 1987.

[5] T. Cover and J. Thomas, *Elements of Information Theory*, Wiley, 1991.

[6] G. R. Cross and A. K. Jain, "Markov Random Field Texture Models," *IEEE Trans. Patt. Anal. Mach. Intell.*, Vol. 5, pp. 25–39, Jan 1983.

[7] H. Derin and H. Elliot, "Modeling and Segmentation of Noisy and Textured Images Using Gibbs Random Field," *IEEE Trans. Patt. Anal. Mach. Intell.*, Vol. 9, pp. 39–55, Jan. 1987.

[8] D. Geiger and F. Girosi, "Parallel and Deterministic Algorithms for MRFs: Surface Reconstruction," *IEEE Trans. Patt. Anal. Mach. Intell.*, Vol. 13, pp. 401–413, May 1991.

[9] S. Geman and D. Geman, "Stochastic Relaxation, Gibbs Distribution and the Bayesian Restoration of Images," *IEEE Trans. Patt. Anal. Mach. Intell.*, Vol. 6, pp. 721–741, Nov 1984.

[10] B. Gidas, "A Renormalization Group Approach to Image Processing," *IEEE Trans. Patt. Anal. Mach. Intell.*, Vol. 11, No. 2, pp. 164–180, 1989.

[11] R. Haralick, "Statistical and Structural Approaches to Texture," *Proc. IEEE*, Vol. 67, pp. 610–621, May 1979.

[12] F. C. Jeng, "Subsampling of Markov Random Fields," *Jour. of Visual Communication and Image Representation*, Vol. 3, pp. 225–229, Sep. 1992.

[13] R. L. Kashyap, "Analysis and Synthesis of Image Patterns by Spatial Interaction Models," in *Progress in Pattern Recognition* (L. N. Kanal and A. Rosenfeld, eds.), North-Holland, Amsterdam, 1981.

[14] S. Krishnamachari, *Hierarchical Markov Random Field Models for Image Analysis*, Ph.D. dissertation, University of Maryland, College Park, 1995.

[15] S. Lakshmanan and H. Derin, "Gaussian Markov Random Fields at Multiple Resolutions," in *Markov Random Fields: Theory and Applications* (R. Chellappa, ed.), pp. 131–157, Academic Press, 1993.

[16] M. R. Luettgen, W. C. Karl, A. S. Willsky, and R. R. Tenney, "Multiscale Representations of Markov Random Fields," *IEEE Trans. on Signal Processing*, Vol. 41, pp. 3377–3397, Dec. 1993.

[17] B. S. Manjunath and R. Chellappa, "A Note on Unsupervised Texture Segmentation," *IEEE Trans. Patt. Anal. Mach. Intell.*, Vol. 13, pp. 478–483, May 1991.

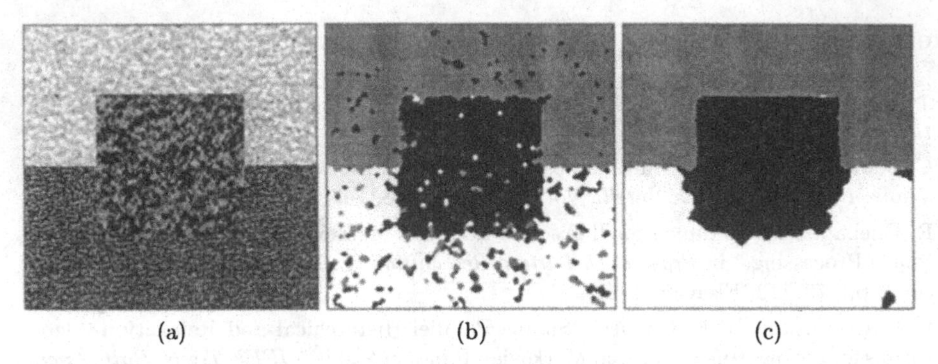

Figure 1: (a) Synthetic texture image, (b) Single resolution segmentation result, (c) Multiresolution segmentation result.

[18] B. S. Manjunath, T. Simchony, and R. Chellappa, "Stochastic and Deterministic Networks for Texture Segmentation," *IEEE Trans. on Acoustics, Speech, and Signal Processing*, Vol. 38, pp. 1039–1049, June 1990.

[19] A. Papoulis, *Probability, Random Variables, and Stochastic Processes*, McGraw-Hill Book Company, 1965.

[20] A. Rosenfeld, "Visual Texture Analysis," Tech. Rep. 70-116, University of Maryland, June 1970.

[21] A. Rosenfeld, "Some Useful Properties of Pyramids," in *Multiresolution Image Processing and Analysis* (A. Rosenfeld, ed.), Springer-Verlag, 1984.

[22] D. Terzopoulos, "Image Analysis using Multigrid Relaxation Methods," *IEEE Trans. Patt. Anal. Mach. Intell.*, Vol. 8, pp. 129–139, March 1986.

[23] C. H. Wu and P. C. Doerschuk, "Tree Approximations to Markov Random Fields," *IEEE Trans. Patt. Anal. Mach. Intell.*, Vol. 17, pp. 391–343, Apr. 1995.

[24] J. Zhang, "The Mean Field Theory in EM Procedures for Markov Random Fields," *IEEE Trans. on Signal Processing*, Vol. 40, pp. 2570–2583, Oct. 1992.

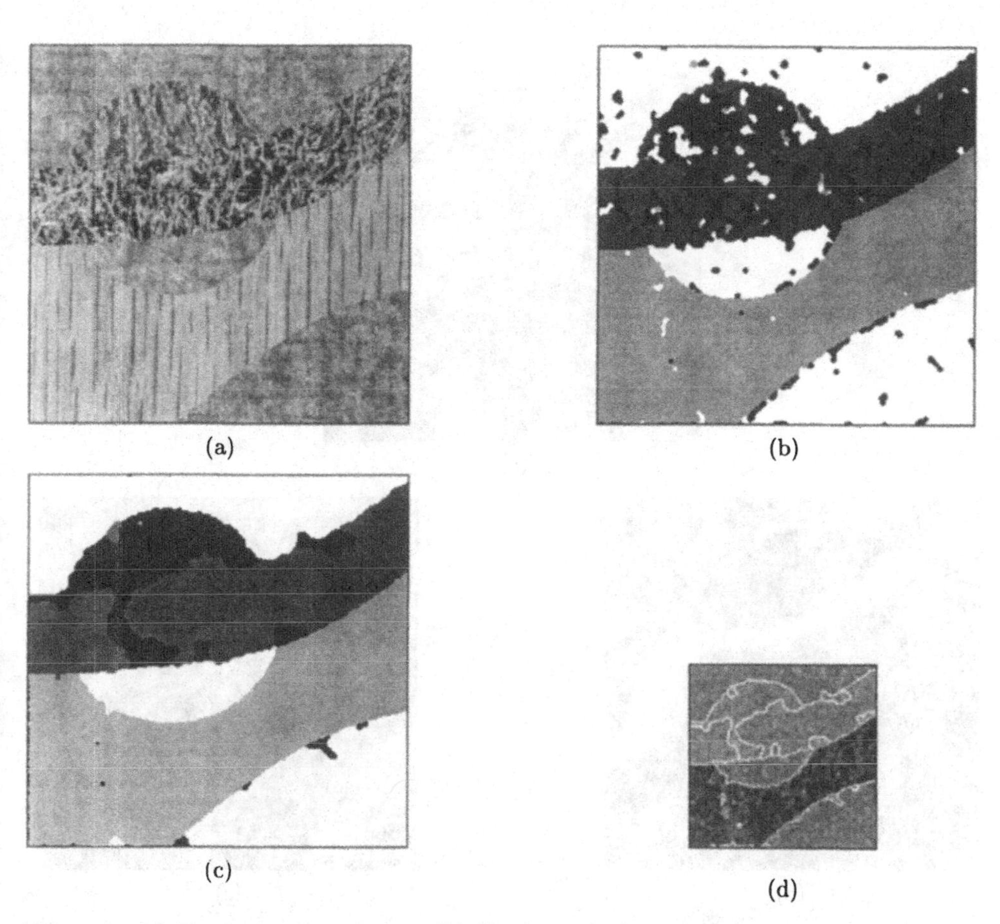

(a)

(b)

(c)

(d)

Figure 2: (a) Brodatz texture image, (b) Single resolution segmentation result, (c) Multiresolution segmentation result, (d) Confidence measures.

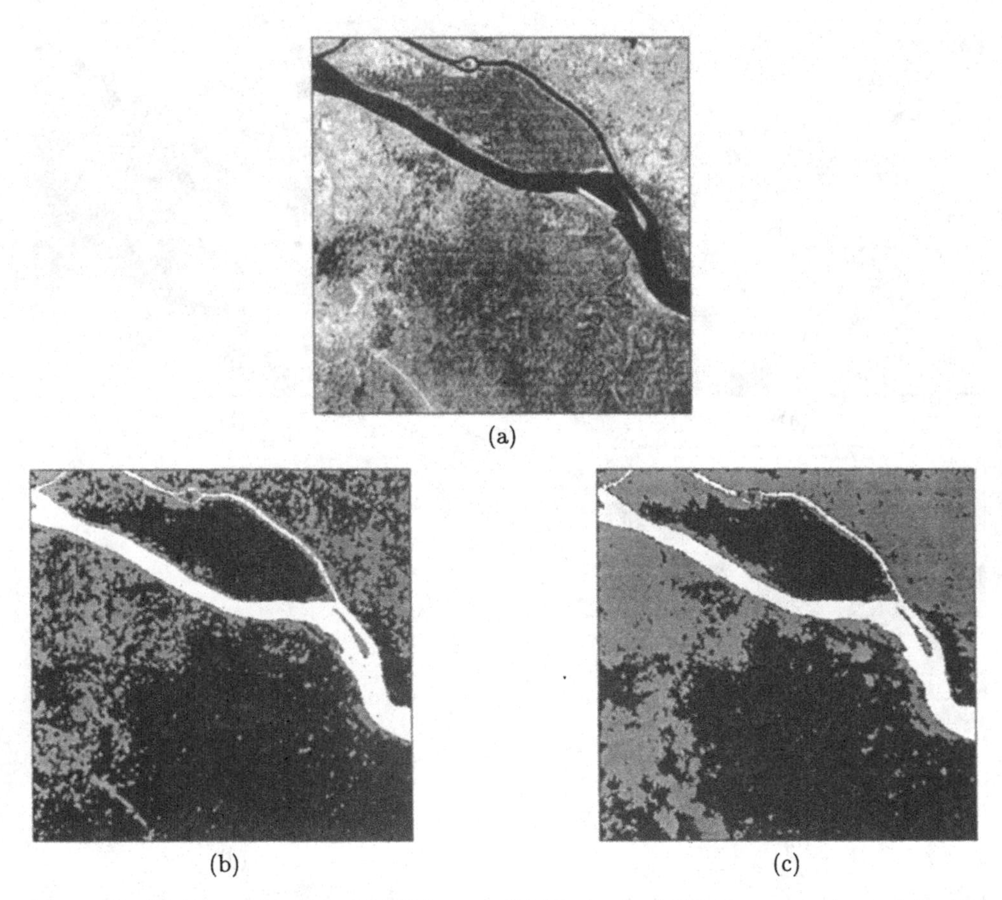

Figure 3: (a) Remotely sensed MSS image, (b) Single resolution segmentation result, (c) Multiresolution segmentation result.

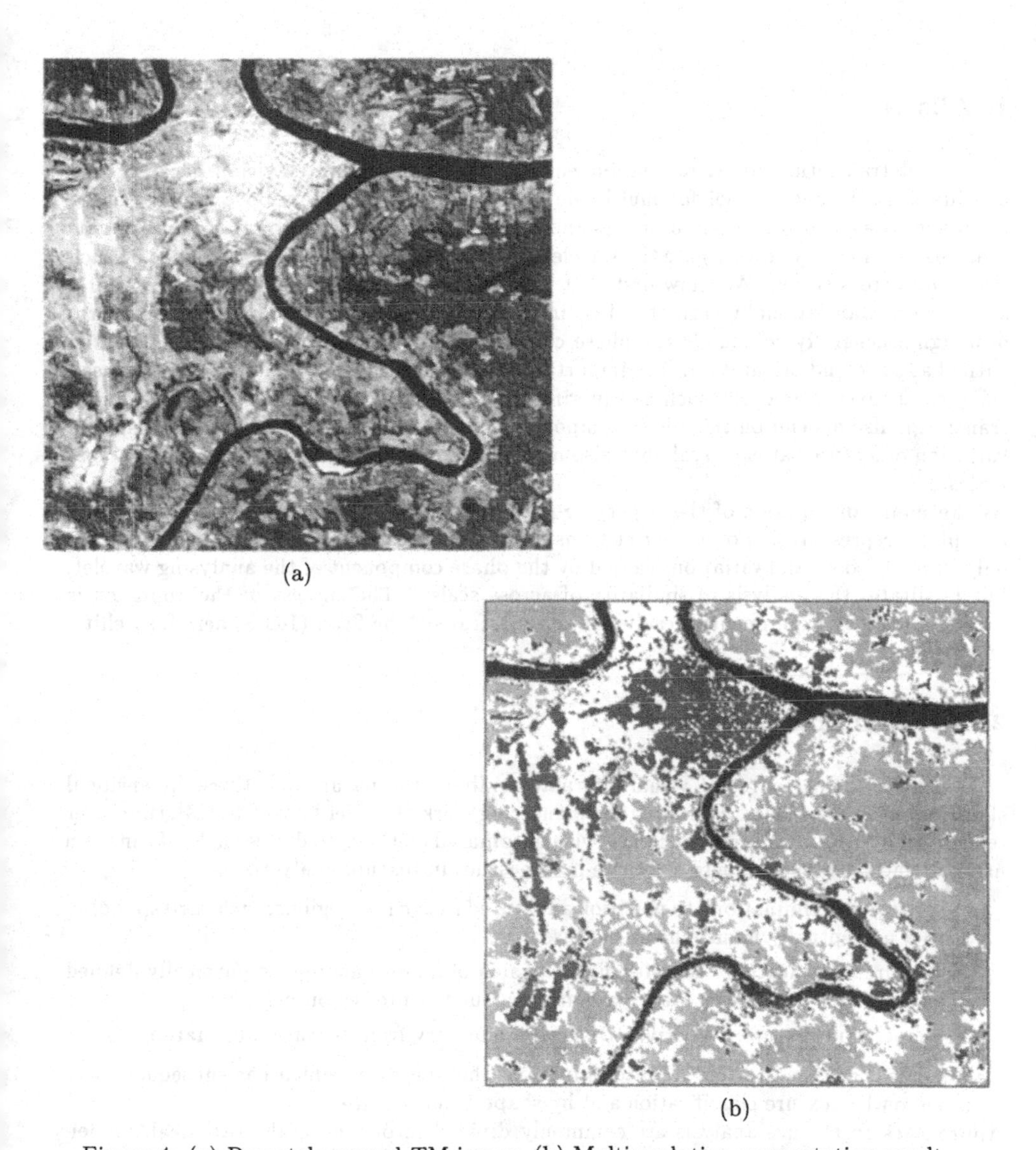

Figure 4: (a) Remotely sensed TM image, (b) Multiresolution segmentation result.

Feature Selection for Texture Segmentation

Michael Brady and Zhi-Yan Xie

1: Abstract

Wavelet transforms are attracting increasing interest in computer vision because they provide a mathematical tool for multiscale image analysis. We develop a wavelet-based approach to segmenting textured images and demonstrate its application to a range of natural images. First, we investigate the wavelet transform representation of an image at each scale and across scales. We show first that the subsampled wavelet multiresolution representation is translationally variant. More importantly, we show that a wavelet transform of a signal generally confounds the phase component of the analysing wavelet associated with that scale and orientation. The importance of this observation is that commonly used features in texture analysis, such as squaring, or half-, full-wave rectification of a wavelet transform, also depend on this phase component. This not only causes unnecessary spatial variation of features at each scale but also makes it more difficult to match features across scales.

As the main contribution of the paper, we propose a complete 2D decoupled local energy and phase representation of a wavelet transform. As a texture feature, local energy is not only immune to spatial variations caused by the phase component of the analysing wavelet, but facilitates the analysis of similarity of across scales. The success of the approach is demonstrated by experimental results for aerial Infrared Line Scan (IRLS) aerial, satellite, and Brodatz images.

2: Introduction

Texture is a rich source of visual information about the nature and three-dimensional shape of physical surfaces. Following the pioneering work of Azriel Rosenfeld [24], computer texture analysis is ultimately concerned with automated methods to derive such information using artificial systems. There are three major issues in texture analysis:

1. Texture discrimination - to partition a textured image into regions, each corresponding to a perceptually homogeneous texture;
2. Texture classification - to determine to which of a finite number of physically defined classes, such as wood or water, a homogeneous texture region belongs;
3. Shape from texture: to derive 3D surface geometry from texture information.

This paper is confined to texture discrimination- the first stage which has subsequently to be relied on by texture classification and by shape from texture.

Approaches to texture analysis are commonly divided into structural, statistical, model-based and transform methods.

Structural approaches [24, 1] represent textures by well-defined primitives (microtexture) and a hierarchy of spatial arrangements (macrotexture) of those primitives. Although there

0-8186-7644-2 $5.00 © 1996 IEEE

is much benefit to be had from abstract descriptions of textures, they can still be ill-defined for natural textures because the variability of both microtexture and macrotexture means there is no clear distinction between them.

In contrast to structural methods, statistical approaches do not attempt to understand "explicitly" the hierarchical structure of a texture, such as the primitives used in structural approaches. Instead they represent a texture indirectly by the nondeterministic properties which govern the distributions and relationships between the grey levels of an image. Methods based on second-order statistics [3, 4, 5, 13] (i.e. the statistics given by pairs of pixels) have been shown to achieve higher discrimination rates than the power spectrum and structural methods [25, 23]. However, such grey level statistics are of very limited descriptive power and have not given satisfactory results in practice.

Model based image texture analysis [9, 11, 20], using stochastic and fractal models, attempt to interpret an image texture by use of a generative image model or stochastic model. The parameters of the model are estimated and used for texture analysis. In practice, the computational complexity arising in the estimation of stochastic model parameters and the difficulties of handling non-stationary textures are the primary problems. Although such methods have had some success in supervised texture segmentation and classification, in general they work poorly for nonsupervised texture segmentation. The fractal model has been shown to be useful for modelling naturally occurring textures. However, although it has been shown to be useful for texture segmentation [22, 21, 27], it lacks orientation selectivity and is incomplete for describing local image structures.

Transform methods of texture analysis, such as the Fourier [2], Gabor [12, 10, 6, 15, 17], and wavelet transforms [19, 8, 16], represent an image in a coordinate system that has an interpretation that is closely related to the characteristics of a texture, such as frequency or size. Methods based on the Fourier transform perform poorly in practice due to its lack of spatial localisation. Although the Gabor transform can overcome the spatial localisation problem to some extent, it is limited in practice because there is usually not a single resolution at which one can localise spatial structures that form natural textured images. For this reason, the wavelet transform offers the best hope.

A continuous wavelet transform (CWT) is defined by

$$W_\psi f(a,b) = \frac{1}{\sqrt{a}} \int_{-\infty}^{\infty} f(x)\overline{\psi}(\frac{x-b}{a})\, dx \tag{1}$$

where $a \in \mathbf{R}^+, b \in \mathbf{R}$ are scale and translation parameters and the wavelet $\psi(x)$ can be any function that satisfies the invertibility condition, i.e. it has to be bandpass. By varying the scale a, $W_\psi f(a,b)$ can capture not only large (low frequency) but also localised (high frequency) components of spatial structures.

Compared with the Gabor transform, the wavelet transforms enjoy several advantages: i) varying the spatial resolution allows it to represent spatial structures at the most suitable scale; ii) there is a wide range of choices for the wavelet function $\psi(x)$; and iii) the flexibility of being able to choose a single (or a set of) wavelet function best suited to the image texture in a specific application. All of these make the wavelet transform particularly attractive for texture segmentation. We begin in Sections 3, 4 by pointing out two theoretical and practical difficulties of using wavelets. Then in Section 5 we define a local energy model that solves these problems. We show typical results on real textured images in Section 6.

3: Translation Invariant Wavelet Multiresolution representation

If the scale and space parameters of CWT are sampled at $\{a = 2^{-j};\ b = 2^{-j}n;\ j, n \in \mathbf{Z}\}$, and if the wavelet $\psi(x)$ is bi-orthogonal, then the wavelet transform is given by

$$\begin{aligned} W_\psi f(j,n) &= 2^{\frac{j}{2}} \int_{-\infty}^{\infty} f(x)\overline{\psi}(2^j x - n)\, dx \\ &\triangleq < f(x), \psi_{j,n}(x) > \end{aligned} \tag{2}$$

where $\psi_{j,n}(x) = 2^{\frac{j}{2}}\psi(2^j x - n)$. $W_\psi f(j,n)$ defines a class of discrete wavelet transforms, the so-called wavelet multiresolution representation (WMR), which can be computed by recursive filtering.

3.1: Translation Invariant WMR

The WMR has been shown useful in image compression because of its completeness and compactness. However, $W_\psi f(j,n)$ is translationally variant:

$$\begin{aligned} W_\psi f(j,n) &= 2^{\frac{j}{2}} < f(x - x_0), \psi_{j,n}(x) > \\ &= 2^{\frac{j}{2}} < f(x), \psi_{j,n-2^j x_0}(x) > \\ &\neq 2^{\frac{j}{2}} < f(x), \psi_{j,n-m}(x) > \end{aligned}$$

for an arbitrary translation x_0, $x_0 = 2^{-j}m;\ m \in \mathbf{Z}$ won't in general be satisfied. In other words, if two identical signals were to appear in different positions, their wavelet transform representations can be very different which is unacceptable for most signal and image processing applications, in particular it badly affects texture segmentation.
One way to overcome this problem is to avoid the scale-dependent subsampling $b = 2^{-j}n$ by setting $b = n$, that is by keeping the same number of samples as the original signal. More precisely, for a signal $f(n)$ the wavelet transform is redefined to be

$$W_\psi f(j,n) = 2^j \sum_k f(k)\overline{\psi}(2^j(k - n))$$

In this way, $W_\psi f(j,n)$ becomes translationally *invariant*.
The 1D wavelet multiresolution representation has been extended to 2D [19] which is defined by

$$\begin{aligned} A^j(m,n) &= < f(x,y), 2^{2j}\phi(2^j x - m)\phi(2^j y - n) > \\ D_x^j(m,n) &= < f(x,y), 2^j\phi(2^j x - m)2^j\psi(2^j y - n) > \\ D_y^j(m,n) &= < f(x,y), 2^j\psi(2^j x - m)2^j\phi(2^j y - n) > \\ D_d^j(m,n) &= < f(x,y), 2^j\psi(2^j x - m)2^j\psi(2^j y - n) > \end{aligned} \tag{3}$$

The mother wavelets associated with $D_x^j(m,n), D_y^j(m,n), D_d^j(m,n)$ are

$$\Psi_x(x,y) = \phi(x)\psi(y) \tag{4}$$
$$\Psi_y(x,y) = \psi(x)\phi(y) \tag{5}$$
$$\Psi_d(x,y) = \psi(x)\psi(y) \tag{6}$$

where ϕ and ψ are the 1D scaling (low-pass) and wavelet function (high-pass), respectively. D_x, D_y, D_d are called the horizontal, vertical and diagonal detail images (channels), respectively, which give strong responses to spatial structures in the horizontal, vertical and diagonal directions[1].
Similarly, it can be shown that A, D_x, D_y, D_d are translationally *variant*. To overcome this problem, we change scale-dependent sampling to uniform sampling as in 1D, but in both the x and y directions. More precisely, an *oversampled* 2D wavelet multiresolution representation is defined by

$$\begin{aligned}
A^j(m,n) &= < f(x,y), 2^j\phi(2^j(x-m))2^j\phi(2^j(y-n)) > \\
D^j_x(m,n) &= < f(x,y), 2^{2j}\Psi_x(2^j(x-m), 2^j(y-n)) > \\
D^j_y(m,n) &= < f(x,y), 2^{2j}\Psi_y(2^j(x-m), 2^j(y-n)) > \\
D^j_d(m,n) &= < f(x,y), 2^{2j}\Psi_d(2^j(x-m), 2^j(y-n)) > \qquad (7)
\end{aligned}$$

in which case A, D_x, D_y, and D_d become translationally *invariant*.
Although the representation given in Eqn. 7 sacrifices the compactness compared with the representation given by Eqn. 3, they provide translational invariance which is essential for texture segmentation, and many other image processing tasks.

4: Phase Dependence of The Wavelet Transform

Considered as a linear convolution with a set of bandpass filters, the wavelet transform enables the image "gradient" to be computed at multiple spatial scales. However the wavelet transform coefficients themselves are not suitable as texture features because they are always zero mean at each scale since $\int_{-\infty}^{\infty} \psi(x)\, dx = 0$. Therefore some non-linear operation is necessary [18].
The most commonly used features are full- and half-wave rectification and the square power of the wavelet detail signals. We have shown [28], however, that in general those features are coupled with the local phase component that depends not only on the analysed signal but also on the analysing wavelet at that scale. This dependency causes two problems that greatly affect practical texture segmentation: "spurious" spatial variations of features at each scale; and the difficulty of matching features across scales.
To illustrate, consider the oversampled 2D wavelet transform of a synthetic images shown in Figure 1. The surfaces of $D^j_x(m,n), D^j_y(m,n), D^j_d(m,n)$ oscillate in space depending on the shape of the wavelets. Patches of moduli $|D^j_d(m,n)|$ at different scales are shown in Figure 2 (top row), and they are clearly affected by the oscillation of the wavelet. Such performance is unacceptable for texture analysis because one wants a uniform feature response in those regions of the image which have uniform texture, while a wavelet is typically an oscillating, wave-like function. Hence, some other nonlinear operation must be found to derive features which can be invariant to *the phase component at each scale* and can also be *matched from one scale to another*.
In 1D, we have previously [26, 28] developed a decoupled local energy and phase representation of a real-valued wavelet transform using the Hilbert transform [28]. In the following section, we show how to extend this representation to 2D.

[1] The orientation of the spatial structure is defined as perpendicular to the direction of maximum gradient.

5: Decoupled 2D Local Energy and Phase Representation

The principal theoretical difficulty in extending the local energy and phase representation of the 1D wavelet transform to 2D is that there does not exist a universal 2D Hilbert transform. Nevertheless, we propose a definition for the local energy and local phase of a 2D wavelet transform which not only provides a complete representation of a 2D wavelet transform in scale-space, but also facilitates the local energy to be independent of the phase components of the analysing wavelets. Moreover, the relationship between a 2D wavelet transform and its local energy is established both in *scale-space* and in *frequency* space.

5.1: Horizontal and Vertical Channels

Recall the mother wavelets associated with the horizontal and vertical channels given in Eqns: 4,5. $D_x^j(m,n)$ can be considered as a 1D wavelet transform with respect to $\psi(y)$ for each column (y axis) after first smoothing each row (x axis) with $\phi(x)$. Similarly, $D_y^j(m,n)$ can be considered as a 1D wavelet transform for each row after first smoothing each column. In these case, the local energy and phase can be defined as 1D [28]:

Definition 5.1 *For a real valued ψ and $f(x,y) \in \mathbf{L}^2(\mathbf{Z}^2)$ the local energy ρ_x, ρ_y and the local phase φ_x, φ_y of $D_x^j(m,n)$ and $D_y^j(m,n)$ are given by*

$$\begin{aligned}
\rho_x^j(m,n) &= \sqrt{[D_x^j(m,n)]^2 + [H_y\{D_x^j(m,n)\}]^2} \\
\rho_y^j(m,n) &= \sqrt{[D_y^j(m,n)]^2 + [H_x\{D_y^j(m,n)\}]^2} \\
\varphi_x^j(m,n) &= Atan2\frac{H_y\{D_x^j(m,n)\}}{D_x^j(m,n)} \\
\varphi_y^j(m,n) &= Atan2\frac{H_x\{D_y^j(m,n)\}}{D_y^j(m,n)}
\end{aligned}$$

where $H_x\{.\}$ ($H_y\{.\}$) denotes the Hilbert transform of the 1D function $D(m,n)$ when n (m) is fixed.

5.2: Diagonal Channel

The mother wavelet associated with the diagonal channel $D_d^j(m,n)$ is $\Psi_d(x,y) = \psi(x)\psi(y)$. In this case, if ψ_H is the Hilbert transform of ψ, we construct four complex functions as follows

$$G_1(x,y) = [\psi(x) + i\psi_H(x)][\psi(y) + i\psi_H(y)] \tag{8}$$
$$G_2(x,y) = [\psi(x) - i\psi_H(x)][\psi(y) - i\psi_H(y)] \tag{9}$$
$$G_3(x,y) = [\psi(x) + i\psi_H(x)][\psi(y) - i\psi_H(y)] \tag{10}$$
$$G_4(x,y) = [\psi(x) - i\psi_H(x)][\psi(y) + i\psi_H(y)] \tag{11}$$

Noticing the conjugacy relationships

$$G_1^j(m,n) = \overline{G_2^j}(m,n) \qquad G_3^j(m,n) = \overline{G_4^j}(m,n)$$

only one pair $\{G_k^j(m,n); k = 1,3\}$ or $\{G_k^j(m,n); k = 2,4\}$ needs to be considered. In the following, we use the first pair. Substituting $\Psi_d(x,y)$ with $\{G_k(x,y); k = 1,3\}$ in Eqn. 7, we generate two complex images by:

$$DG_k^j(m,n) \stackrel{\text{def}}{=} < f(x,y), 2^{2j}G_k(2^j(x-m), 2^j(y-n)) > ;\ k = 1,3 \quad (12)$$
$$= (f(x,y) * 2^{2j}\overline{G}_k(-2^jx, -2^jy)) \quad (13)$$

The properties of functions $\{DG_k^j(m,n); k = 1,3\}$ are essential for deriving the decoupled local energy and local phase representation of the diagonal channel. We present them in the following lemma.

Lemma 5.1 *For each scale j The functions $DG_1^j(m,n)$ and $DG_3^j(m,n)$ can be represented by $D_d^j(m,n)$ as*

$$DG_1^j(m,n) = D_d^j(m,n) - H_y\{H_x\{D_d^j(m,n)\}\} + i(H_x\{D_d^j(m,n)\} + H_y\{D_d^j(m,n)\}) \quad (14)$$
$$DG_3^j(m,n) = D_d^j(m,n) + H_y\{H_x\{D_d^j(m,n)\}\} + i(H_x\{D_d^j(m,n)\} - H_y\{D_d^j(m,n)\}) \quad (15)$$

and they give a strong response to spatial structures at, or close to $\frac{1\pi}{4}$ and $\frac{3\pi}{4}$, respectively.

Proofs: see Appendix.

Now we are in the position to define the local energy and local phase of the diagonal channel.

Definition 5.2 *The functions $DG_1^j(m,n)$, $DG_3^j(m,n)$ are complex functions and can be written as*

$$DG_1^j(m,n) = \rho_{x+y}^j(m,n)e^{i\varphi_{x+y}^j(m,n)}$$

$$DG_3^j(m,n) = \rho_{x-y}^j(m,n)e^{i\varphi_{x-y}^j(m,n)}$$

The $[\rho_{x+y}^j(m,n)]^2$ and $\varphi_{x+y}^j(m,n)$ are called the local energy and the local phase of $D_d^j(m,n)$,respectiv at or close to $\frac{\pi}{4}$. The $[\rho_{x-y}^j(m,n)]^2$ and $\varphi_{x-y}^j(m,n)$ are called the local energy and the local phase of $D_d^j(m,n)$,respectively, at or close to $\frac{3\pi}{4}$.

Now we have defined four local energy channels for each scale j, denoted by $\rho_x^j(m,n)$, $\rho_y^j(m,n)$, $\rho_{x+y}^j(m,n)$, $\rho_{x-y}^j(m,n)$, which are oriented in the horizontal, vertical and $\frac{\pi}{4}$, $\frac{3\pi}{4}$ directions, respectively. Comparing the local energy and the wavelet detail images shown in Figure 1, the phase dependency embedded in the wavelet detail images has been removed in the local energy representations. Moreover, the local energy images at different scales become comparable in terms of shape similarity as indicated by the correlation of the images adjacent scales shown in Figure 2 (bottom row).

5.3: Properties of The Local Energy and Local Phase

The following theorem shows that the local energy and local phase defined above provides a complete representation of the 2D wavelet transform. The local energy and the wavelet transform are equivalent in the frequency domain (conserve energy), but they are very different in scale-space.

Theorem 5.1 *For a real valued $\psi(x)$ and $f(x,y) \in \mathbf{L^2(Z^2)}$,*

1. *The wavelet transform $D_x^j(m,n)$, $D_y^j(m,n)$ and $D_d^j(m,n)$ can be represented completely by the local energies and local phases and are given by*

$$D_x^j(m,n) = \rho_x^j(m,n)\cos\varphi_x^j(m,n) \tag{16}$$
$$D_y^j(m,n) = \rho_y^j(m,n)\cos\varphi_y^j(m,n) \tag{17}$$
$$D_d^j(m,n) = \frac{1}{2}(\rho_{x+y}^j(m,n)\cos\varphi_{x+y}^j(m,n) + \rho_{x-y}^j(m,n)\cos\varphi_{x-y}^j(m,n)) \tag{18}$$

2. *For each scale $j < 0$,*

$$\sum_m\sum_n[D_x^j(m,n)]^2 = \frac{1}{2}\sum_m\sum_n[\rho_x^j(m,n)]^2 \tag{19}$$
$$\sum_m\sum_n[D_y^j(m,n)]^2 = \frac{1}{2}\sum_m\sum_n[\rho_y^j(m,n)]^2 \tag{20}$$
$$\sum_m\sum_n[D_d^j(m,n)]^2 = \frac{1}{8}\sum_m\sum_n([\rho_{x+y}^j(m,n)]^2 + [\rho_{x-y}^j(m,n)]^2) \tag{21}$$

Proofs: see Appendix.
The wavelet detail images and their associated local energy images are very different in scale-space: the former confounds the phase component, the latter does not. Further, from Eqns. 16, 17, 18, it is clear that full-, half-wave rectification or squaring of the wavelet transform also confound the phase component. The difference between the modulus and the local energy of the wavelet transform are shown in Figure 2.

6: Application to Texture Segmentation

To overcome the phase dependency and spatial localisation problems, we suggest that the local energies be used as local features. Using these features, we develop a computation scheme that is used for texture segmentation. The scheme is depicted in Figure 3.

At the **first level**, the 2D **oversampled wavelet transform** is applied to an image. This transform decomposes an image into a stack of images, each of which is given by an oversampled wavelet detail image denoted by $D(\theta, j, x, y)$ at sampled orientation $\theta = \{\theta_1, \cdots, \theta_n\}$ and sampled scale $a = \{2^{-j};\ j = -1, -2, \cdots, -J\}$. For a 2D separable wavelet transform, an image is decomposed into a pile of images $\{D_x^j(x,y), D_y^j(x,y), D_d^j(x,y); j = -1, -2, \cdots, -J\}$.
The **second level** is a nonlinear operation to remove the phase dependency from each image $D(\theta, j, x, y)$, to obtain a pile of local energy images $\rho(\theta, j, x, y)$ given by

$$\rho(\theta,j,x,y) = \begin{cases} \rho_x^j(x,y) & \theta = 0 \\ \rho_y^j(x,y) & \theta = \frac{\pi}{2} \\ \rho_{x+y}^j(x,y) & \theta = \frac{\pi}{4} \\ \rho_{x-y}^j(x,y) & \theta = \frac{3\pi}{4} \end{cases} \tag{22}$$

This level operates only within a single scale, hence it is also called **intra-scale nonlinear fusion**. The local energy images of a Infrared line scan (IRLS) aerial images are shown in Figure 4.
The **third level** derives two texture features in wavelet scale-space, i.e. a multi-scale

orientational measure $\alpha(j,x,y)$ and an energy measure $F(j,x,y)$. This level is composed of two sub-processes, namely **inter-scale clustering** and **inter-orientation fusion**. The inter-scale clustering is denoted as a c within a circle in Figure 3 because this process is actually implemented by correlation, and it is designed to associate the local energy descriptors $\rho(\theta,j,x,y)$ across scales such that the spatial localisation problem is minimised globally. More precisely, for each $\rho(\theta,j,x,y)$, a new feature image $\rho'(\theta,j,x,y)$ is computed by

$$\rho'(\theta,j,x,y)=\begin{cases}\rho^j_x(x-tx1[j],y-ty1[j]) & \theta=0\\ \rho^j_y(x-tx2[j],y-ty2[j]) & \theta=\frac{\pi}{2}\\ \rho^j_{x+y}(x-tx3[j],y-ty3[j]) & \theta=\frac{\pi}{4}\\ \rho^j_{x-y}(x-tx4[j],y-ty4[j]) & \theta=\frac{3\pi}{4}\end{cases} \tag{23}$$

where $tx1,ty1$ is a translation vector which is determined by a correlation process such that $\rho'(\theta=0,j,x,y)$ across scales are better aligned in space according to the image structure giving rise to the descriptors. Similarly, the vectors of $tx2,ty2,tx3,ty3$ and $tx4,ty4$ can be determined for the orientations $\pi/2,\pi/4$ and $3\pi/4$, respectively.
Unlike the other levels given above, the inter-orientation fusion is not universal. It is specific to each application and to the meaning of different orientation channels. Currently, a simple formula is used to combine four oriented local energy images into quantitative and orientational measures of local energy denoted as $F(j,x,y)$ and $\alpha(j,x,y)$ which are defined by

$$\begin{aligned}F(j,x,y) &= \sqrt{[\rho'(\theta,j,x,y)|_{\theta=0}]^2+[\rho'(\theta,j,x,y)|_{\theta=\frac{\pi}{2}}]^2}\\ &\quad +c*\sqrt{[\rho'(\theta,j,x,y)|_{\theta=\frac{\pi}{4}}]^2+[\rho'(\theta,j,x,y)]^2|_{\theta=\frac{3\pi}{4}}} \qquad (24)\\ \alpha(j,x,y) &= \arg\left(\frac{\rho'(\theta,j,x,y)|_{\theta=\frac{\pi}{2}}}{\rho'(\theta,j,x,y)|_{\theta=0}}\right) \qquad (25)\end{aligned}$$

In other words, the total local energy $F(j,x,y)$ is given by a weighted sum of two parts, each part determined by an orthogonal pair of local energy measures. The combined local energy images $F(j,x,y)$ are shown in Figure 4.
The **fourth level** is segmentation, which is carried out in four steps. The first step is filtering $F(j,x,y)$ by Gaussian smoothing, which allows texture density in a local neighbourhood to be computed. The resultant images are then input for clustering at the second step. This exploits the observation that in the feature space, a well-chosen set of features induces well separated clusters corresponding to different classes. Minimisation of the Kullback information distance [14] is applied at each scale and the value of the threshold is determined automatically. One problem of this simple thresholding is that intensity edges are sometimes misclassified as urban textures in our experiments. This is because the Kullback clustering method only uses simple global statistics (i.e. histogram) and limited texture classes (only two classes). Currently, this effect is minimised by applying post-processing. It can be removed by exploiting the fact that the local energy for isolated intensity edges and intensity edge surrounded by many other edges in its neighbourhood (i.e. typical urban texture) behave very differently across scales [26]. The third step is post-processing. Morphological opening and closing is used to remove intensity edges, isolated small patches and holes. Finally, texture boundaries are detected by finding the points of discontinuity in the image.

6.1: Results

The texture segmentation scheme given above is implemented and has been tested on more than 30 real aerial and satellite images. Typical results are shown in Figure 5. Figure 5 (a), (b) show typical IRLS images taken from a low flying aircraft. The goal (part of a system under development for matching images on successive fly-pasts and matching/constructing a map) is to segment rural and urban areas. The patches in Figure 5 (b) correspond to parks within the surrounding urban area. Figure 5 (c) shows the segmentation of a satellite image taken over Plymouth area, the segmentation result is matched quite well with the map over same area. Finally Figure 5 (d) shows the segmentation of two Brodatz textures [7] (cotton canvas and woolen cloth).

7: Conclusions

In this paper, we have introduced an oversampled wavelet multiresolution representation to achieve translation invariance. Then, we developed a complete, decoupled local energy and phase representation of a 2D oversampled wavelet transform. This representation not only provides a better understanding of the wavelet transform in space at each single scale, but also facilitates matching across scales. The usefulness of this decoupled local energy and phase representation is demonstrated by its application segment textures in several classes of natural images.

8: Proofs

Lemma: 5.1 For $k = 1$, Eqn. 12 can also be written by

$$\begin{aligned} DG_1^j(m,n) &= < f(x,y), 2^{2j}G_1(2^j(x-m), 2^j(y-n)) > \\ &= < f(x,y), 2^{2j}\Psi_d(2^j(x-m), 2^j(y-n)) > \\ &- < f(x,y), 2^{2j}H_x\{H_y\{\Psi_d(2^j(x-m), 2^j(y-n))\}\} > \\ &+ i < f(x,y), 2^{2j}H_x\{\Psi_d(2^j(x-m), 2^j(y-n))\} > \\ &+ i < f(x,y), 2^{2j}H_y\{\Psi_d(2^j(x-m), 2^j(y-n))\} > \end{aligned}$$

by noticing $H_y\{H_x\{D_d^j(m,n)\}\} =< f(m,n), 2^{2j}H_x\{H_y\{\Psi_d(2^j(x-m), 2^j(y-n))\}\} >$, we have

$$DG_1^j(m,n) = D_d^j(m,n) - H_y\{H_x\{D_d^j(m,n)\}\} + i(H_x\{D_d^j(m,n)\} + H_y\{D_d^j(m,n)\})$$

To study the orientation selectivity of $DG_1^j(m,n)$, we note that

$$\hat{DG}_1^j(u,v) = \begin{cases} 4\hat{D}_d^j(u,v) & \text{if } u \leq \frac{N}{2} \text{ and } v \leq \frac{N}{2} \\ 0 & \text{otherwise} \end{cases}$$

where we adopt the convention that $u, v \leq \frac{N}{2}$ for positive frequency and $u, v > \frac{N}{2}$ for negative frequency. However, it is known that the distribution of $|\hat{D}_d^j(u,v)|$ is dominant at or close to the $\frac{\pi}{4}$ and $\frac{3\pi}{4}$ axes. Hence the distribution of $|\hat{DG}_1^j(u,v)|$ is only possible to be dominant at or close to $\frac{\pi}{4}$ axis. In other words, $DG_k^1(m,n)$ only gives strong response to spatial structures in or close to the direction[2] $\frac{\pi}{4}$. $DG_3^j(m,n)$ can be proved in a similar

[2] In other words, it gives strong response to spatial structures whose maximum gradient is in, or close to $\frac{3\pi}{4}$.

way.

□

Theorem: 5.1

1. Using Eqn. 14, 15, we have

$$Re\{DG_1^j(m,n)\} = D_d^j(m,n) - H_y\{H_x\{D_d^j(m,n)\}\}$$
$$Re\{DG_3^j(m,n)\} = D_d^j(m,n) + H_y\{H_x\{D_d^j(m,n)\}\}$$

Adding them together,

$$D_d^j(m,n) = \frac{1}{2}(Re\{DG_1^j(m,n)\} + \{DG_3^j(m,n)\})$$
$$= \frac{1}{2}(\rho_{x+y}\cos(\varphi_{x+y}) + \rho_{x-y}\cos(\varphi_{x-y}))$$

2. For each scale j, applying Parseval's theorem, we have

$$\sum_m\sum_n |D_d^j(m,n)|^2 = \sum_u\sum_v |\hat{D_d}^j(u,v)|^2$$
$$= \sum_u\sum_v |\hat{f}(u,v)\bar{\hat{\Psi}}_d(2^{-j}u, 2^{-j}v)|^2$$
$$= \sum_u\sum_v |\hat{f}(u,v)|^2 |\hat{\Psi}_d(2^{-j}u, 2^{-j}v)|^2 \qquad (26)$$

Considering the Fourier transform of the wavelet Ψ_d, we have

$$|\hat{\Psi}_d(2^{-j}u, 2^{-j}v)|^2 = \frac{1}{16}|\sum_{k=1}^{4} \hat{G}_k(2^{-j}u, 2^{-j}v)|^2$$
$$= \frac{1}{16}(\sum_{k=1}^{2} |\hat{G}_k(2^{-j}u, 2^{-j}v)|^2 + \sum_{k=3}^{4} |\hat{G}_k(2^{-j}u, 2^{-j}v)|^2$$
$$= \frac{1}{8}(|G_1(2^{-j}u, 2^{-j}v)|^2 + |G_3(2^{-j}u, 2^{-j}v)|^2) \qquad (27)$$

Substituting Eqn. 27 to Eqn. 26,then

$$\sum_m\sum_n |D_d^j(m,n)|^2 = \frac{1}{8}\sum_u\sum_v |\hat{f}(u,v)\hat{G}_1(2^{-j}u, 2^{-j}v)|^2 + |\hat{f}(u,v)\hat{G}_3(2^{-j}u, 2^{-j}v)|^2)$$
$$= \frac{1}{8}\sum_m\sum_n (|DG_1^j(m,n)|^2 + |DG_3^j(m,n)|^2)$$
$$= \frac{1}{8}\sum_m\sum_n ((\rho_{x+y}^j(m,n))^2 + (\rho_{x-y}^j(m,n))^2)$$

□

9: Acknowledgement

This work was supported by a strategic research programme of the Defence Research Agency. We acknowledge many interactions with Margaret Varga, Paul Ducksbury, Phil

Kent. and John Radford. Michael Brady thanks INRIA for hosting his sabbatical, in particular Nicholas Ayache, Marc Berthod, Oliver Faugeras, and Zhengyou Zhang. Thanks also to Robyn Owens and Andrew Zisserman for comments on the ideas presented here. We really acknowledge the inspiration we have had from Azriel Rosenfeld's work over the past quarter century.

References

[1] N. Ahuja and A. Rosenfeld. Mosaic models for textures. *IEEE PAMI*, PAMI-3:1–11, 1981.

[2] R. Bajcsy and L. Lieberman. Computer identification of visual surfaces. *Computer Graphics and Image Processing*, pages 118–130, 1973.

[3] J. Beck. Similarity grouping and peripheral discriminability under uncertainty. *Am. J. Psychol.*, pages 1–19, 1972.

[4] J. Beck et al., editors. *Organization and Representation in Perception, ch. Textural Segmentation*. Erlbaum, Hillsdale, NJ, 1982.

[5] J. Beck, J. Hope, and B. Rosenfeld. *Human and Machine Vision*. Academic Press, 1983.

[6] A. C. Bovik, M. Clark, and W. S. Giesler. Multichannel texture analysis using localized spatial filters. *IEEE PAMI*, 12:55–73, 1990.

[7] Phil Brodatz. *Textures*. Dover Publications, Mineola NY, 1966.

[8] P. H. Carter. Texture discrimination using wavelets. In *SPIE applications of digital image processing XIV*, volume 1567, pages 432–438, 1991.

[9] R. Chellappa and S. Chatterjee. Classification of textures using gaussian markov random fields. *IEEE Trans. on Acoustic Speech Signal Processing*, 33(4):959–963, 1985.

[10] J. G. Daugman. Uncertainty relation for resolution in space, spatial frequency, and orientation optimized by two-dimensional visual cortical filters. *J. Opt. Soc. Am.*, 2:1160–1169, 1985.

[11] H. Derin and H. Elliot. Modeling and segmentation of noisy and textured images using gibbs random fields. *IEEE Trans. on Pattern Analysis and Machine Intelligence*, 9(1):39–55, 1987.

[12] D. Gabor. Theory of communication. *Journal of IEE*, 93:429–457, 1946.

[13] Robert M. Haralick, K.Shanmugam, and I. Dinstein. Textural features for image classification. *IEEE SMC*, SMC-3(6):610–621, 1973.

[14] Robert M. Haralick and Linda Shapiro. *Computer and Robot Vision*. Addison-Wesley Publishing Company, 1992.

[15] A. K. Jain and F. Farrokhnia. Unsupervised texture segmentation using gabor filters. *Pattern Recognition*, 24:1167–1186, 1991.

[16] A. Laine and J. Fan. Texture classification by wavelet packet signatures. *IEEE PAMI*, 15(11):1186–1190, 1993.

[17] J. Malik and P. Perona. A computational model for texture segmentation. In *IEEE Proceeding of Computer Vision and Pattern Recognition*, 1989.

[18] Jitendra Malik and Pietro Perona. Preattentive texture discrimination with early vision mechanisms. *Journal of the Optical Society of America A*, 7(5):923–932, 1990.

[19] Stephane G. Mallat. Multifrequency channel decompositions of images and wavelet models. *IEEE ASSP*, 37:2091–2110, 1989.

[20] B. S. Manjunath and R. Chellappa. Unsupervised texture segmentation using markov random field models. *PAMI*, 13(5):478–482, 1991.

[21] Uwe Mussigmann. Homogeneous fractals and their application in texture analysis. *Fractals in the Fundamental and Applied Sciences, edited by H.-O. Peitgen, J.M. Henriques and L.F. Penedo*, pages 269–281, 1991.

[22] Alex P. Pentland. Fractal-based description of natural scenes. *IEEE PAMI*, PAMI-6(6):661–674, 1984.

[23] A. Rosenfeld and A. C. Kak. *Digital Picture Processing*. Academic Press, 1982.

[24] A. Rosenfeld and E. Troy. Visual texture analysis. *Technical Report 70-116, University of Maryland, College Park*, 1970.

[25] J.S. Weszka, C. R. Deya, and A. Rosenfeld. A comparative study of texture measures for terrain classification. *IEEE Trans. on Systems, Man and Cybernetics*, SMC-6:269–285, 1976.

[26] Z. Xie. *Multi-scale Analysis and Texture Segmentation*. PhD thesis, University of Oxford, 1994.

[27] Zhi-Yan Xie and Michael Brady. Fractal dimension image for texture segmentation. In *Proceedings of 2nd International Conference on Automation, Robotics and Computer Vision*, volume 1, pages CV–4.3.1 to CV–4.3.5, 1992.

[28] Zhi-Yan Xie and Michael Brady. A decoupled local energy and phase representation of a wavelet transform. In *VCIP'95 (Visual Communications and Image Processing)*, 1995.

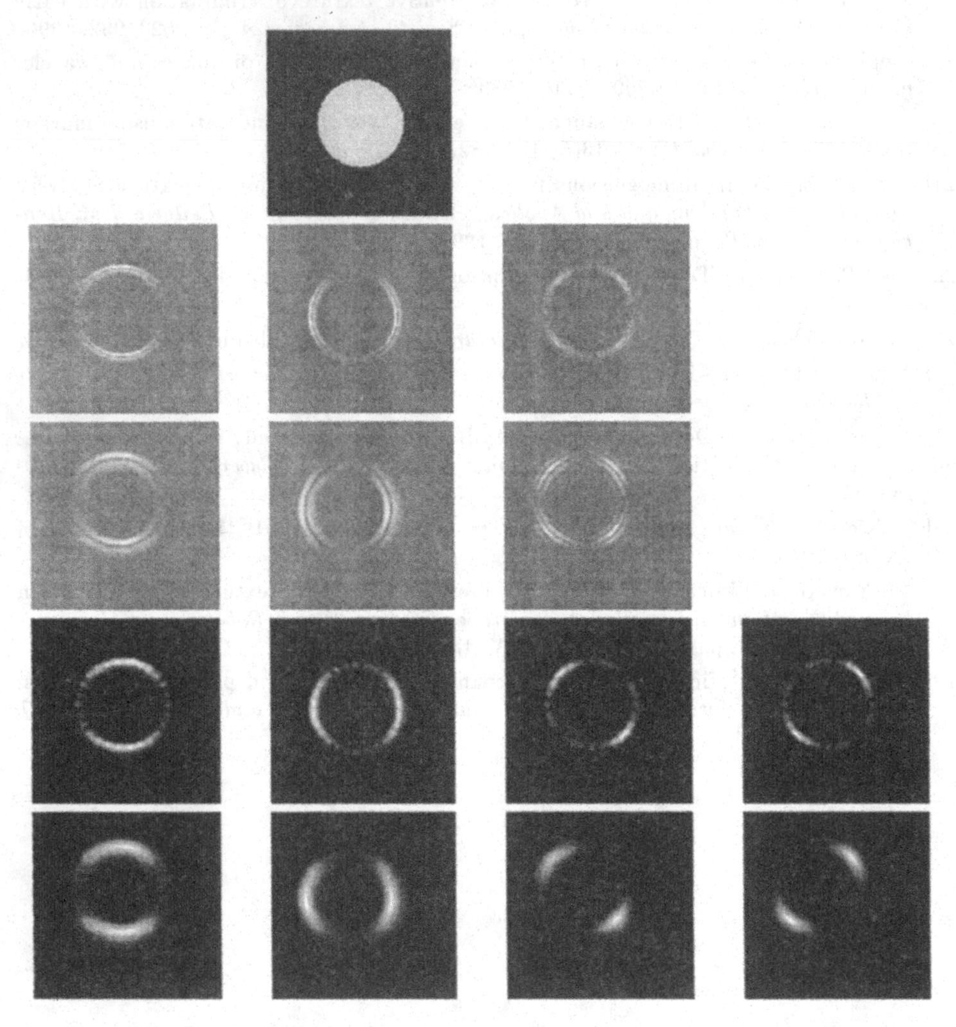

Figure 1. The wavelet decompositions of a synthetic image. top to bottom (from left to right): original synthetic image; the detail images at $j = -2, j = -3$; the local energy images at scale $j = -2, -3$, respectively. The Daubechies wavelet of length 4 is used throughout.
ordinary paragraph.

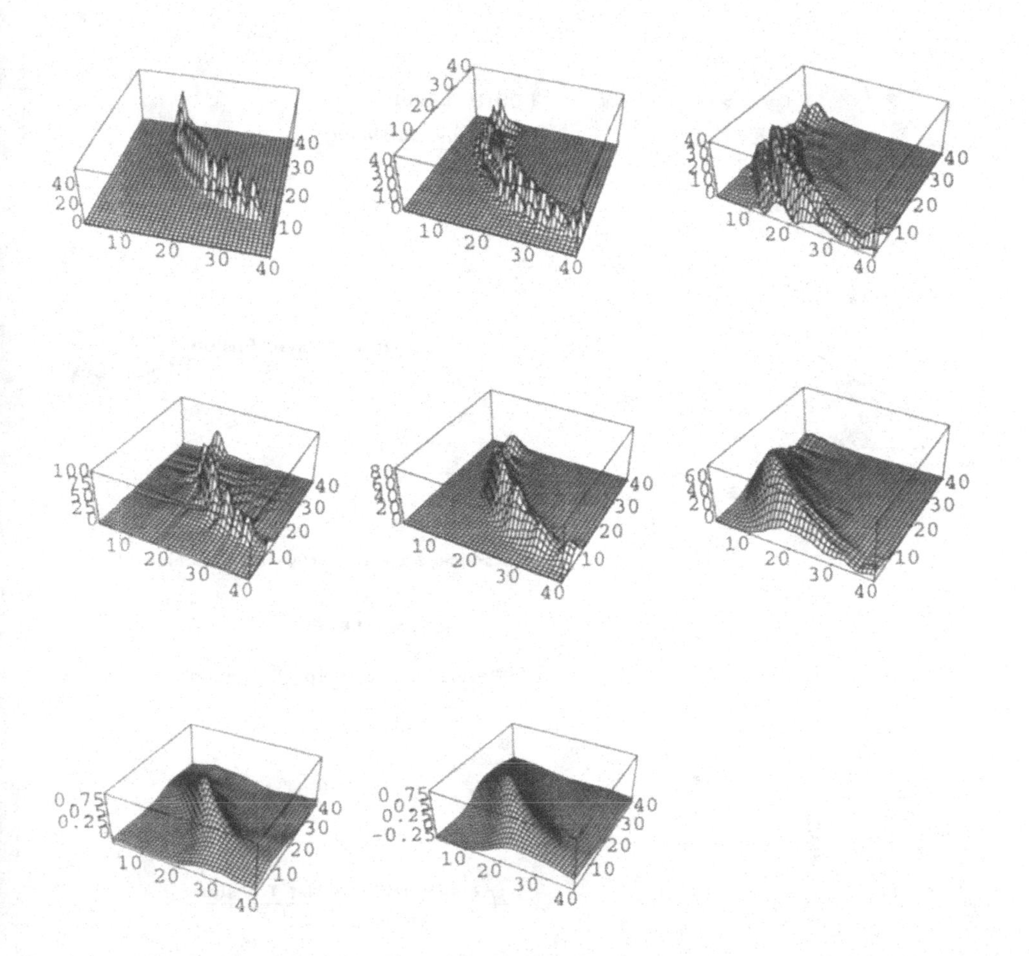

Figure 2. Comparison between the modulus and the local energy of the wavelet transform. Top row (left to right): the plot of patches of $|D^j_d|$ of the synthetic image at scales $j = -1, -2, -3$, respectively; middle row: the plot of the local energy ρ^j_{x-y} for the same patch at scale $j = -1, -2, -3$, respectively; bottom row: the plot of the linear correlation coefficients of two local energy images at adjacent scales, $j = -1, -2$ and $j = -2, -3$ respectively.
ordinary paragraph.

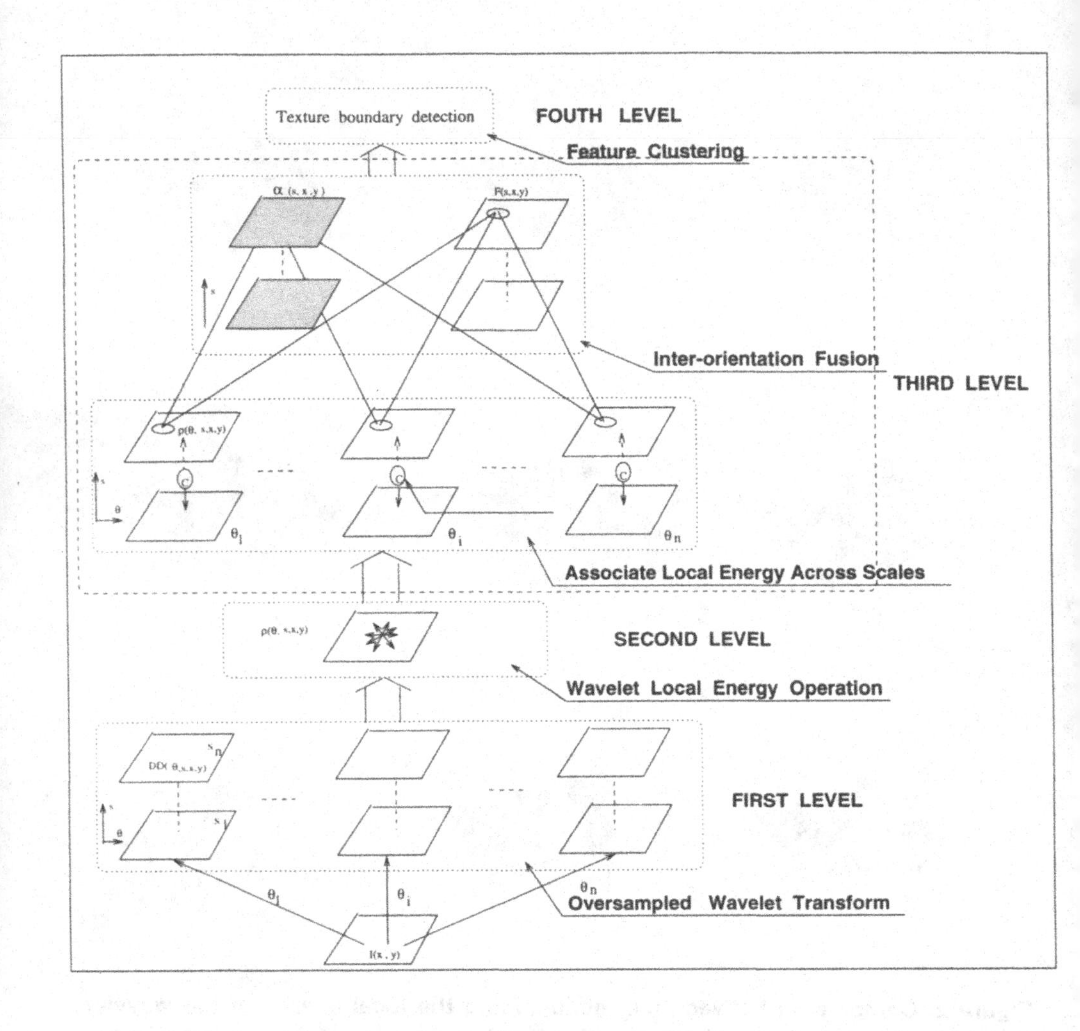

Figure 3. A schematic view of texture segmentation system in wavelet scale-space. The 1st level is to represent an image by a set of wavelet detail images $D(\theta, s, x, y)$, where each column and row is corresponding to the wavelet representation at different orientation and scale, respectively; the 2nd level is to derive a set of local energy images $\rho(\theta, s, x, y)$ by applying the local energy operation to each of the images; the 3rd level is to derive two feature images $F(s, x, y)$ and $\alpha(s, x, y)$ at different scales; the 4th level is to detect texture boundaries.
ordinary paragraph.

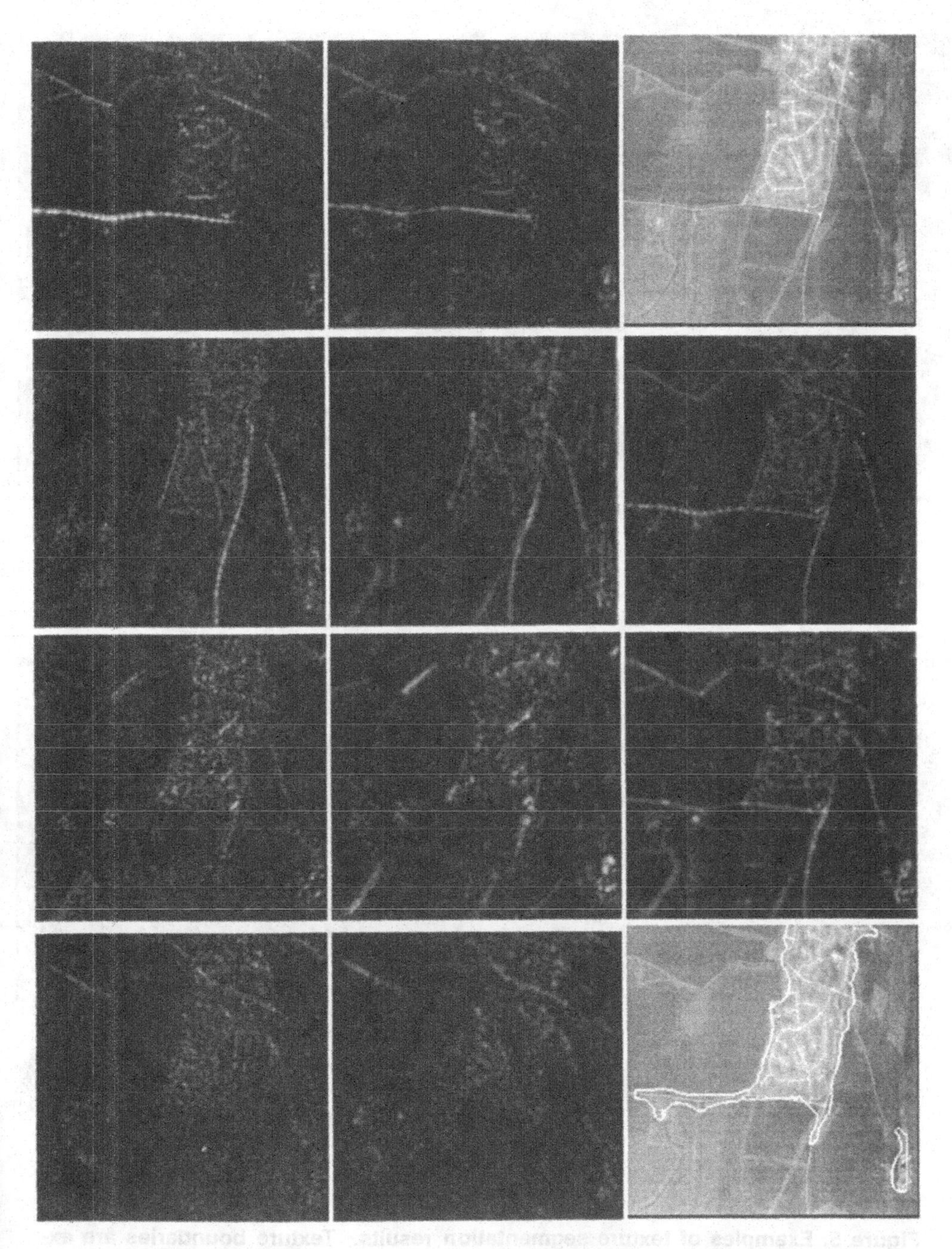

Figure 4. The left (middle) column from top to bottom: the local energy images $\rho(\theta, s, x, y)$ at scale $j = -1$ ($j = -2$) in horizontal, vertical, $\frac{\pi}{4}$ and $\frac{3\pi}{4}$ directions; the right column (top to bottom): original image 256×256, the combined local energy image $F(s, x, y)$ at scale $s = 2^{-j}, j = -1, -2$, respectively, where $c = 3.0$ and the segmentation result.
ordinary paragraph.

Figure 5. Examples of texture segmentation results. Texture boundaries are extracted and superimposed on their original images. Top row: Urban regions have been extracted for real IRLS aerial images; bottom row (left to right): Urban regions have been extracted for a satellite image, cotton canvas (in right bottom corner) have been picked up from woolen cloth background for a Brodatz montage image. ordinary paragraph.

A Transform For Multiscale Image Segmentation

Narendra Ahuja
Beckman Institute and Coordinated Science Laboratory
University of Illinois at Urbana-Champaign

Abstract

This paper describes a new transform to extract image regions at all geometric and photometric scales. It is argued that linear approaches such as convolution and matching have the fundamental shortcoming that they require *a priori* models of edge geometry. The proposed transform avoids this limitation by letting the structure emerge, bottom-up, from interactions among pixels, in analogy with statistical mechanics and particle physics. The transform involves global computations on pairs of pixels followed by vector integration of the results, rather than scalar and local linear processing. An attraction force field is computed over the image in which pixels belonging to the same region are mutually attracted and the region is characterized by a convergent flow. It is shown that the transform possesses properties that allow multiscale segmentation, or extraction of original, unblurred structure at all different geometric and photometric scales present in the image. This is in contrast with much of the previous work wherein multiscale structure is viewed as the smoothed structure in a multiscale decomposition of image signal. Scale is an integral parameter of the force computation, and the number and values of scale parameters associated with the image can be estimated automatically. Regions are detected at all *a priori* unknown scales resulting in automatic construction of a segmentation hierarchy, in which each pixel is annotated with descriptions of all the regions it belongs to. Although some of the analytical properties of the transform are presented for piecewise constant images, it is shown that the results hold for more general images, e.g., those containing noise and shading. Thus the proposed method is intended as a general approach to multiscale, integrated edge and region detection, or low-level image segmentation.

1 Introduction

This paper is concerned with the problem of low level image segmentation, or partitioning of an image into homogeneous regions, that represent low level image structure. A region can be characterized as possessing a certain degree of interior homogeneity and a contrast with the surround which is large compared to the interior variation. This is a satisfactory characterization from both perceptual and quantitative viewpoints. The type of homogeneity (constancy, smoothness) and the magnitude of the contrast may vary, and the regions may have arbitrary size and shape. Different values of allowed homogeneity and contrast lead to different partitioning of image into regions. As greater homogeneity and lower contrast values are allowed, new, smaller regions emerge in the partition. Thus, a decrease in the minimum acceptable contrast leads to increased, hierarchical decomposition of the image, which culminates in constant-value regions at the bottom of the hierarchy. The depth of the hierarchy and the values of region homogeneity, contrast, shape and size parameters associated with the different levels vary across the image. Features such as

0-8186-7644-2 $5.00

depth and branching factor of the tree thus defined are unrelated across subtrees, each solely determined by the image and therefore *a priori* unknown. The homogeneity and contrast parameters associated with different image regions will be said to form the set of photometric scales present in the image, while the region shapes and sizes will be said to define the geometric scales present.

Finding a solution of the low level segmentation problem poses two main challenges. First, a valid image region must be detected regardless of its shape, size, degree of homogeneity, and contrast. Second, all geometric and photometric scales at which regions happen to occur across an image must be identified. If these two problems are solved, the result will be a segmentation tree representing the multiscale, low level, image structure. To obtain such a tree for an arbitrary image using intensity based homogeneity and contrast is the objective of image segmentation pursued in this paper.

Limited work has been done to meet both of the above challenges. Much of the previous work on multiscale analysis is concerned with a scale-space decomposition of the image signal, determined by a single scale parameter. The decomposition amounts to a blurring of the image to different degrees. The image structure is present across this scale-space continuum and the extraction of image regions of different sizes and contrasts from this continuum is not addressed. Further, even if they were extracted, the regions in the different decompositions would be correspondingly smoothed. Automatic estimation of scale parameters is typically not addressed. Even at a given scale, robust detection of a region continues to be an area of active investigation, mainly through the work on edge detection. Region detection such that the detected boundaries are closed and coincident with the true region boundaries regardless of region parameters is not a solved problem. Most methods are linear and often use restrictive region models, e.g., allowed geometric and photometric complexity of edges. Although these models simplify processing, they cause fundamental limitations in the detection accuracy and sensitivity achieved which is partly why the problem of region and edge detection continues to evade a satisfactory solution.

A central theme of this paper is to achieve both of the above goals of accurate region detection and automatic scale estimation. This is accomplished by introducing a new transform which converts an image into another image having two major properties. First, for a specific pair of scale values, the transform leads to well-defined signatures of corresponding image regions which are easy to detect. The transform definition incorporates the duality of interior- and edge-based descriptions of regions. Thus, the transform performs integrated edge and region detection, and can be viewed as a multiscale blob and edge detector at the same time. Second, all scale values for which regions occur in the image can be computed by analyzing the results of the transform as the scale values used are varied.

A major motivation for the transform comes from a desire to develop a segmentation method which is not rigidly bound to specific models of regions, e.g., models of spatial variation. It is desired that groupings of pixels that reflect a smooth photometric variation and stand in relative contrast to their surround be detected as regions, regardless of the exact nature of variation and contrast values. This appears necessary to deal with the variety characteristic of real imagery. To achieve such performance, the transform computes affinities among image *points* or *pixels* for grouping with other pixels, letting the structure emerge bottom-up from "interactions" among the pixels instead of imposing *a priori* chosen models of region edges and interior. As one consequence of this, the emergent region geometry is not restricted, since pixels can group together to form any connected set.

The transform computes a family of force fields for a given image where the force vector at a point denotes its affinity to the rest of the image. On either side of a region boundary, the pixels have high affinities, but there is little affinity between pixels across regions.

The strength of interaction between pixels, and consequently their affinities, depend upon their distances and contrasts, and this allows association of the computed affinities and segmentation with spatial and photometric scales. Since the transform allows interaction between a pixel and all other pixels, it can be viewed as collecting globally distributed evidence for image structure and making it available locally, e.g. at the locations of region edges and medial axes (The medial axis of a region is defined as the locus of points inside the region which are equidistant from two or more points on the region boundary [6, 17]). The regions are encoded in the force field via distinct signatures amenable to robust, local identification. In this sense, the transform performs Gestalt analysis.

This paper introduces the transform and shows how it can be used for segmentation (the basic idea of the transform can be found in [1]). It does not present specific segmentation algorithms. The segmentation is intended to represent low-level image structure at all scales, thus with applicability to textured as well as smooth images. To analyze and illustrate the basic properties of the transform, we model regions, whenever necessary, as possessing uniform gray levels and step edges. However, the transform properties and segmentation results are shown to apply to images containing general types of regions as discussed above, e.g., having shading and noise. Section 2 discusses some basic desired characteristics of segmentation and how they motivate the proposed approach. Section 3 describes the transform, describes some of its properties of interest, and shows how these properties facilitate multiscale segmentation. Section 4 analyzes the segmentation performance of the transform. Section 5 presents concluding remarks.

2 Background and Objectives

In this section, we first discuss past work on the two major subproblems of image segmentation: structure detection at a single scale and multiscale analysis (Sec. 2.1). This leads us to formulate the characteristics desired in a satisfactory segmentation (Sec. 2.2).

2.1 Two Aspects of Image Segmentation

We will first review the past work on photometric and geometric models of a region used for segmentation at a single scale. [26] estimates a 2D functional that minimizes a cost function comprised of the difference between the images and estimated intensity values, the length of detected edges, and the variation in the functional away from edges, which are combined using *a priori* chosen relative weights. Morphological methods are used in [23] to detect regions as intensity hills in grayscale landscape. Although, a region can be detected by identifying either its interior pixels or edges, the latter method has been investigated more extensively. An edge separates two different regions and thus two different types of gray level populations. Edge detection methods use different models of edge geometry, and gray level variation along edge as well as within region. These models are fitted to local pixel populations to determine if an edge is present or not. Such local responses are then combined to derive a more global segmentation. Clearly, the validity of the models of the edge as well the gray level populations are critical factors in achieving a valid segmentation. We will now review some models used in the previous work. It is common to treat the problem of edge detection as mainly that of selecting a point along the intensity profile across edge, assuming such a profile can be extracted from the image. Accordingly, a model of the intensity profile is used to precisely define an edge and to optimally detect its location. Different types of intensity models of an edge have been proposed, according to the nature of the two populations and the spatial profile of the transition from one to the other across the edge [30, 29, 5, 11, 14]. To meet the assumption that edge profile through a pixel can be identified, it is common in edge detection work to implicitly or explicitly use a model of edge curvature. The use of such geometric model constrains the number of possible, different subdivisions of the pixel neighborhood (e.g., a 3x3 or a 5x5) into two regions which must be analyzed to detect the presence of an edge in the neighborhood. For example, the assumption of local straightness of edge is common which makes it very easy to select neighborhoods on the two sides of the edge. [22] assumes that the edge is locally straight (and that the intensity changes linearly along a direction parallel to the edge.) Nalwa and Binford [27] assume straightness to extract a sample edge profile. Even the computation of gradient which is common to many edge detectors, e.g., the diffusion based methods [21, 34],romeney), implicitly assumes local edge straightness. The same can be said about Laplacian based edge detectors. The use of straightness is very explicit in the different types of discrete edge masks each of which is meant to detect a different edge orientation [31]. To detect intensity facets meeting at an edge [13], a model of edge geometry is required so candidate neighborhoods from each side of the edge can be identified. The work on optimal edge detection (e.g., [9]) is also subject to the validity of the assumed model of the edge geometry. In short, the work on edge detection has lead to different approaches to estimate edge location and orientation for edges having some (implicitly or explicitly) assumed local curvature properties. Image edges do not always conform to these assumptions, and deviations lead to detection errors. Examples of such errors incurred using the Laplacian-of-Gaussian operator for different edge geometries can be found in [4]. To avoid some of these problems, [12] uses the Markov random field model to obtain an estimate similar to that in [26] but allows for end points, corners and junctions in the edge models used. Another example of an approach that avoids the dependence on geometric models of edge is given in [28] where interpixel correlations in spatio-temporal space are considered instead of interwindow correlations.

The second major aspect of segmentation is related to scale. As we stated earlier, scale as pursued in this paper is associated with both geometric and photometric sensitivity to detail. Thus, a pixel may simultaneously belong to different regions each having a different contrast value and size, giving rise to the tree representation mentioned earlier. Large regions may be said to have a coarse spatial scale while smaller sizes may be said to be associated with finer spatial scales. Analogously, an edge contour which separates two regions of a given contrast scale may not be detected at a higher scale associated with a larger contrast. The exact number and parameters of scales for a given image are *a priori* unknown. Therefore, multiscale segmentation must automatically estimate these parameters and detect the corresponding regions. Although the general notion of multiscale operators has been examined for a long time [32], there has been limited work on definition, analysis and automatic estimation of multiscale image structure as pursued in this paper. Our objective here is to separate *original* (unsmoothed) image structure at different scales (regions with different sizes and contrasts), as well as identify the spatial and topological relationships among the regions. We obtain a multiscale structural decomposition of the image, and not a multiscale decomposition of the image signal as performed in much of the past work [18, 35, 36, 20] where coarse scale structure is detected from blurred images and is therefore a smoothd version of the original image structure. In addition to accruing edge displacement error [4], the latter leads to artifcats due to multiscale analysis, e.g., phantom edges [10]. Among other approaches to multiscale segmentation, multiscale blob detection using morphological methods is described in [7, 19], and computation of multiscale medial axis representation is discussed in [25, 3, 17].

2.2 Desired Characteristics and Objectives

The above discussion leads us to the following desired characteristics of multiscale structure detection and segmentation.

A. Shape and Topology Invariance: The regions should be correctly detected regardless of their shapes and relative placement. For example, an edge point must be detected at only one and the correct location, regardless of whether the edge in the vicinity of the point is straight, curved or even contains a corner or a vertex where multiple regions meet.
B. Photometric Scaling: It should be possible to detect all regions which are in contrast to their surround, regardless of the actual degree of within-region homogeneity and the value of the contrast. Regions having large contrast may be associated with higher scales.
C. Spatial Scaling: It should be possible to detect all regions regardless of their shapes and sizes. Higher scales may be associated with larger regions.
D. Stability and Automatic Scale Selection: Image structures associated with different scales correspond to segmentations that are locally invariant to changes in geometric and contrast sensitivities. Since the contrasts and sizes of regions contained in an arbitrary image are *a priori* unknown, they should be identified automatically.

The transform presented in this paper has been motivated by the objective of achieving these desired characteristics. Specifically, the objective is to derive multiscale segmentation of the image and represent it through a hierarchical, tree structure in which the different image segments, their parameters, and their spatial interrelationships are made explicit. The bottom (leaf) nodes of the hierarchy correspond to regions consisting of individual pixels or connected components of constant gray level, and the path from a leaf to the root node specifies how the leaf regions recursively merge with adjacent regions to form larger regions each of which is homogeneous with respect to its surround and is characterized by its own contrast. Alternate representations of the same image structure and contrast information are also possible, e.g., by ordering regions according to contrast. In this paper, we will not dwell on the different possible data structures that could be used for representation. Rather, we will demonstrate how the transform extracts information about the *a priori* unknown region geometries, homogeneities and contrasts, and associates this structural information with each image pixel. Any specific image representation may be constructed from such annotated pixel array.

3 The Transform and Image Segmentation

In this section we first discuss how the problems with the previous methods and the desired characteristics of the segmentation motivate an approach such as that underlies the proposed transform (Sec. 3.1). We then introduce the transform (Sec. 3.2) and present its properties (Sec. 3.3) that demonstrate how the transform makes explicit the image structure and facilitates image segmentation, and why the resulting segmentation possesses the desired characteristics listed in Section 2.2. Section 3.4 describes how a given image region appears in the vector field computed by the transform. In Sec. 3.5, we discuss the estimation of the unknown scale parameters associated with an image, which are required to extract the unknown structures present in different parts of the image. Sec. 3.6 describes the hierarchical image structure that is extracted as the final result of segmentation. Whenever necessary, we will consider constant-value regions to analytically and qualitatively describe the transform behavior. However, the properties of the transform responsible for its segmentation capability remain valid for images containing more general types of regions, such as those having noise (statistical constancy) and shading (smoothly varying values or higher-order homogeneity) as will be explained in Sec. 4. Thus, the transform is proposed for use in segmentation of piecewise smooth images.

3.1 Overcoming Limitations of Linear Processing via the Transform

In the literature, different models of edge profile (step, ramp, roof) and their validity have been investigated. However, the limitations and impact of the assumptions made about edge geometry have received limited attention. Since any convolution kernel for edge detection must incorporate a template for the expected edge geometry, no linear, convolution based approach would avoid the limitations resulting from the use of geometric models of edge. In the digital case, one could attempt to circumvent the problem by enumerating all possible edge geometries in a neighborhood. But the number of resulting kernels will fast increase with neighborhood size and will be prohibitively large for any reasonable size neighborhood.

The inspiration for the proposed solution comes from physics where microscopic homogeneity of physical properties leads to islands of, say, similar particles or molecules. An island shape is congruent with the space occupied by a set of contiguous, similar particles, whatever the complexity of the boundary! The particles group together and coalesce into regions based on the similarity of their intrinsic properties only, regardless of their relative locations. The common property of particles then characterizes the region they form. As an alternate analogy, the grouping process is like the alignment of microscopic domains over an area of ferromagnetic material. The key process is that of interaction among particles which leads to bindings among similar particles.

The problem of segmentation has similarities to the above physical process. The goal is to find a partition of the image, regardless of the boundary complexity, such that each cell of the partition has a characteristic property, say, homogeneity of gray level. This analogy suggests a formulation of the segmentation process in terms of a suitably defined method of interpoint interaction - one that would group, bottom-up, each set of points of the same property to form a region having a boundary of any complexity. Being a parameter of the grouping process, different acceptable degrees of the presence of the property within a region would yield groupings over different regions, making scale an integral part of structure detection.

In the next section, we introduce a transform which achieves the above grouping.

3.2 The Transform

The transform converts the image I into a vector field $\mathbf{F}$. The vector $\mathbf{F_p}$ at an image location $\mathbf{p}$ is defined as

$$\mathbf{F_p} = \int_{\mathbf{q}\neq\mathbf{p}} d_s(\mathbf{r_{pq}}, \sigma_s(\mathbf{p})) d_g(\Delta I, \sigma_g(\mathbf{p})) \hat{\mathbf{r}}_{\mathbf{pq}} \mathbf{dq} \tag{1}$$

where

$\hat{\mathbf{r}}_{\mathbf{pq}}$ = unit vector in the direction from $\mathbf{p}$ to another image location $\mathbf{q}$;

$\sigma_s(\mathbf{p})$ = spatial scale parameter at $\mathbf{p}$; related to the shortest distance to region boundary; all valid $\sigma_s(\mathbf{p})$ values are computed automatically;

$\sigma_g(\mathbf{p})$ = photometric scale parameter at $\mathbf{p}$; denotes contrast of region with surround; all valid $\sigma_g(\mathbf{p})$ values are computed automatically;

ΔI = absolute gray level difference between image points under consideration;

$d_s(\mathbf{a}, b)$ = A nonnegative, nonincreasing function of $||\mathbf{a}||$, not identically 0 for $||\mathbf{a}|| \leq b$, and 0 for $||\mathbf{a}|| > b$, and

$d_g(a,b) =$ A nonnegative, nonincreasing and symmetric function of a, not identically 0 for $a \leq b$, and 0 for $a > b$.

Since $d_g(a,b)$ as defined above cannot be a linear function of a for unrestricted values of a, the transform is a function of image input.

3.2.1 Formulation of the Transform

Having stated the definition of the transform, we will now explain the motivation behind this formulation, and construct the definition from basic principles.

Consider a region R and a point P inside it. It is desired that the similarity of P to all the other points within R, relative to those outside R, be recognized, regardless of the distance to and curvature of the nearest region boundary. To do this, the transform defines a neighborhood around P which is sufficiently small so that the pixels within the neighborhood which have the most influence on the computation at P are within R. The similarity is estimated by comparing the gray level at P with those of the points in the neighborhood, rather than testing some (position independent) gray level statistics of points withing the neighborhood, or comparing the statistics of different sets of pixels near P. An estimate of the local shape of R near P, specifically of the nearest region border, is computed. This is achieved by a vector integration of the results of pairwise comparisons of points at different orientations, instead of computing a scalar, weighted average at P as is done in linear methods. Contiguous points having compatible estimates of local shape of R are grouped together. This results in the detection of an arbitrarily large region of arbitrary shape.

To capture the local region geometry, the transform computes an attraction-force field over the image wherein the force at each point denotes its affinity to the rest of the image. The force vector points in the direction in which the point experiences a net attraction from the points in the rest of the image. For example, a point inside a region would experience a force towards the interior of the region. This force is computed as the resultant of attraction-forces due to all other image points. If $F(\mathbf{p},\mathbf{q})$ denotes the magnitude of the force vector $\mathbf{F}(\mathbf{p},\mathbf{q})$ with which a pixel P at location $\mathbf{p}$ is attracted by another pixel Q at location $\mathbf{q}$, then the transform is given by

$$\mathbf{F}(\mathbf{p},\mathbf{q}) = F(\mathbf{p},\mathbf{q})\hat{\mathbf{r}}_{\mathbf{pq}},$$

where $\hat{\mathbf{r}}_{\mathbf{pq}}$ denotes the unit vector in the direction from P to Q, i.e.,

$$\hat{\mathbf{r}}_{\mathbf{pq}} = \frac{\mathbf{q}-\mathbf{p}}{\|\mathbf{q}-\mathbf{p}\|}$$

In the real image plane, an image is transformed into a continuous vector field. The vector $\mathbf{F_p}$ at point P is given by

$$\mathbf{F_p} = \int_{\mathbf{q}\neq\mathbf{p}} F(\mathbf{p},\mathbf{q})\hat{\mathbf{r}}_{\mathbf{pq}}d\mathbf{q} \tag{2}$$

where $\mathbf{q}$ can be any image location other than $\mathbf{p}$. In the discrete case,

$$\mathbf{F_p} = \sum_{\mathbf{q}\neq\mathbf{p}} F(\mathbf{p},\mathbf{q})\hat{\mathbf{r}}_{\mathbf{pq}}$$

We need to specify what forms the force function $F(\mathbf{p},\mathbf{q})$ could take. We will do so by identifying the characteristics that any such function must possess, to yield the correct segmentation for a given pair of spatial and photometric scales as well as exhibit appropriate

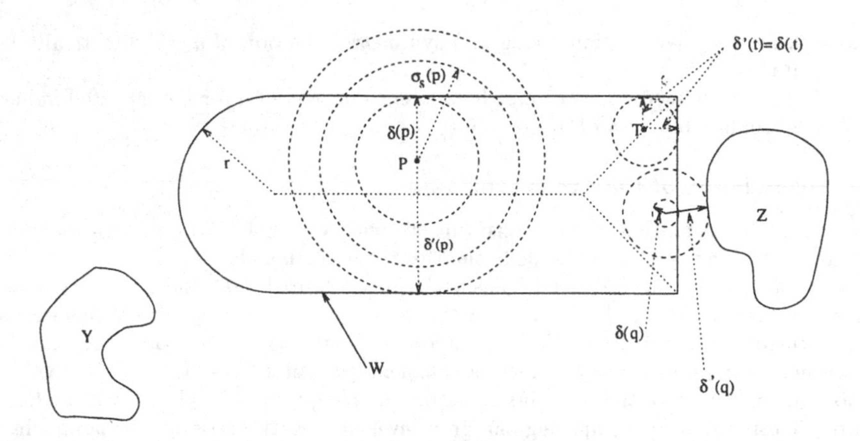

Figure 1. The medial axis of the region W is the locus of points equidistant from multiple points on W's border, shown here by dotted lines. A point P's distance to the nearest border point is given by $\delta(\mathrm{p})$. A disk D of radius $> \delta'$ may intersect with W in more than one connected components (e.g., point P), or it may intersect with other regions as well (e.g., point Q). For points along the medial axis, $\delta' = \delta$ (e.g., point T). The spatial scale parameter σ_s at any point is defined as a value between δ and δ' (radius of the middle circle for point P).

behavior across multiple scales. Specific choices of force functions having these characteristics will then define different instances of the transform.

Since the presence of an edge of a region at any given spatial scale must be determined by its adjoining regions rather than by distant points across other intervening regions, the force F exerted on a given pixel P by another pixel Q should be a nonincreasing function of the distance between P and Q. Further, a pixel should be attracted more to a pixel within its own region than to one in a different region, and the attraction between two pixels should depend only on the magnitude of their gray level difference and not its sign. This is accomplished by making F to be a nonincreasing and symmetric function of the difference between the gray levels of P and Q.

Since both spatial and photometric vicinities are relative to the scales of interest, let us now consider how to integrate the scale information in the computation of F. Consider a region W which exists amidst many other regions in an image (Fig. 1). To detect any structural characteristics of W, the computation of F at a point must not extend to nonlocal parts of W or to other regions, since otherwise the result will depend upon nonlocal structure of W or multiregion structure. To be specific, first consider the situation where a point P is inside W but not on its medial axis. Let $\delta(\mathbf{p})$ denote the distance to the unique border point (or border segment) of W closest to P, and let D denote a disk of radius $r = \delta(\mathbf{p})$ centered at P. If r is increased, for some value $r = \delta'(\mathbf{p}) > \delta(\mathbf{p})$, D's intersection with W will begin to consist of multiple disconnected regions or D will begin to intersect with other regions near W (e.g., regions Y and Z in Fig. 1). Now consider the second situation where the point P is along the medial axis of W. For any such point, $\delta'(\mathbf{p}) = \delta(\mathbf{p})$ since a disk of radius $> \delta(\mathbf{p})$ will intersect W in multiple disconnected components, and possibly other regions. In order that $\mathbf{F_p}$ reflects the structure of W surrounding P, it must be ensured that the contribution to F at $\mathbf{p}$ from points at a distance $\geq \delta'(\mathbf{p})$ is negligible. Such spatial locality of computation is enforced by including a spatial (or geometric) scale parameter σ_s in the definition of F. $\sigma_s(\mathbf{p})$ associates a "cut-off distance" with $\mathbf{p}$ such that the points farther than this distance make negligible contribution to F value at $\mathbf{p}$. In particular, we define the spatial scale parameter $\sigma_s(\mathbf{p})$ at each point $\mathbf{p}$ in some region R as having a value given by $\sigma_s(\mathbf{p}) = \delta(\mathbf{p}) + \alpha(\delta'(\mathbf{p}) - \delta(\mathbf{p}))$, where $0 < \alpha < 1$. Thus, $\delta(\mathbf{p}) < \sigma_s(\mathbf{p}) < \delta'(\mathbf{p})$ at all points inside R except along the medial axis where $\sigma_s(\mathbf{p}) = \delta(\mathbf{p}) = \delta'(\mathbf{p})$. To achieve the desired monotone dependence of F on the spatial scale as well as on the distance as discussed in the previous paragraph, we make it proportional to a function $d_s(\mathbf{r_{pq}}, \sigma_s(\mathbf{p}))$, where $d_s(\mathbf{r_{pq}}, \sigma_s(\mathbf{p}))$ is a nonnegative, nonincreasing function of $\|\mathbf{r}\|$, not identically 0 for $\|\mathbf{r_{pq}}\| \leq \sigma_s(\mathbf{p})$, and 0 for $\|\mathbf{r_{pq}}\| > \sigma_s(\mathbf{p})$.

Just as the spatial scale parameter is chosen to ensure that F at $\mathbf{p}$ depends on spatial structure in the neighborhood of $\mathbf{p}$, the photometric (contrast, or gray level) scale parameter at P is determined by the degree of gray level homogeneity and contrast of a specific region of interest containing P. It ensures that F at $\mathbf{p}$ is determined by points within the region, i.e., the point experiences negligible attraction to another point having gray level difference larger than a cut-off value characteristic of within-region gray level variability. To achieve this, we make F proportional to the function $d_g(\Delta I, \sigma_g(\mathbf{p}))$, where ΔI denotes the absolute gray level difference between points P and Q, and $\sigma_g(\mathbf{p})$ is the photometric scale parameter. $d_g(\Delta I, \sigma_g(\mathbf{p}))$ is a nonnegative, nonincreasing and symmetric function of ΔI, not identically 0 for $\Delta I \leq \sigma_g(\mathbf{p})$, and 0 for $\Delta I > \sigma_g(\mathbf{p})$.

Thus, $\mathbf{F_p}$ can be written as defined earlier in Equation (1), namely,

$$\mathbf{F_p} = \int_{\mathbf{q} \neq \mathbf{p}} d_s(\mathbf{r_{pq}}, \sigma_s(\mathbf{p})) d_g(\Delta I, \sigma_g(\mathbf{p})) \hat{\mathbf{r}}_{\mathbf{pq}} \mathbf{dq}$$

Two observations follow from the above definition of F. First, the scale parameters at different points within a region R are mutually dependent since they are all determined by the structure of R. In particular, σ_s at a point P inside a region R depends on P's location relative to R's boundary, and all points inside R have the same σ_g value which corresponds to the contrast of R with its surround. For example, in a piecewise constant image, σ_s varies continuously and σ_g is piecewise constant. Second, since the region structure is to be determined in the first place, σ_s and σ_g are *a priori* unknown to the segmentation algorithm and must be computed at each point. Since a point in general belongs to many different regions at different scales, σ_s and σ_g will have multiple values at a point.

Note that when a homogeneous region (or background) is at least partly enclosed by the image border, the computation of F will be undefined at those image points P whose nearest border point is on image boundary. This is because the computation of $\sigma_s(\mathbf{p})$ involves points at distances $> \delta(\mathbf{p})$ from $\mathbf{p}$ (Fig. 1) and some such points are outside the image. To resolve this problem, we will treat the entire image as surrounded by a hypothetical, constant-value region whose contrast relative to the given image is infinite. Accordingly, for computational purposes, the points outside the image will be assumed to be accessible but having a gray level of infinity. This would yield the given, finite size image as the largest and least homogeneous region within the hierarchical segmentation of the hypothetical, infinitely large image.

3.3 Properties

In this section, we present some properties of $\mathbf{F}$ which collectively describe the relationship between the spatial structure of $\mathbf{F}$ and the image structure, and consequently, suggest $\mathbf{F}$ as a means of image segmentation. For brevity, the proofs are outlined for only some of the properties; the rest can be found in [2].

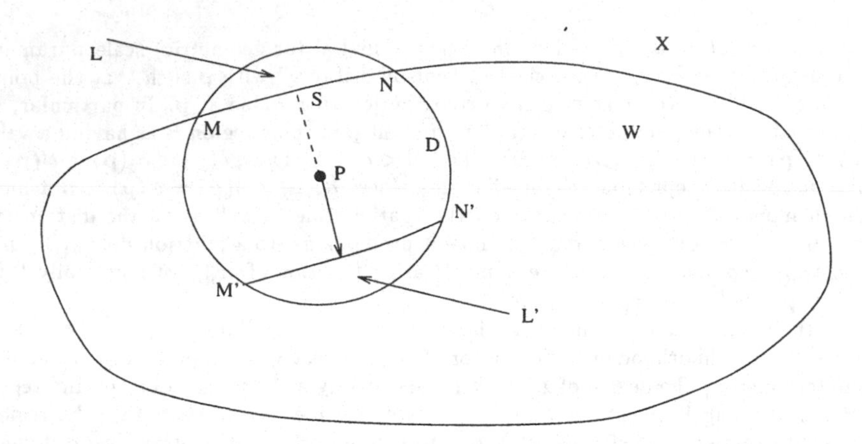

Figure 2. The force at a point P inside a homogeneous region points inward (Property 2).

1. Null Response: Suppose an image contains a constant-value disk, an arbitrary number of arbitrarily located other regions not intersecting with the disk, and a constant-value background. Then the value of $\mathbf{F}$ is zero at the disk center.

2. Inward Attraction: Let P be a point inside a homogeneous region W. Let D be the disk of radius $\sigma_s(\mathbf{p})$ centered at P which intersects with W and its border with a homogeneous region X of contrast C. Let M and N be the two points where the borders of D and W intersect (there will be exactly two such points by definition of σ_s). Then, for $\sigma_g < C$, there exists a point S along W's border within D and between the points M and N such that the direction of $\mathbf{F_p}$ at P is given by the vector from S to P, i.e., $\mathbf{F_p} = k\hat{\mathbf{r}}_{\mathbf{sp}}$, where k is a positive constant and $\hat{\mathbf{r}}_{\mathbf{sp}}$ denotes the unit vector from S to P. Thus $\mathbf{F_p}$ points inward.

Proof: Let L denote the region of intersection between D and X (Fig. 2). If we construct another region L' such that L and L' are symmetric about P, with the segment M'N' being symmetric to MN, then the force at P due to the neighborhood (D-L-L') will be 0 by symmetry (see Property 1). Therefore, the force $\mathbf{F_p}$ at $\mathbf{p}$ is due to the points in the regions L∪L' and is given by,

$$\mathbf{F_p} = \mathbf{F_{p}}_L + \mathbf{F_{p}}_{L'} \tag{3}$$

where $\mathbf{F_p}_L$ and $\mathbf{F_p}_{L'}$ denote the forces at $\mathbf{p}$ due to regions L and L', respectively. Now for all $\mathbf{q}_1 \in L$, it is given that $\Delta I(\mathbf{p}, \mathbf{q}_1) > \sigma_g$. Therefore, $d_g(\Delta I(\mathbf{p}, \mathbf{q}_1), \sigma_g) = 0$ from the definition of d_g. For all $\mathbf{q}_2 \in L'$, $\Delta I(\mathbf{p}, \mathbf{q}_2) = 0 < \sigma_g$. Since d_g is a nonincreasing function of ΔI not identically 0, and $\Delta I(\mathbf{p}, \mathbf{q}_2 \in L') = 0 < \sigma_g < \Delta I(\mathbf{p}, \mathbf{q}_1 \in L)$, we have

$$d_g(\Delta I(\mathbf{p}, \mathbf{q}_2), \sigma_g) > 0 = d_g(\Delta I(\mathbf{p}, \mathbf{q}_1), \sigma_g). \tag{4}$$

Thus $\mathbf{F_p}_L = 0$, and consequently, $\mathbf{F_p} = \mathbf{F_p}_{L'}$. Since d_s is not identically 0 and nonnegative, it follows from Equations (1) (3) and (4) that $\mathbf{F_p}_{L'} > 0$. Therefore, $\mathbf{F_p} = \mathbf{F_p}_{L'} > 0$. Since $\mathbf{F_p}_{L'}$ is the net force on P due to all points $\mathbf{q_2} \in$ L', the direction of $\mathbf{F_p}_{L'}$ must be given by the unit vector $\hat{\mathbf{r}}_{\mathbf{ps}''}$ for some point S" ∈ L'. If S' denotes a point of intersection of the line joining P and S" with the segment M'N', then the direction of $\mathbf{F_p}$ is given by

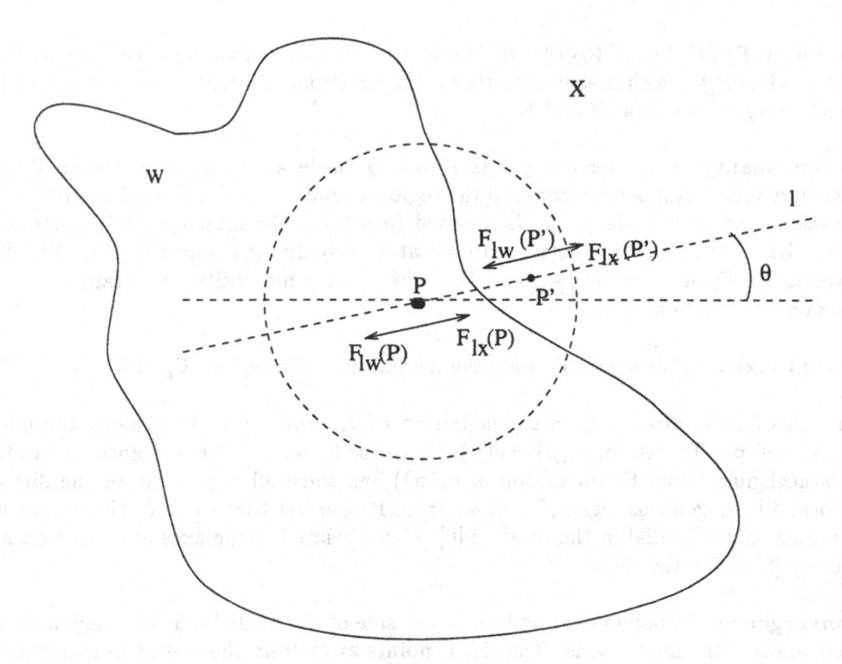

Figure 3. Divergence. If a point P in region W and another point P' in region X are infinitesimally far from each other, and hence also from the boundary between W and X, then the force vectors at P and P' are equal and opposite (Property 3).

the unit vector $\hat{\mathbf{r}}_{\mathbf{ps}'}$. Therefore, $\mathbf{F_p} = k\hat{\mathbf{r}}_{\mathbf{ps}'}$, where k is a constant. But for each point $\mathbf{s}'$ on segment M'N', there is another point $\mathbf{s}$ on segment MN where the line PS intersects the segment MN, and $\hat{\mathbf{r}}_{\mathbf{ps}'} = \hat{\mathbf{r}}_{\mathbf{sp}}$. Therefore, $\mathbf{F_p} = k\hat{\mathbf{r}}_{\mathbf{sp}}$.

3. Divergence: Consider a pair of points P and P' inside and outside, respectively, a homogeneous region W, and infinitesimally close to W's boundary. Then $\mathbf{F_p X F_{p'}} = \mathbf{0}$, i.e., the force vector undergoes direction reversal across the boundary, regardless of the shape of the boundary.

Proof: Let P' belong to region X (Fig. 3). Consider the line l through P and P', and the forces at P and P' due to points along l and within a distance $\sigma_s(\mathbf{p})(= \sigma_s(\mathbf{p}'))$. Let $F_{lW}(P)$ denote the total attraction force at P along l due to and towards the points inside W. Let $F_{lX}(P)$ denote the total attraction force at P along l due to and towards the points inside X. Analogously, let F_{lX} (P') and F_{lW} (P') denote the total attraction forces on P' along l due to and towards the points inside X and W, respectively. Then the total attraction force $\mathbf{F}_{lW}$ (P) at P towards W is given by $\mathbf{F}_{lW}(P) = (F_{lW}(P) - F_{lX}(P))\hat{\mathbf{w}}$ where $\hat{\mathbf{w}}$ is the unit vector along l and towards W. Similarly, the total attraction force $\mathbf{F}_{lX}$ (P') at P' towards X is given by $\mathbf{F}_{lX}(P') = (F_{lX}(P') - F_{lW}(P'))\hat{\mathbf{x}}$ where $\hat{\mathbf{x}}$ is the unit vector along l and towards X. Since each of W and X is a constant-gray-level region, we have $F_{lW}(P) = F_{lX}(P')$ and further since $\mathbf{F}$ is an even function of gray level difference, $F_{lX}(P) = F_{lW}(P')$. Therefore, $\mathbf{F}_{lW}(P) = -\mathbf{F}_{lX}(P')$. Now the total force $\mathbf{F}_W(P)$ at P towards W is given by $\mathbf{F}_W(P) = \int_0^\pi \mathbf{F}_{lW}(P)d\theta$ and the total force $\mathbf{F}_X$ (P') at P' towards

X is given by $\mathbf{F}_X(P') = \int_0^\pi \mathbf{F}_{lX}(P')d\theta$. From the previous three equations we see that, $\mathbf{F}_W(P) = -\mathbf{F}_X(P')$, which means that there is a directional discontinuity of magnitude π between the force vectors at P and P'.

4. Orthogonality: Consider two points P and Q inside a homogeneous region W such that the unit vector $\hat{\mathbf{r}}_{\mathbf{pq}}$ is orthogonal to the region boundary at the point S where the line PQ intersects region boundary, and is directed into W. If the intercept of W's boundary with the disk D of radius $r = \sigma_s(\mathbf{p})$ centered at P is symmteric about the line PQ, then the direction of $\mathbf{F}_\mathbf{p}$ is given by $\hat{\mathbf{r}}_{\mathbf{pq}}$, i.e., $\mathbf{F}_\mathbf{p} = k\hat{\mathbf{r}}_{\mathbf{pq}}$ for some positive constant k, i.e., $\mathbf{F}_\mathbf{p}$ is orthogonal to the boundary.

5. Medial Axis: At all points P along the medial axis of a region, $\mathbf{F}_\mathbf{p} = \mathbf{0}$.

Proof: This follows directly from the definition of σ_s. For any point P along the medial axis, $\delta(\mathbf{p}) = \delta'(\mathbf{p})$. Therefore, $\sigma_s(\mathbf{p}) = \delta(\mathbf{p})$. Since the force at P due to points farther than $\sigma_s(\mathbf{p})$ is negligible (from the definition of $\sigma_s(\mathbf{p})$) and since all points within the distance $\sigma_s(\mathbf{p})$ form a homogeneous region, it follows from Property 1 that $\mathbf{F}_\mathbf{p} = 0$. (Note that here the detected axis is medial in the sense of [6]. Other related shape axes are based on local symmetery [8] or inertia [33]).

6. Convergence: At points near and on either side of the medial axis of a region R, **F** is directed towards the medial axis. That is, **F** points away from the closest border segment in the sense described in Property 2.

7. Smoothness: If $d_g(\Delta I, \sigma_g)$ is a continuous function of ΔI, and $d_s(\mathbf{r}, \sigma_s)$ is a continuous function of $\mathbf{r}$ and σ_s, then **F** is a spatially continuous function at all nonboundary points of a region.

Proof: Consider a point P inside a completely homogenous region R, and another point T inside R and arbitrarily close to P. Then,

$$\mathbf{F}_\mathbf{p} = \int_{\mathbf{q}\neq\mathbf{p}} d_s(\mathbf{r}_{\mathbf{pq}}, \sigma_s(\mathbf{p}))d_g(\Delta I(\mathbf{p}, \mathbf{t}), \sigma_g(\mathbf{p}))\hat{\mathbf{r}}_{\mathbf{pq}}d\mathbf{q}$$

and

$$\mathbf{F}_\mathbf{t} = \int_{\mathbf{q}\neq\mathbf{t}} d_s(\mathbf{r}_{\mathbf{tq}}, \sigma_s(\mathbf{t}))d_g(\Delta I(\mathbf{t}, \mathbf{q}), \sigma_g(\mathbf{t}))\hat{\mathbf{r}}_{\mathbf{tq}}d\mathbf{q}$$

Now for any third point Q, $\Delta I(\mathbf{p}, \mathbf{q}) = \Delta I(\mathbf{t}, \mathbf{q})$ since P and T have the same gray level. Further, since P and T are arbitrarily close, $(\mathbf{r}_{\mathbf{tq}} - \mathbf{r}_{\mathbf{pq}})$ is arbitrarily small. Also, $\sigma_s(\mathbf{t}) - \sigma_s(\mathbf{p})$ is arbitrarily small from the definition of σ_s. Therefore, for any given choice of $\sigma_g(\mathbf{p})(= \sigma_g(\mathbf{t}))$, $(\mathbf{F}_\mathbf{p} - \mathbf{F}_\mathbf{t})$ is arbitrarily small. Consequently, $\mathbf{F}_\mathbf{p}$ is continuous everywhere within a region.

8. Closure: For any piecewise constant image, the contours along which **F** exhibits divergence are closed.

3.4 A Region's Signatures

The above properties of the transform collectively suggest the following structure of **F** associated with an image region R, or the signatures of R in the **F**-field. When a σ_g value corresponding to R's contrast is used to compute $\mathbf{F}$, it leads to much attraction between pixels within R and little attraction between pixels across R's boundary. Further, if the σ_s value at each point is chosen corresponding to R, then R is characterized by a spatially smooth and inward force flow, where the force lines emanate from the region border and converge at R's medial axis given by $\mathbf{F} = 0$. The direction of $\mathbf{F}$ undergoes a divergent discontinuity of magnitude π across the entire region border except at those border points where $\mathbf{F} = 0$, i.e., border points which also lie on the medial axis such as corner points. Thus $\mathbf{F}$'s magnitude along a region's border varies but is 0 at corners.

A test of $\mathbf{F}$'s validity can be performed by considering the nature of $\mathbf{F_P}$ at a point P inside an arbitrary shaped region R as R shrinks in size uniformly and vanishes. Let us first consider the case where P belongs to the medial axis of R. Here, $\delta(\mathbf{p}) = \sigma_s(\mathbf{p}) = \delta'(\mathbf{p})$. Since $\mathbf{F_P}$ is computed over a constant-value disk of radius $\sigma_s(\mathbf{p})$, $\mathbf{F_P} = 0$. Now consider, a point P off the medial axis of R. Then, $\delta(\mathbf{p}) < \sigma_s(\mathbf{p}) < \delta'(\mathbf{p})$. However, as R continues to shrink, the area of the region L' (= area of region L) in Fig. 2 approaches 0. Since the net force at P is that due to L' (see the proof of Property 2),

$$\lim_{Area(R)\to 0} \mathbf{F_P} = 0. \quad (5)$$

Thus, the influence on the $\mathbf{F}$ field caused by an image region vanishes as the area of the region vanishes, as is to be expected.

3.5 Estimation of Scale Parameters

Recall that image regions are in general recursively embedded, each standing in contrast with its surround and characterized by its own gray level homogeneity. An image point is associated with multiple regions, and therefore with multiple degrees of homogeneity and contrast scales. Since in the proposed transform an increasing value of σ_g corresponds to increasingly nonuniform regions, σ_g comprises one index into the structural hierarchy. Further, since each region has its own shape and size, and the spatial scale parameter at a point depends on its location with respect to region boundary, each image point is also associated with a number of spatial scales (σ_s values). Therefore, each image point is associated with a number of (σ_s, σ_g) pairs, corresponding to the different regions that it belongs to. In real images, a point is contained in only a small number of regions, and is therefore characterized by a small number of scale pairs. $\mathbf{F}$ makes explicit at each image location regions corresponding to all scales present at that location. A particular selection of regions across the image corresponds to a specific cutset of the segmentation hierarchy.

The signatures of a region in the $\mathbf{F}$ field, as described in the previous subsection, are obtained assuming that $\mathbf{F}$ at each point is computed using the appropriate pair of (σ_s, σ_g) values. We will now explain how these unknown scales can be estimated to yield the multiscale image structure. For estimation of these values, we will treat σ_s and σ_g as variables and identify their values that correspond to image regions. Suppose that for all images of interest, $(\sigma_s)_{\min} \le \sigma_s \le (\sigma_s)_{\max}$ and $(\sigma_g)_{\min} \le \sigma_g \le (\sigma_g)_{\max}$. That is, the ranges of sizes and contrasts to be encountered in images have known bounds. If no specific information is available for images to be processed, the image size and the maximum gray level can be used as $(\sigma_s)_{\max}$ and $(\sigma_g)_{\max}$, respectively, while $(\sigma_s)_{\min} = (\sigma_g)_{\min} = 1$ can be used as the minimum size and contrast a region could have. Suppose the transform is used to compute the force at each point for $(\sigma_s)_{\min} \le \sigma_s \le (\sigma_s)_{\max}$, and $(\sigma_g)_{\min} \le \sigma_g \le (\sigma_g)_{\max}$. Suppose that at a point P, the pair of values $(\sigma_{s1}, \sigma_{g1})$ and $(\sigma_{s2}, \sigma_{g2})$ correspond to two regions R1 and R2 at two adjacent scales which contain P, with R2 containing (>) R1, and $\sigma_{g2} > \sigma_{g1}$. That is, there is no other region R3 such that R1 < R3 < R2. By our definition of scale given in Sec. 1, R1 and R2 represent two adjacent natural scales occuring in the image at P. Let $\mathbf{V_1}$ and $\mathbf{V_2}$ denote, respectively, the values of **F** corresponding to $(\sigma_{s1}, \sigma_{g1})$ and $(\sigma_{s2}, \sigma_{g2})$. As the value of the variable σ_s is increased from σ_{s1}, the image area responsible for a nonzero value of **F** at P (area L in Fig. 2) will increase, resulting in an increase in the magnitude of **F**. Now from Property 2, as the value of σ_s at a point P increases beyond $\sigma_s(\mathbf{p})$, the shape of L changes gradually and **p** moves along the subregion boundary, resulting in a gradual change in $\hat{\mathbf{r}}_{\mathbf{sp}}$ A similar gradual change in **F** will also be associated with an increase in σ_g. Thus, **F** will slowly deviate from $\mathbf{V_1}$ as (σ_s, σ_g) values increase. However, for a sufficiently large value of σ_s, i.e., $\sigma_s > \delta'(\mathbf{p})$, the value of **F** will begin to depend on multiple disconnected components of R1 which are subregions of R2. Since the direction $\hat{\mathbf{r}}_{\mathbf{sp}}$ is in general different for subregions R1 and R2, $\mathbf{V_1}$ and $\mathbf{V_2}$ are in general different. Similarly, for $\sigma_{g1} << \sigma_g << \sigma_{g2}$, the subregions of R2 will make significant contribution to **F** value at P. Consequently, **F** will change with (σ_s, σ_g). As (σ_s, σ_g) approach $(\sigma_{s2}, \sigma_{g2})$, **F** will assume the relatively stable value of $\mathbf{V_2}$. Therefore, in the $\sigma_s\sigma_g$-space, the locations where **F** is stable will be scattered, associated with structures at different pairs of scale values. Somewhere between each pair of nearby locations of stable points in the $\sigma_s\sigma_g$-space corresponding to R1 and R2, **F** will make a sharper transition from the value $\mathbf{V_1}$ to $\mathbf{V_2}$. By traversing the $\sigma_s\sigma_g$-space, computing **F** using all σ_s, σ_g values at all image points, and identifying those parts in the $\sigma_s\sigma_g$-space where **F** is locally stable (has locally minimal variation), we can determine all the scales associated with P. The scale values at all image points can be estimated jointly, because together the values comprise the signatures of regions as explained in the previous subsection. Such scale estimation is robust for two reasons. First, at each point only qualitative changes in **F** are detected. Second, the qualitative changes at different points are analyzed jointly to detect the spatial signatures of a region. This further suppresses any noise in **F** which is already low because of the large neighborhoods used in the computation of **F**.

The point pattern in $\sigma_s\sigma_g$-space defined by the locations of the actual scale values corresponding to any image point is unique for the point and the image, and represents the *a priori* unknown multiscale structure determined by the algorithm at the point. This structure is restricted for common images because an image point does not have multiple contrasts associated with the same spatial scale although it may be contained in multiple regions of the same contrast. Therefore, it would suffice to perform a linear sweep of the $\sigma_s\sigma_g$-space in the σ_g direction, while identifying all (if any) σ_s values which yield locally stable **F** for each σ_g.

3.6 Region Detection

For each region in the image, occuring at any scale, the automatic scale estimation process computes at each pixel in and around the region a $\sigma_s - \sigma_g$ pair of values which correspond to the region's shape (σ_s) and contrast (σ_g). When the **F** field is computed using these $\sigma_s - \sigma_g$ values, the field contains the signatures of the region. The detection of regions would thus require partitioning of the **F** field such that each cell has the signatures of a region.

A simple approach to finding candidate regions is to locate contours of force divergence. This is easy and the result robust since the directional discontinuity across such contours is known to be π. All characteristics comprising the region's signature may then be matched jointly with the local **F**-field around the candidate regions to test the region hypothesis. The details of such hypothesis formation and testing are outside the scope of this paper.

3.7 Segmentation Hierarchy

The properties of the transform and the capability of automatic estimation of the scale parameters discussed in the previous subsections allow the construction of a hierarchical representation of segmentation. The computed scale values for nearby pixels are mutually compatible in that for all points within a region, σ_s values are continuous and σ_g values are constant.

The regions of a single gray level form the smallest regions. These may be defined as the leaf nodes of the hierarchy. Homogeneous regions having larger sizes or gray level variations are used to define higher levels. For example, the hierarchy may be based on spatial containment relationship; thus, the regions containing a particular leaf region are arranged in increasing order of size to define the path from the leaf node to the root. Nearby pixels merge into increasingly large regions as the path to the root is traversed. The subtree below any node in the hierarchy is unrelated to any other disjoint subtree, i.e., the structure, path lengths and σ_s-σ_g values associated with the nodes are unrelated across the subtrees. They reflect the *a priori* unknown spatial structure within an image. In this sense, the hierarchy define a recursive partition of the image into arbitrarily shaped regions, analogous to the irregular pyramid representation of [24]. Specific algorithms to compute the scale parameters and to obtain different hierarchical representations are not within the scope of this paper; these will be reported in subsequent publications.

4 Performance Analysis

Let us first review the overall performance of the transform with respect to the desired characteristics (A-D) listed in Section 2.2. Characteristics A, invariance to local edge geometry and topology, serves as a key motivation for proposing the transform, and is central to its design. The discussion in Sec. 3 makes it clear how this characteristic is possessed by **F**. For example, the capability of multiscale segmentation holds even if more than two regions share a border point since the properties of the transform leading to region signatures are not affected by shape and adjacency characteristics of the regions. With regard to desired characteristic B, the scale parameter σ_g provides a mechanism to accomplish contrast scaling. As σ_g increases, adjacent regions may merge. This is because the attraction of a point in one region from another point across region boundary may increase sufficiently so that the directional discontinuity in **F** responsible for the edge may vanish. Thus changing σ_g achieves the same result as contrast based split-and-merge of regions [15, 19], and therefore, the desired contrast scaling. Analogously, for any given σ_g, scale parameter σ_s helps achieve geometric scaling (desired characteristic C). Larger σ_s values at a point correspond to more global structures having a given contrast σ_g, which results in the capability to detect different spatial scales. The desired characteristic D is of course met since scales can be automatically estimated as explained in Sec. 3.5.

In the rest of this section, we will divide the performance of the transform into two types. The first type is concerned with the capabilities to detect off-axis signatures of a region, specifically those represented by Properties 2, 3, 7 and 8. The second type consists of region signatures related to the medial axis, represented by Properties 5 and 6. Properties 1 and 4 do not directly contribute to the region signatures. Sections 4.1 and 4.2 examine the type 1 and type 2 performance, respectively. In discussing each type of performance, we consider two types of deviations from the model of piecewise constancy that was often used in deriving the transform's properties. These deviations better characterize real images as stated earlier. We investigate the effect these deviations have on the particular aspects of the region signatures, i.e. on the different properties. We then examine the impact of different choices of the function F, i.e., the functions d_s and d_g, on the transform's performance. No detailed proofs are given for these claims as doing so will require the use of specific models of the deviations and further analysis which are beyond the scope of this paper. Results of further, experimental evaluation of the performance can be found in [2].

4.1 Off-Axis Signatures

In this section, we examine the first type of performance of the transform, namely, the impact on inward attraction, divergence, smoothness and closure properties.

4.1.1 Deviation from Piecewise Constancy

In the discussion so far, we have assumed that the regions have a constant gray level. We will now explain how the segmentation performance of the transform extends to regions having other types of smooth variations. If the region does not have a constant intensity then the force at a point P in the region will include additional components due to differential rates of change of intensities in different directions away from P (unlike the case for Property 1). For the simple case of an intensity ramp, the changes in intensity around P are antisymmetric. Since force depends on the absolute intensity difference, P still experiences equal and opposite forces from radially symmteric locations within the region resulting in zero net force from these locations.

Regions in real images often contain shading which is more complex spatial variation of intensity than represented by the ramp considered above, e.g., given by a polynomial in image coordinates x and y. Consider a point P within such a region R and another point Q in R within a neighborhood of radius $\sigma_s(\mathbf{p})$ centered at P. Then, $d_g(\Delta I(\mathbf{p}, \mathbf{q}))$ will vary for different points Q, unlike was the case for constant-value regions. This variation will in general be nonlinear, partly due to the nonlinear variation in $\Delta I(\mathbf{p}, \mathbf{q})$. Now consider another point T within the neighborhood but not within R. If the range of $\Delta I(\mathbf{p}, \mathbf{q})$ values is sufficiently small compared to $\Delta I(\mathbf{p}, \mathbf{t})$ for all choices of Q and T, then many of the properties of the transform may still hold. The proofs of Properties 2, 3, 7 and 8 given earlier suggest that the boundaries of the regions may still be detected as before. Accordingly, at a region boundary, there will still be directional discontinuities because of the large gray level discontinuity; and in the process of finding the scale parameter values for the region, stable response of $\mathbf{F}$ will be found for the same values of σ_s and σ_g as if the region were homogeneous. Any directional discontinuities found at locations other than the region border, due to nonlinear variation in gray level, will not persist if σ_g is varied. Since the gray levels vary smoothly within the region, Property 7 suggests that F will still be continuous within the region although it will exhibit differences in F values from the piecewise constant case. Therefore, regions with shading but in contrast with the surround should still have signatures similar to those for the piecewise constant case. Verification of the above extrapolation of the properties and the exact restatement of these and other properties for the general case would require exact models of within-region intensity variation, and will be omitted here.

4.1.2 Intensity Noise

We will now consider sensitivity to noise in intensity values. First, suppose that the regions have constant values but contain independently distributed, zero-mean, additive noise having a distribution which is symmetric with respect to the mean. Consider a point P and any other point Q within a neighborhood of radius $\sigma_s(\mathbf{p})$ around P. Then, $d_g(\Delta I(\mathbf{p},\mathbf{q}))$ will have the same mean as for the case when the noise is absent. Since $d_g(\Delta I(\mathbf{p},\mathbf{q}))$ is a symmetric function of $\Delta I(\mathbf{p},\mathbf{q}))$, the expected value of $\mathbf{F}(\mathbf{p},\mathbf{q})$ will remain unchanged compared to the case without noise. Therefore, the expected value of $\mathbf{F_p}$ due to all points $\mathbf{q}$ in the image is the same with or without noise. Now suppose that the regions exhibit ramp-like intensity variation which is contaminated by independent, zero-mean, additive noise. Again, because of gray level antisymmetry about P, and the symmetry of $d_g(\Delta I)$ with respect to ΔI, the region boundary will remain unchanged assuming the region contrast with the surround is high compared to within region variation.

For shaded regions, the noise effects will be anisotropic because the region intensities are asymmetric. For a given point P, consider two other points Q and T within a neighborhood of radius $\sigma_s(\mathbf{p})$ around P. Q is within the region R but T is across R's border. If the range of $\Delta I(\mathbf{p},\mathbf{q})$ values is sufficiently small compared to the range of $\Delta I(\mathbf{p},\mathbf{t})$ values, then resulting $\mathbf{F}$ will have limited differences relative to the noiseless case. That is, boundaries of regions with shading but in contrast with the surround will still coincide with direction discontinuities in $\mathbf{F}$ and will therefore still be detected. However, as for noiseless shaded regions, exact analysis is necessary to obtain the true characterization of the region signature and its dependence on noise which we will again omit in this paper.

4.1.3 Choices of F

While defining the transform (Sec. 3), we stated that $d_s(\mathbf{r_{pq}},\sigma_s(\mathbf{p}))$ should be a nonincreasing function of the magnitude of $\mathbf{r_{pq}}$, and $d_g(\Delta I,\sigma_g(\mathbf{p}))$ should be a nonincreasing and symmetric function of ΔI. A variety of such functions could be used including pulse (box-car), Gaussian, exponential, and linear functions. For example, we may use a Gaussian for d_s as well as d_g, having standard deviations of $k_s\sigma_\mathbf{s}$ and $k_g\sigma_\mathbf{g}$, resepectively, where k_s and k_g are normalization constants. The choice of Gaussian for d_s and d_g results in optimal localization properties in both spatial and transform domains, in addition to others such as separability in computation. Then, the transform at image location $\mathbf{p}$ is given by

$$\mathbf{F_p} = \int_{\mathbf{q}\neq\mathbf{p}} e^{-\frac{\|\mathbf{r_{pq}}\|^2}{2(\mathbf{k_s}\sigma_\mathbf{s}(\mathbf{p}))^2}} e^{-\frac{\Delta I^2(\mathbf{p},\mathbf{q})}{2(\mathbf{k_g}\sigma_\mathbf{g}(\mathbf{p}))^2}} \hat{\mathbf{r}}_\mathbf{pq} \mathbf{dq} \tag{6}$$

The properties of the transform given in Sec. 3 hold for any choices of such functions. Although the exact values of the force vectors and the computational speeds depend on specific choices of the functions, the region signatures and segmentation hierarchy remain unchanged. We have verified empirically that this in fact is the case for the four choices of box-car, Gaussian, exponential and linear functions [2].

4.2 Axial Signatures

This section discusses the impact of the two kinds of deviations from piecewise constancy and the choices of F on the medial axis related signatures. There are two ways in which the transform yields multiscale description of region shape. First, of course, is through the detection of region boundaries which may be used to estimate the medial axis following its definition, or using existing algorithms [6, 31]. The reliability of the region axis detected by this method directly depends on the corresponding reliability of detected region boundaries which we have discussed above. The second, more direct way in which the transform extracts multiscale region shape information is by making explicit the location of the medial axis in the field signatures of the region. We will now consider the performance of this method of medial axis detection with respect to shading and noise. As stated in Property 5, for constant-value regions the $\mathbf{F} = \mathbf{0}$ curve within the region where $\mathbf{F}$ directions converge represents the medial axis. Now in the vicinity of locations where $\mathbf{F} = \mathbf{0}$, the magnitude of $\mathbf{F}$ will change at a rate determined by how fast the neighborhood of radius σ_s begins to have significant intersection with an adjacent region, and thus on the the local region shape. The steeper the $\mathbf{F}$ variation, the more accurate will be the detection of the $\mathbf{F} = \mathbf{0}$ locations, and therefore, the medial axis. However, if the intensity value within the region is not constant, then the disk of radius $\sigma_s(\mathbf{p})$ centered at a point on the medial axis will not in general have isotropic intensity distribution about the point. The force at the point will not be 0, and the points where $\mathbf{F} = \mathbf{0}$ may be off the medial axis. Therefore, when shading is present, the detection of medial axis is less reliable than the detection of region borders. With regard to noise, there will be no expected change in results for noisy piecewise constant regions for reasons analogous to those given for off-axis signatures. However, in shaded regions, noise will further increase the medial axis deviation beyond that already present due to shading alone. This is because the function d_g in the integral defining the transform is in general nonlinear, and therefore, in the presence of shading, uniform noise distribution around the point will cause unequal deviations in the $\mathbf{F}$ value in different directions, which will lead to deviation in the location of the $\mathbf{F} = \mathbf{0}$ curve. Finally, let us consider the effect of different choices of F. For the piecewise constant case, changes in F will not cause any deviation in the mdeial axis because of symmetry. However, when shading is present different choices of the nonlinear function d_g will in general result in different distributions of force magnitudes and hence deviation in the $\mathbf{F} = \mathbf{0}$ curve.

5 Summary

We have introduced a transform for integrated detection of image edges and regions at all natural scales, and thus for general purpose low-level image segmentation. Our objective here is to present the transform and its capabilities; details and different applications are left to future work. The transform computes a force vector at each image point, such that the spatial distribution of vectors makes explicit region edges and medial axes at all scales. The points within and on the boundary of a region are attracted towards its interior, thus causing a directional discontinuity across the boundary. The medial axes also exhibit similar discontinuities in the force direction. An important property of the transform is that the transform space is the same as image space, unlike for example the Fourier transform. The resulting multiscale representation is directly useful for analysis by humans, e.g., image browsing, manipulation and retrieval. The use of scale parameters and the detection of structure from qualitative characteristics of -field makes the use of the transform robust and free of critical thresholds. We have presented the properties of the transform. Experimental results, not reported here, show that the segmentations provided by the transform have few errors even for images with large noise, intricate geometric structure, and shading. The transform could be used in a variety of applications where determination of perceptually salient image structure is critical or advantageous, such as image compression, motion analysis, texture analysis and perceptual grouping. We plan to report on some of these applications in the future. We have not given algorithms for using the transform to automatically estimate the scales and identify region signatures. These will also be reported later along with the applications. Since no assumptions specific to the sensing modality are made in defining the transform, it could be applied to other types of data such as range images and synthetics aperture radar images. It will be interesting to compare the transform with force based clustering algorithms [16] which we plan to do.

Acknowledgements

This support of the Advanced Research Projects Agency under grant N00014-93-1-1167 administered by the Office of Naval Research, and the National Science Foundation under grant IRI 93-19038 is gratefully acknowledged. Thanks are also due to the anonymous reviewers of another, journal version of this paper whose comments were very helpful.

References

[1] N. Ahuja. A transform for detection of multiscale image structure. In *Proceedings Computer Vision and Pattern Recognition*, pages 780–781, New York, 1993.

[2] N. Ahuja. A transform for multiscale image segmentation. Technical Report 96-01, Beckman Institute Technical Note, July 1995.

[3] N. Ahuja and J. Chuang. Shape representation using a generalized potential field model. *IEEE Transactions on Pattern Analysis and Machine Intelligence*, 1996. to appear.

[4] V. Berzins. Accuracy of lapacian edge detectors. *Computer Vision, Graphics and Image Processing*, pages 195–210, 1984.

[5] T. Binford. Inferring surfaces from images. *Artificial Intelligence*, pages 205–244.

[6] H. Blum. A transformation for extracting new descriptors of shape. In Walthen Dunn, editor, *Models for the Perception of Speech and Visual Form*, pages 362–380. MIT Press, Cambridge, MA, 1967.

[7] R. Boomgaard and A. Smeulders. Towards a morphological scale-space theory. In *Shape in Picture: Mathematical Descriptions of Shape in Grey-Level Images*, pages 631–640. Springer-Verlag, New York, 1992.

[8] M. Brady and H. Asada. Smoothed local symmetries and their implementation. *International Journal of Robotics Research*, 3(3):36–61, 1984.

[9] J.F. Canny. A computational approach to edge detection. *IEEE Transactions on Pattern Analysis and Machine Intelligence*, pages 679–698, 1986.

[10] J. J. Clark. Authenticating edges produced by zero-crossing algorithms. *IEEE Transactions on Pattern Analysis and Machine Intelligence*, pages 43–57, January 1989.

[11] L.S. Davis. A survey of edge detection techniques. *Computer Graphics and Image Processing*, pages 248–270, 1975.

[12] S. Geman and D. Geman. Stochastic relaxation, gibbs distributions, and the bayesian restoration of images. *IEEE Transactions Pattern Analysis and Machine Intelligence*, PAMI-6(11):721–741, 1984.

[13] R. M. Haralick. Digital step edges from zerocrossings of second directional derivative. *IEEE Transactions on Pattern Analysis and Machine Intelligence*, pages 58–68, 1984.

[14] R. M. Haralick and L. Shapiro. A survey of image segmentation algorithms. *Computer Vision, Graphics and Image Processing*, 1492.

[15] S.L. Horowitz and T. Pavlidis. Picture segmentation by a directed split and merge procedure. In *Proc. Second Intl. Joint Conf. on Pattern Recognition*, pages 424–433, 1974.

[16] A. K. Jain and R. C. Dubes. *Algorithms for Clustering Data*. Prentice-Hall, 1983.

[17] J. A. Koenderink. *Solid Shape*. MIT Press, 1990.

[18] T. Lindeberg. *Scale-Space Theory in Computer Vision.* Kluwer Academic, Boston, 1994.

[19] T. Lindeberg and J. O. Eklundh. Scale-space primal sketch: Construction and experiments. In *Image and Vision Computing*, pages 3–18. January 1992.

[20] Y. Lu and R.C. Jain. Behavior of edges in scale space. *IEEE Transactions on Pattern Analysis and Machine Intelligence*, pages 337–356, 1989.

[21] J. Malik and P. Perona. Edge detection by diffusion. Technical report, University of California, Berkeley, 1986.

[22] D. Marr and E. Hildreth. A theory of edge detection. In *Royal Society of London*, volume B-207, 1980.

[23] F. Meyer and S. Beucher. Morphological segmentation. *Vis. Communication Image Representation*, 1(1):21–46, 1990.

[24] A. Montavert, P. Meer, and A. Rosenfeld. Hierarchical image analysis using irregular tessellations. *IEEE Transactions on Pattern Analysis and Machine Intelligence*, pages 307–316, 1991.

[25] B. S. Morse, S. M. Pizer, and A. Liu. Multiscale medial analyis of medical images. In H. Barrett and A. Gmitro, editors, *Proceedings 14th International Conference on Information Processing in Medical Imaging.* Springer-Verlag, 1993.

[26] D. Mumford and J. Shah. Boundary detection by minimizing functionals, i. In *IEEE Conference on Computer Vision and Pattern Recognition*, pages 22–26, 1985.

[27] V. Nalwa and T. Binford. On detecting edges. *IEEE Transactions on Pattern Analysis and Machine Intelligence*, pages 699–714, 1986.

[28] V. S. Nalwa. Experiments with a spatiotemporal correlator. In *Proceedings IEEE Conference on Computer Vision and Pattern Recognition*, pages 712–716, 1992.

[29] P. Perona and J. Malik. Detecting and localizing edges composed of steps, peaks and roofs. In *Third Intl. Conf. Computer Vision*, pages 52–57, Osaka, Japan, Dec. 1990.

[30] J. Ponce and M. Brady. Toward a surface primal sketch. Technical Report MIT-AI-TR-824, MIT, 1985.

[31] A. Rosenfeld and A. Kak. *Digital Picture Processing*, volume 2. Academic Press, 1981.

[32] A. Rosenfeld and M. Thurston. Edge and curve detection for visual scene analysis. *IEEE Transactions on Computers*, pages 562–569, 1971.

[33] J. Brian Subirana-Vilanova. Curved inertia frames and the skeleton sketch: Finding salient frames of reference. In *Proceedings IEEE International Conference on Computer Vision*, pages 702–708, 1990.

[34] R. Whitaker and S. Pizer. Geometry-based image segmentation using anisotropic diffusion. In *Shape in Picture: Mathematical Descriptions of Shape in Grey-Level Images*, pages 641–650. Springer-Verlag, New York, 1992.

[35] A. Witkin. Scale space filtering. In *International Joint Conference on Artificial Intelligence*, 1983.

[36] A. Yuille and T. Poggio. Scaling theorems for zero crossings. *IEEE Transactions on Pattern Analysis and Machine Intelligence*, pages 15–25, 1986.

Image Segmentation Using Clustering

Anil K. Jain
Department of Computer Science
Michigan State University
East Lansing, MI 48824
jain@cps.msu.edu

Patrick J. Flynn
School of Electrical Engineering and Computer Science
Washington State University
Pullman, WA 99164
flynn@eecs.wsu.edu

Abstract

Pattern Recognition methodology has influenced many areas of computer vision research. This chapter addresses one of the most direct and long-lived applications of classification to image analysis, namely the use of unsupervised classification (cluster analysis) for image segmentation. The use of clustering for image segmentation dates back to the late 1960s and many of the techniques developed then are still in popular use today. This chapter outlines the segmentation and clustering problems and their close relationship, and surveys the use of clustering for segmentation in a variety of different domains (*e.g.*, intensity images, textured images, multispectral images, and range images).

1: Introduction

An *image segmentation* is a fundamental component in many high-level computer vision applications, and is typically viewed as an unsupervised classification task [1]. The segmentation of the image(s) presented to an image analysis system is critically dependent on the scene to be sensed, the imaging geometry, configuration, and sensor used to transduce the scene into a digital image, and ultimately the desired output (goal) of the system. Image segmentation procedures are often labeled as *edge-based* or *region-based*, referring in each case to a dominant computational philosophy of identifying homogeneous regions or contours of local inhomogeneity, respectively. An important region-based technique is *split-and-merge* segmentation [10]. Some attempts [18, 19] have also been made to merge the results of edge-based and region-based methods to improve segmentation results.

Clustering is a generic label for a variety of procedures designed to find "natural" groupings in *unlabeled* data presented as a set of *patterns*, each a list of *features*. Clustering is often referred to as *unsupervised classification* to reinforce the notion that the goal is to discover

0-8186-7644-2 $5.00 © 1996 IEEE

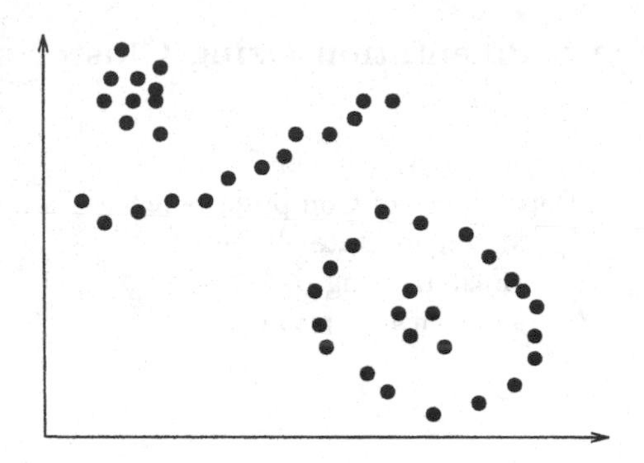

Figure 1. Cluster structures. Clusters can be compact or elongated, well-separated or poorly separated from other clusters.

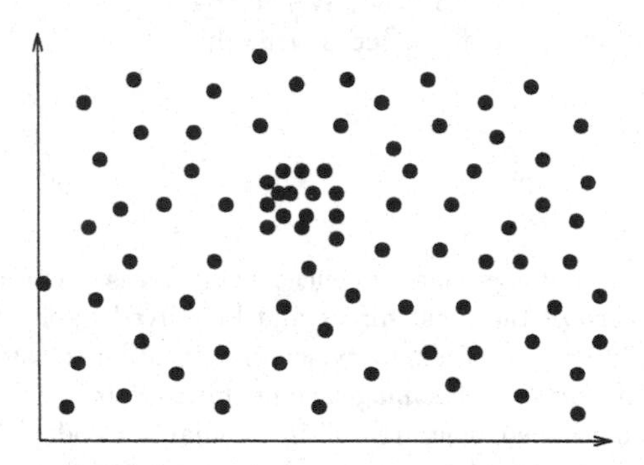

Figure 2. A "dot" cluster.

structure (classes or categories) in a multivariate data set without an explicit model for each pattern class, an oracle to supply correct or true classifications for representative samples, or an external procedure to validate the groupings produced by the clustering algorithm. Clustering is a difficult problem because "natural groupings" in multidimensional data can have very different properties, as illustrated in Figure 1. Another example of clustering (addressed by Sher and Rosenfeld [3]) is the detection and delineation of regions containing a high density of patterns compared to their background; Figure 2 shows such an image.

The applicability of clustering methodology to the image segmentation problem was recognized over three decades ago, and the paradigms underlying the initial pioneering efforts are still in use today. A recurring theme is to define feature vectors at every image location (pixel) composed of both functions of image intensity and functions of the pixel location itself. This basic idea has been successfully used for intensity images (with or without texture), range (depth) images and multispectral images. Figure 3 depicts the scheme. Contrary to many applications of pattern recognition, the output of a clustering-based segmenter is easily visualized as a labeling of the input image with (for example) different colors or gray scale shades representing different label values.

The goals of this chapter are as follows:

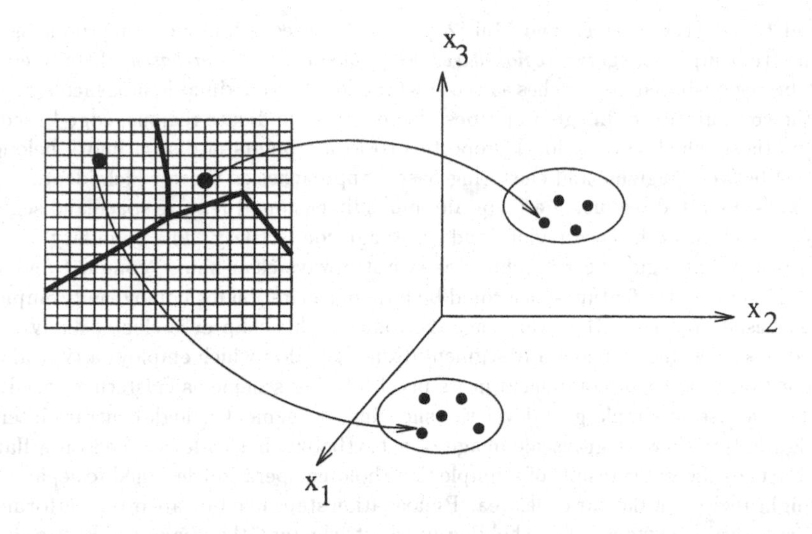

Figure 3. Feature representation for clustering. Image measurements and positions are transformed to features. Clusters in feature space correspond to image segments.

1. review image segmentation and the major paradigms used to obtain segmentations,
2. survey clustering methodology including its placement in the taxonomy of pattern recognition techniques,
3. motivate the application of clustering to segmentation in general,
4. recognize the work of those who pioneered this technique, and
5. describe selected applications of clustering to specific image segmentation domains over the last thirty years.

We make no effort to be exhaustive in this survey, but strive to mention those approaches that laid the foundation for the work of others or achieved significant milestones in the application of cluster analysis to image segmentation.

2: Segmentation: Definition, Intuition, and Major Paradigms

An *image segmentation* is typically defined as an exhaustive partitioning of an input image into regions, each of which is considered to be homogeneous with respect to some image property of interest (*e.g.*, intensity, color, or texture) [11]. If

$$\mathcal{I} = \{x_{ij}, i = 1 \ldots N_r, j = 1 \ldots N_c\}$$

is the input image with N_r rows and N_c columns and measurement value x_{ij} at pixel (i, j), then the segmentation can be expressed as $\mathcal{S} = \{S_1, \ldots S_k\}$, with the lth segment

$$S_l = \{(i_{l_1}, j_{l_1}), \ldots (i_{l_{N_l}}, j_{l_{N_l}})\}$$

consisting of a connected subset of the pixel coordinates. No two segments share any pixel locations ($S_i \cap S_j = \emptyset \quad \forall i \neq j$), and the union of all segments covers the entire image ($\cup_{i=1}^{k} S_i = \{1 \ldots N_r\} \times \{1 \ldots N_c\}$).

Jain and Dubes [12], after Fu and Mui [21], identify three techniques for producing segmentations from input imagery: *region-based*, *edge-based*, or *cluster-based*. Intuitively, we can describe region-based approaches as those which identify maximal homogeneous regions through the computation of image properties defined over candidate regions, edge-based approaches as those which employ local properties to locate points of discontinuity belonging to the edges between regions, and clustering-based approaches as those which identify compact and well-separated regions in a (typically multidimensional) feature space and segment the images based on both these regions and on image connectivity. The boundary between clustering-based and region-based techniques is not always clear since "compact and well-separated clusters" in the feature space could be part of a criterion for homogeneity employed in a region-based approach. However, since the focus of this chapter is cluster analysis and its applications we would consider any segmentation algorithm which employs a typical clustering algorithm as a major component in its pixel labeling scheme as clustering-based.

Consider the use of simple gray level thresholding to segment a high-contrast intensity image. Figure 4(a) shows a grayscale image of a textbook's bar code scanned on a flatbed scanner. Part (b) shows the results of a simple thresholding operation designed to separate the dark and light regions in the bar code area. Binarization steps like this are often performed in character recognition systems. Thresholding in effect 'clusters' the image pixels into groups based on the one-dimensional intensity measurement [2, 4]). A postprocessing step separates the classes into connected regions. While simple gray level thresholding is adequate in some carefully controlled image acquisition environments and much research has been devoted to appropriate methods for thresholding [22, 23], complex images require more elaborate segmentation techniques. For example, Ohlander [20] used multidimensional histograms for segmentation.

Segmentation, like many other vision tasks, is defined by its ultimate use, and the diversity of image types and noise sources has made image segmentation a fertile research area. This diversity has also established segmentation as a *difficult* research area since in many cases techniques learned in one application domain do not transfer with ease (or at all) to another domain. The evaluation of a segmented image is a topic that has not been addressed with regularity by researchers. To the extent that segmentation performance is subjectively measured, a methodological gap exists and should be addressed [23]. The nature and quality of the input imagery, as well as expectations on the characteristics of the output regions, dictate the set of valid segmentation strategies that should be attempted and perhaps even the details of their design and implementation.

A critical issue associated with segmentation is the image representation. This reduces to the problem of defining the feature set, its dimensionality, and its type and scale. In the segmentation techniques surveyed in this chapter, the image measurement can be intensity, range, or features derived from intensity or range (*e.g.*, texture energy at a specified scale and orientation, or an estimated normal to a surface), its dimensionality ranges from one to twenty-eight, its type can be discrete or continuous, and its scale can be ordinal or ratio (for a discussion of data type and scale in the context of clustering methodology, see Jain and Dubes [12]). Many segmenters use measurements which are both *spectral* (*e.g.*, the multispectral scanner used in remote sensing) and *spatial* (based on the pixel's location in the image plane). The measurement at each pixel will be viewed henceforth as a point in a multidimensional space (a *feature space*), where the nature of the space is described when needed. Figure 3 shows a three-dimensional feature space (x_1, x_2, x_3) corresponding to an image measurement.

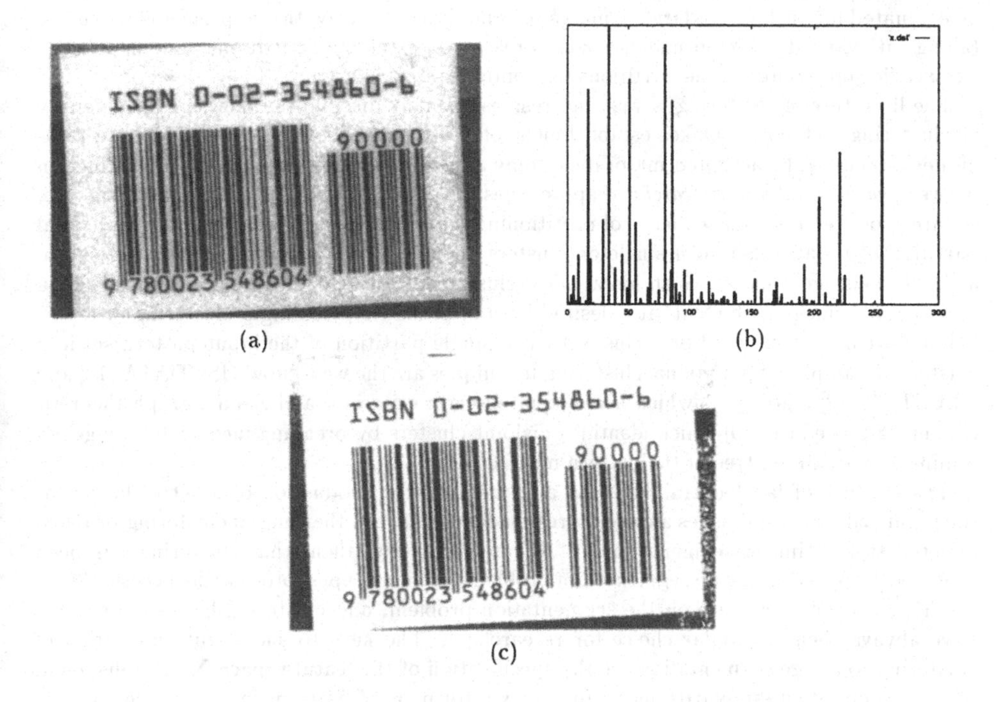

Figure 4. Binarization via thresholding. (a): Original grayscale image. (b): Gray-level histogram. (c): Results of thresholding.

3: Clustering Procedures: Context and Taxonomy

The essential elements of a clustering problem are a feature space $\mathbf{X}$ (typically multidimensional), samples (*patterns*) $X = \{\mathbf{x}_1, \ldots \mathbf{x}_n\}$ from that space, presumed cluster structure in the samples, and the absence of accessible class information for the samples. A clustering procedure is expected to (i) identify the number of clusters K present in the available samples, and (ii) obtain a label set $\mathbf{L} = \{l_1, \ldots l_n\} (l_i \in \{1 \ldots K\}, \forall i)$ for the samples. These two tasks may be performed simultaneously or sequentially; in some techniques K is specified to the algorithm by the user.

The combinatorics of the clustering problem make simple enumerative schemes impractical for all reasonably-sized problems. The number $S(n, K)$ of distinct clusterings of n objects into K clusters is a Stirling number of the second kind [12]:

$$S(n, K) = \frac{1}{K!} \sum_{i=1}^{K} (-1)^{K-i} \binom{K}{i} i^n$$

For example, while 34,105 distinct partitions of ten objects into four clusters exist, there are 11,259,666,000 partitions of nineteen objects into four clusters. If the number of clusters is unknown but can be bounded above by (say) K_{max}, the number of possible partitionings is $\sum_{K=1}^{K_{max}} S(n, K)$. Moreover, in order to select the best clustering a criterion would need to be evaluated for each of the large numbers of candidates. Clearly, this approach is infeasible (along with variants which prune the space of reasonable candidate partitions), and in practice we sacrifice optimality of the partitions for computational efficiency.

The literature of clustering is vast and resides in many disciplines; likewise, the diversity of clustering techniques makes establishment of a comprehensive taxonomy of these techniques difficult. The accepted major dichotomy among clustering methods is the distinction between *partitional* and *hierarchical* approaches. In essence, a hierarchical clustering procedure produces a nested sequence of partitionings covering the spectrum between the trivial partition of n patterns into n singleton clusters and the trivial partition of n patterns into a single n-element cluster. Such hierarchical clusterings are often depicted graphically by a *dendrogram* which can be 'cut' at a desired level of proximity, yielding a clustering or a partition. In contrast, a partitional clustering is a single partition of the input pattern set into clusters. Examples of partitional clustering techniques are the well-known ISODATA [14] and CLUSTER [12] algorithms which use a squared-error criterion, and Zahn's graph-theoretic clustering procedure [49] which identifies disjoint clusters by breaking inconsistent edges in a minimum spanning tree of the pattern matrix.

The absence of labeled training data and the need for production of a set of labels for the input patterns establishes a close correspondence between the general clustering problem and the task of image segmentation. It is not surprising, then, that clustering has been repeatedly applied to image segmentation problems of many types, often with success. There is a rich body of literature on the segmentation problem, and clustering-based approaches have always been a popular choice for researchers. The keys to successful application of clustering to image segmentation are the specification of the feature space $\mathbf{X}$, establishment of an appropriate dissimilarity measure $d(\mathbf{x}_i, \mathbf{x}_j)$ for pairs of patterns in that space, and the choice of an appropriate algorithm for clustering this type of data.

The difficult issue of evaluation of segmentation results has received some attention over the years. The parallel issue in cluster analysis is *cluster validity*; while all clustering algorithms will produce a clustering of an input pattern set (and in general those clusterings will differ), how should the quality of a clustering be measured? Cluster validity has been extensively studied and offers some lessons to those interested in validating segmentations. A related issue that is not as germane to segmentation as validity is *cluster tendency*: does an input pattern set indeed contain cluster structure? The source of images in a segmentation application is typically contrived to avoid circumstances where only one segment (or homogeneous region) is present, and image structure (*e.g.*, image plane connectivity) makes the assignment of each pattern to its own cluster unreasonable.

Not all clustering problems contain compact and well-separated clusters. Yet hyperellipsoidal clusters are assumed to be present by many clustering techniques. A variety of methods for "cleaning" the input pattern set have been proposed over the years. Jolion and Rosenfeld [6] proposed to associate with each input pattern a weight that depends on the empirical density observed in that pattern's vicinity. Cluster statistics computed during clustering use these empirical weights and were demonstrated to reduce the bias caused by the inappropriate linkage of 'background noise' patterns into clusters.

Although clustering methodology is a mature field with many well-understood and powerful techniques to offer prospective users, research into new clustering techniques continues. For example, Mao and Jain [48] recently developed a self-organizing network for data clustering, and applied it to both traditional data sets from the statistical literature, and the examples from the texture segmentation problem (described below). The expectation maximization

(EM) technique [17, 55] for resolving mixture models yields a clustering algorithm which can be applied to image segmentation.

3.1: Squared Error Clustering

Squared-error clustering procedures are partitional clustering methods which attempt to minimize a squared-error objective function over the space of possible partitioning of the pattern set. The squared error for a clustering of n patterns into K clusters is

$$E_K^2 = \sum_{j=1}^{K} \sum_{i=1}^{n_j} \|\mathbf{x}_i^{(j)} - \mathbf{m}_j\|^2,$$

where $\mathbf{x}_i^{(j)}$ is the ith pattern belonging to cluster j and

$$\mathbf{m}_j = \frac{1}{n_j} \sum_{i=1}^{n_j} \mathbf{x}_i^{(j)}$$

is the mean vector of cluster j (formed from the n_j patterns assigned to it). The goal of squared error clustering algorithms is to find the partition which (for a given K) minimizes E_K^2. Intuitively, we can minimize E_K^2 (which is sometimes called the ***within-cluster variation***) by obtaining compact and well-separated clusters.

Many squared-error algorithms proceed iteratively, beginning with an initial clustering chosen randomly or 'sensibly', with a goal of decreasing the squared error at each iteration. The initial partition affects the final clustering, and simple minimization procedures can become trapped in local minima of the squared-error objective function; a standard technique for overcoming this problem is to run the clustering program several times with different initial conditions on each run, and save that clustering which yields the minimum value of E_K^2. Additional heuristics allow the ability to adjust the number of clusters during the algorithm's operation; these adjustments typically split clusters with large variance and merge clusters which have close mean vectors. The behavior of the splitting and merging heuristics is controlled by user-settable thresholds.

Even with special treatment to avoid local minima, convergence of a squared-error partitional clustering algorithm does not necessarily yield a configuration with the globally minimal E_K^2. Convergence criteria used in many clustering programs include a maximum number of relabeling iterations, a lack of significant decreases in E_K^2, and a minimal number of label changes during an iteration.

We now present an outline of the CLUSTER algorithm, based on the description of Jain and Dubes [12]. This algorithm has been used successfully in many image segmentation problems. The objective of CLUSTER is to minimize E_K^2. In its most popular version it has a single user-specified parameter: K_{max}, the maximum number of clusters to consider. CLUSTER will produce a sequence of clusterings containing $2, 3, \ldots K_{max}$ clusters. During an initialization phase, a candidate set of $K_{max} - 1$ clusterings is created as follows. Cluster centers for the 2-cluster solution are chosen to be the centroid of the pattern set and the pattern farthest from that centroid. The clustering itself is obtained by minimum distance classification with respect to these two centers. The pattern farthest from its cluster center is chosen as the third cluster center. This process repeats until there are K_{max} distinct cluster centers.

The main portion of CLUSTER contains two phases which are repeated in sequence until a pass through both phases does not decrease E_K^2. The first phase is a classical K-means pass which adjusts cluster memberships in each clustering to decrease E_K^2; this K-means pass is repeated until no class labels change or a maximum number of iterations is performed. The second phase of the algorithm is a forcing pass which merges clusters in pairs to see if a better clustering can be achieved. After the K-means and forcing passes complete, the squared error of the clustering is compared with its previous value, and the process repeats if the squared error decreased. When CLUSTER completes, a number of summary statistics are produced to aid in the (difficult) task of choosing the appropriate number of clusters.

3.2: Clustering Applications in Image Analysis

While the focus of this chapter is on the use of clustering techniques for the task of image segmentation, clustering has seen widespread use for other problems in image analysis. Perceptual grouping tasks (usually motivated by Gestalt principles and often viewed as an *intermediate-level* visual module) often make use of either traditional or *ad hoc* clustering procedures. Rosenfeld and Lee [8] proposed a heuristic for deriving connectivity graphs from line drawings. Scher *et al.* [7] developed a clustering procedure to group small collinear line segments into larger segments. The analysis of dot patterns (and the identification and representation of clusters of dots in digital imagery) was addressed by Velasco and Rosenfeld [9] and Ahuja and Tuceryan [50]; this latter work included an explicit comparison of specialized rules with classical clustering algorithms applied to the same dot patterns.

4: Image Segmentation Via Clustering

How has clustering been applied to image segmentation? In the remainder of this chapter, we will survey, compare, and contrast the variety of segmentation contexts in which clustering has been applied.

4.1: Intensity Image Segmentation

Schachter, Davis, and Rosenfeld [5] applied a local feature clustering approach to segmentation of gray-scale images. This paper emphasized the appropriate selection of features at each pixel rather than the clustering methodology, and proposed the use of image plane coordinates (spatial information) as additional features to be employed in clustering-based segmentation. For example, this paper succinctly shows that intensity should not be supplemented by simple edge features when constructing a feature space for clustering-based segmentation, since the edge features display little global variation. The goal of clustering was to obtain a sequence of hyperellipsoidal clusters starting with cluster centers positioned at maximum density locations in the pattern space, and growing clusters about these centers until a χ^2 test for goodness of fit was violated. A variety of features were discussed and applied to both grayscale and color imagery.

Silverman and Cooper [27] employed an agglomerative clustering algorithm to the problem of unsupervised learning of clusters of *coefficient vectors* for two image models that correspond to image segments. The first image model is a polynomial form for the observed image measurements; the assumption here is that the image is a collection of several adjoining graph surfaces which are sampled on the raster grid to produce the observed image. The algorithm proceeds by obtaining vectors of coefficients of least-squares fits to the data in M disjoint image windows. An agglomerative clustering algorithm merges (at each step) the two clusters which yield a minimum global between-cluster Mahalanobis distance. The same framework was applied to segmentation of textured images, but for such images the polynomial model was inappropriate and a parameterized Markov Random Field model was assumed instead.

Wu and Leahy [32] applied the principles of network flow to unsupervised classification and produced a novel hierarchical algorithm for clustering. In essence, the technique views the unlabeled patterns as nodes in a graph, where the weight of an edge (*i.e.*, its capacity) is a measure of similarity between the corresponding nodes. Clusters are identified by removing edges from the graph to produce connected disjoint subgraphs. In image segmentation, pixels which are 4-neighbors or 8-neighbors in the image plane share edges in the constructed adjacency graph, and the weight of a graph edge is based on the strength of a hypothesized image edge between the pixels involved (this strength is calculated using simple derivative masks). Hence, this segmenter works by finding closed contours in the image and is best labeled edge-based rather than region-based.

Vinod *et al.* [33] developed two neural networks which in combination can be used to perform pattern clustering. A two-layer network operates on a multidimensional histogram of the data to identify 'prototypes' which are used to classify the input patterns into clusters. These prototypes are fed to the classification network, another two-layer network operating on the histogram of the input data, but trained to have differing weights from the prototype selection network. In both networks, the histogram of the image is used to weight the contributions of patterns neighboring the pattern under consideration to the location of prototypes or the ultimate classification; as such, it is likely to be more robust compared to techniques which assume an underlying parametric density function for the pattern classes. This architecture was tested on gray-scale and color segmentation problems.

Zhang and Modestino [34] address the *cluster validation* problem in the context of image segmentation. The correct number of clusters (image segments) is decided using an information-theoretic criterion originally developed for the selection of the proper order and parameters of an autoregressive model for time-series data. This criterion is computed from a maximum likelihood estimate of the model parameters, which in this case were obtained from "homogeneous" regions produced by a K-means clustering of the input data. The proper number of clusters is chosen by iterating K from 1 to a predefined maximum K_{max} and selecting that K which minimizes the criterion. This technique was applied to the segmentation of images assumed to contain regions well approximated by Gaussian random fields. More recently, Zhang *et al.* [60] applied the expectation maximization algorithm to the segmentation of images described by several different models.

Jolion *et al.* [35] extracted clusters sequentially from the input pattern set by identifying hyperellipsoidal regions (bounded by loci of constant Mahalanobis distance) which contain a specified fraction of the unclassified points in the set. The extracted regions are compared against the best-fitting multivariate Gaussian density through a Kolmogorov-Smirnov test, and the fit quality is used as a figure of merit for selecting the 'best' region at each iteration. The process continues until a stopping criterion is satisfied. This procedure was applied to the problems of threshold selection for multithreshold segmentation of intensity imagery and segmentation of range imagery.

4.2: Range Image Segmentation

Clustering techniques have been successfully used for the segmentation of range images, which are a popular source of input data for three-dimensional object recognition systems [13]. Range sensors typically return raster images with the measured value at each pixel being the coordinates of a 3D location in space. Depending on the sensor's configuration, these 3D positions can be understood as the locations where rays emerging from the image plane locations in either a parallel bundle or a perspective cone intersect the objects in front of the sensor.

The local feature clustering concept is particularly attractive for range image segmentation since (unlike intensity measurements) the measurements at each pixel have the same units (length); this would make *ad hoc* transformations or normalizations of the image features unnecessary if their goal is to impose equal scaling on those features. However, range image segmenters often add additional measurements to the feature space, removing this advantage.

Hoffman and Jain [38] described a range image segmentation procedure employing squared-error clustering in a six-dimensional feature space as a source of an "initial" segmentation which is refined (typically by merging segments) into the output segmentation. The procedure was enhanced by Flynn and Jain [41] and used in a recent systematic comparison of range image segmenters [42]; as such, it is probably one of the longest-lived range segmenters which has performed well on a large variety of range images.

This segmenter works as follows. At each pixel (i, j) in the input range image, the corresponding 3D measurement is denoted (x_{ij}, y_{ij}, z_{ij}), where typically x_{ij} is a linear function of j (the column number) and y_{ij} is a linear function of i (the row number). A $k \times k$ neighborhood of (i, j) is used to estimate the 3D surface normal $\mathbf{n}_{ij} = (n^x_{ij}, n^y_{ij}, n^z_{ij})$ at (i, j), typically by finding the least-squares planar fit to the 3D points in the neighborhood. The feature vector for the pixel at (i, j) is the six-dimensional measurement $(x_{ij}, y_{ij}, z_{ij}, n^x_{ij}, n^y_{ij}, n^z_{ij})$, and a candidate segmentation is found by clustering these feature vectors. For practical reasons, not every pixel's feature vector is used in the clustering procedure; typically 1000 feature vectors are chosen by structured or random subsampling (Jain and Hoffman concluded that there was little reason to prefer one sampling technique over the other). Recent work by Judd *et al.* [45] demonstrates that a combination of algorithmic enhancements to the clustering algorithm and distribution of the computations over a network of workstations can allow an entire 512×512 image to be clustered in a few minutes.

The issue of interpattern proximity in this six-dimensional feature space deserves some comments. As noted above, the feature space used here is 'mixed' in that three of the features are ratio-scaled continuous measurements (x, y, z), while the remaining features are ordinal-scaled continuous features measured on the unit sphere (n^x, n^y, n^z). The position and orientation features are not directly comparable, and a pragmatic approach was used to make the computation of proximity between patterns more appropriate. Each of the six features was normalized to zero mean and unit standard deviation before the clustering procedure was executed, and the Euclidean distance measure was used. This set of choices works well in practice.

The CLUSTER algorithm was used to obtain segment labels for each pixel. Hoffman and Jain also experimented with other clustering techniques (*e.g.*, complete-link, single-link, graph-theoretic, and other squared-error algorithms) and found CLUSTER to provide the best combination of performance and accuracy. An additional advantage of CLUSTER is that it produces a sequence of output clusterings (*i.e.*, a 2-cluster solution up through a K_{max}-

cluster solution where K_{max} is specified by the user and is typically 20 or so); each clustering in this sequence yields a clustering statistic which combines between-cluster separation and within-cluster scatter. That clustering which optimizes this statistic is chosen as the best clustering.

Each pixel in the range image is assigned the segment label of the nearest cluster center. This minimum distance classification step is not guaranteed to produce segments which are connected in the image plane; therefore, a connected components labeling algorithm allocates new labels for disjoint regions that were placed in the same cluster. Subsequent operations include surface type tests, merging of adjacent patches using a test for the presence of crease or jump edges between adjacent segments, and surface parameter estimation

Figure 5 shows this processing applied to a range image. Part (a) of the figure shows the input range image; part (b) shows the distribution of surface normals. In part (c), the initial segmentation returned by CLUSTER and modified to guarantee connected segments is shown. Part (d) shows the final segmentation produced by merging adjacent patches which do not have a significant crease edge between them. The final clusters reasonably represent distinct surfaces present in this complex object.

4.3: Texture Image Segmentation

The analysis of textured images has been of interest to researchers for several years. Tuceryan and Jain [46] provide a comprehensive survey of texture definitions, models, and analysis techniques. Texture segmentation techniques have been developed using a variety of texture models and image operations. In this section, we survey some clustering-based approaches to the segmentation of texture images.

Nguyen and Cohen [31] addressed the texture image segmentation problem by modeling the image as a hierarchy of two Markov Random Fields, obtaining some simple statistics from the image texture in each block to form a feature vector, and clustering these blocks using a fuzzy K-means clustering method. The clustering procedure here is modified to jointly estimate the number of clusters as well as the fuzzy membership of each feature vector to the various clusters.

Jain and Farrokhnia [43] developed a system for segmenting texture images. The system uses Gabor filters to obtain a set of 28 orientation- and scale-selective features characterizing the texture in the neighborhood of each pixel. To avoid the "curse of dimensionality" problem associated with a large number of features per pattern, the set of features (ordered by sample variance) which collectively capture a prespecified fraction of the total variance of the original 28 features is used as a reduced feature set. A saturating nonlinear transformation is applied independently to each of the retained feature images, and the feature images are filtered with a uniform kernel. These filtered feature images are then subsampled uniformly to select 1000 or fewer feature vectors, which are then clustered with the CLUSTER program described above. An index statistic proposed by Dubes [44] is used to select the best clustering. Minimum distance classification is used to label each of the original image pixels. This technique was tested on several texture mosaics including the natural Brodatz textures and synthetic images. Figure 6(a) shows an input texture mosaic consisting of four of the popular Brodatz textures [16]. Part (b) shows the segmentation produced when the Gabor filter features are augmented to contain spatial information (pixel coordinates). This Gabor filter technique has proven very powerful and has been extended to the automatic segmentation of text in documents [40] and segmentation of objects in complex background [39].

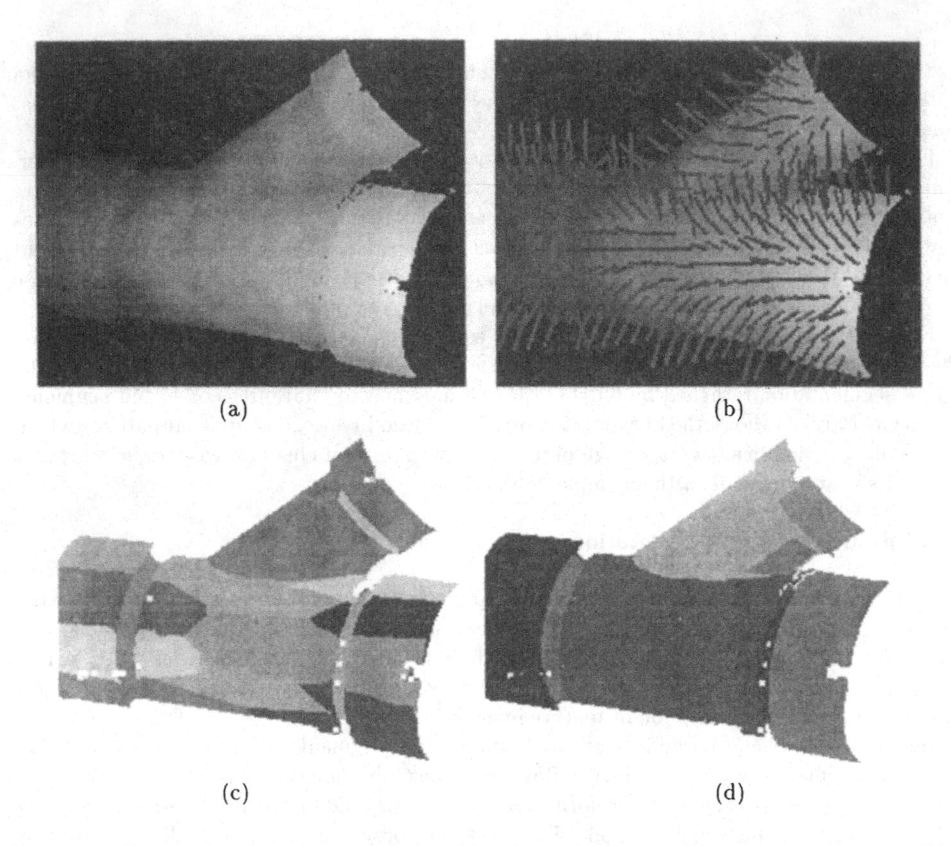

Figure 5. Range image segmentation using clustering. (a): Input range image. (b): Surface normals for selected image pixels. (c): Initial segmentation (19 cluster solution) returned by CLUSTER using 1000 six-dimensional samples from the image as a pattern set. (d): Final segmentation (8 segments) produced by postprocessing.

Mao and Jain [56] combined multiresolution analysis and autoregressive image models in a texture classification and segmentation system. The autoregressive image model is rotation-invariant and admits a simple parameter estimation procedure. Model parameters are computed at multiple image resolutions (constructed using low-pass filtering and subsampling). The performance of this system as a classifier (*i.e.*, identifying input textures as belonging to one of several predefined classes characterized by model parameters) was impressive. The system was modified for use as a texture segmenter. Instead of subsampling the image, multiple parameter estimation windows of different sizes were designed and used, and a squared-error clustering algorithm was used to generate segmentations. A careful evaluation of the ability of features to discriminate between clusters, the assignment of feature weights, and the evaluation of clustering output are notable features of this work.

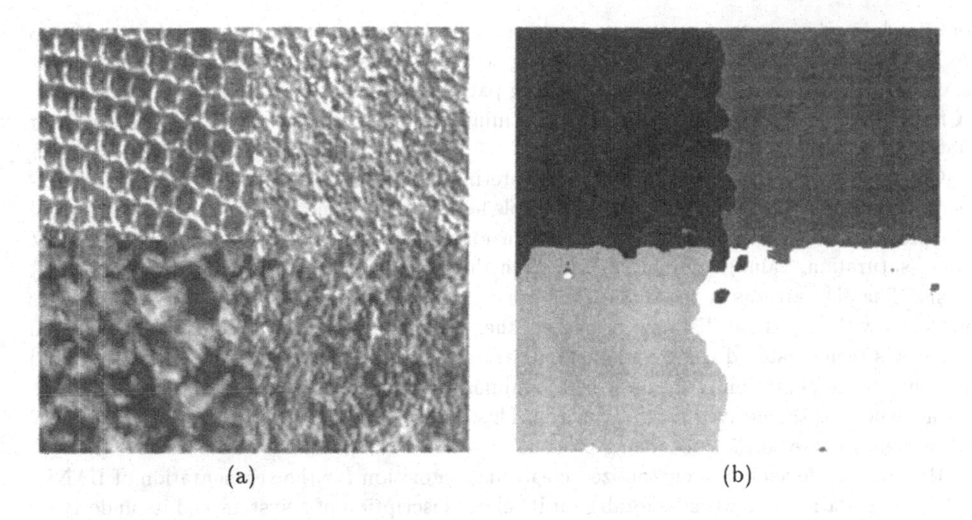

Figure 6. Texture image segmentation results. (a): Four-class texture mosaic. (b): Four-cluster solution produced by CLUSTER with pixel coordinates included in the feature set.

4.4: Color/Multispectral Image Segmentation

Haralick and Kelly [26] proposed two clustering procedures for images (including multispectral images). The *spatial* clustering procedure sequentially builds an image partition starting with a single cluster; at each iteration, a new cluster is defined as a connected subset of pixels with the same measurement value(s), and grown to connected pixels with similar measurement value(s). A *measurement space* clustering procedure iteratively forms connected regions of the measurement space using a probabilistic measure of proximity between subsets in that space. The latter procedure was tested on an aerial image of an urban scene with three measurements per pixel.

Amadasun and King [25] constructed a segmentation system for multispectral imagery employing agglomerative clustering. Image neighborhoods which are uniform according to a mean-value criterion are extracted, summarized by their mean vectors, and fed to an agglomerative clustering procedure to identify different segments. The agglomerative procedure is close to a single-link algorithm, merging the classes of the two closest patterns until the number of output classes equals the (prespecified) number of desired segments. The final image segmentation is obtained from a minimum-distance classification of each (multispectral) pixel with respect to the obtained clusters.

Coleman and Andrews [28] applied a K-means clustering procedure to multispectral image segmentation. Their presentation is rich with the essential and practical details of applying pattern recognition methodology to the image segmentation problem (*e.g.*, the necessity of an eigenvector rotation to obtain decorrelated features, the use of the Bhattacharyya distance measure for feature selection). A variety of features were used in order to classify images with varying amounts of texture; these features included the responses of the Sobel operator to local intensity as well as nearly uniform regions of varying sizes obtained from a nonlinear image filter. The K-means procedure was executed on a subset of the output features obtained by eigenvector rotation and selection, and output clusterings consisting of two to sixteen clusters were produced (beginning with two clusters, a new cluster is formed at each iteration by identifying the most outlying pattern and using it as the new cluster center). A final image segmentation is obtained using minimum distance classification with the set of obtained feature vectors.

Gowda [30] applied the ISODATA [14] clustering algorithm to the segmentation of multispectral images (LANDSAT frames). A notable feature of this method was the use of several transformations to take the original four-dimensional measurement at each pixel into the HSV (hue, saturation, value) color basis, and from there into a pseudo RGB (red, green, blue) basis. The 3D patterns in the RGB space are condensed into a 2D array which summarizes patterns within a small distance of one another by their mean vectors. This set of mean vectors is then clustered using a multistage variant of the popular ISODATA algorithm, and the output segmentation is obtained by a minimum distance classification of the input pixels with respect to the derived cluster centers. This approach to image segmentation uses very little memory due to the condensing procedure.

Bryant [47] developed a customized clustering algorithm for the segmentation of LANDSAT image data. This work is notable for its clear description of a system and its underlying algorithms, and for its comprehensive survey and critique of clustering-based multispectral image segmentation techniques *circa* 1979. Bryant's system has two major phases. In the first phase, simple gradient thresholding is used to identify 'fields' (small uniform regions) in the input data. Samples from these fields are used to form both the subset of data fed to the clustering algorithm and that algorithm's initial set of candidate cluster centers. The clustering algorithm itself works by performing minimum-distance classification with rejection of pixels which are too far from any cluster center. Image-plane cleaning operations (designed to label pixels containing a mixture of class information) are also applied to improve the results. The number of clusters can be automatically adjusted by internally judging the quality of the current set of cluster centers. This approach was tested on several LANDSAT scenes.

Fukuda [29] developed two clustering procedures for segmentation and applied them to color images. The first procedure segments the image by recursively subdividing the image into blocks until each block has a small dispersion about its sample mean in the measurement space, and then grouping blocks with similar mean vectors into connected segments. The second procedure performs one subdivision step (without recursion), obtains a sequence of clusterings by varying a dispersion threshold used as a merging criterion, selects that threshold which yielded the maximum number of clusters, summarizes each cluster by a cluster center, and merges those cluster centers to yield fewer than a prespecified number. In each algorithm, the final image segmentation is employed by a minimum-distance classification of each pixel with respect to the set of output cluster centers.

Uchiyama and Arbib [37] applied competitive learning techniques to the color image segmentation problem. The authors demonstrate an equivalence between clustering and vector quantization, and 'units' which represent cluster centers are sequentially generated. After each generation, the units compete for randomly chosen members of the input pattern set, and the winning unit (which is closest to the pattern) has its weight vector (*i.e.*, the location of the cluster center) updated toward the pattern. Units are created where the local density of patterns is large as evidenced by a large number of 'wins' by a specific unit. In applying this technique to color image segmentation, the authors motivate the use of a feature transformation to improve the segmentation results.

Clustering can be used as a preprocessing stage to identify pattern classes for subsequent

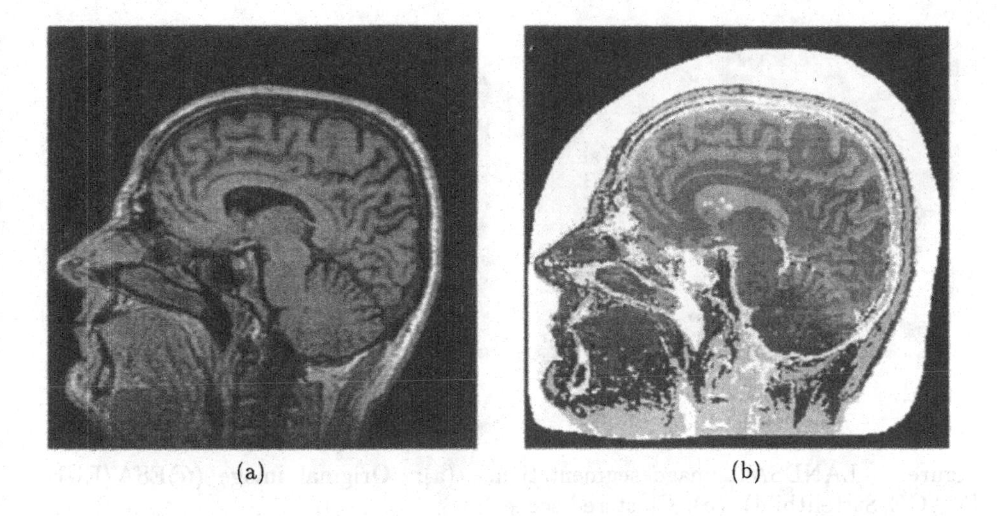

Figure 7. Multispectral Medical Image Segmentation. (a): A single channel of the input image. (b): 9-cluster segmentation.

supervised classification. Taxt and Lundervold [53, 54] employed a partitional clustering algorithm and a manual labeling technique to identify material classes (*e.g.*, cerebrospinal fluid, white matter, striated muscle, tumor) in registered images of a human head imaged at five different magnetic resonance imaging channels (yielding a five-dimensional feature vector at each pixel). A number of clusterings were obtained and combined with domain knowledge (human expertise) to identify the different classes. Decision rules for supervised classification were based on these obtained classes. Figure 7(a) shows one channel of an input multispectral image; part (b) shows the 9-cluster result.

Solberg [62] applied the K-means algorithm to segmentation of LANDSAT imagery. Initial cluster centers were chosen interactively by a trained operator, and correspond to land-use classes such as urban areas, soil (vegetation-free) areas, forest, grassland, and water. Figure 8(a) shows the input image rendered as grayscale; part (b) shows the result of the clustering procedure.

4.5: Multiple Motion Segmentation

Cluster analysis has also been applied to the problem of resolving multiple motions in image sequences. The intuitive model used in current work is that the observed image is composed of *layers*, each with an associated motion, and the segmentation problem is expected to resolve the layers. Jepson and Black [59] used the expectation maximization algorithm [17] to resolve a mixture density model for the motion parameters. Here, the mixture models the presence of several different motion classes. Sawhney *et al.* [57] adopt a similar approach. Wang and Adelson [58] used a K-means clustering procedure to segment different affine motion classes observed in image sequences. Dubuisson and Jain [61] fused motion segmentation, color segmentation based on a split-and-merge paradigm and edge information from the Canny edge detector to extract the contours of moving objects. Figure 9

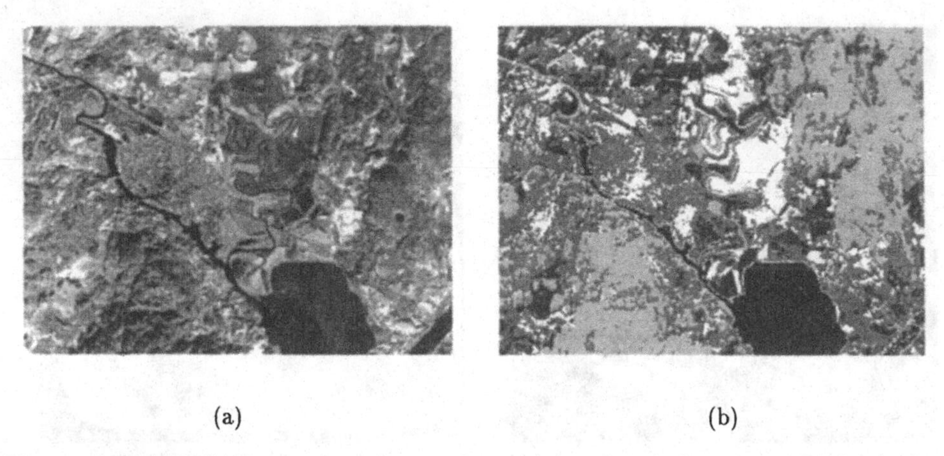

(a) (b)

Figure 8. LANDSAT image segmentation. (a): Original image (©ESA/EUR-IMAGE/Sattelitbild). (b): Clustered scene.

shows the segmentation obtained in a single frame of a motion sequence using this technique.

5: Summary

In this chapter, the application of clustering methodology to image segmentation problems has been motivated and surveyed. The historical record shows that clustering is a powerful tool for obtaining classifications of image pixels. Key issues in the design of any clustering-based segmenter are the choice of pixel measurements (features) and dimensionality of the feature vector (*i.e.*, should the feature vector contain intensities, pixel positions, model parameters, filter outputs, *etc*?), a measure of similarity which is appropriate for the selected features and the application domain, the identification of a clustering algorithm (*e.g.*, squared-error, mode-seeking, graph-theoretic, *etc.*), the development of strategies for feature and data reduction (to avoid the "curse of dimensionality" and the computational burden of classifying large numbers of patterns), and the identification of necessary pre- and post-processing techniques (*e.g.*, image smoothing and minimum distance classification). The use of clustering for segmentation dates back to the 1960s and new variations continue to emerge in the literature. Challenges to the successful use of clustering include the high computational complexity of many clustering algorithms and their incorporation of strong assumptions (often multivariate Gaussian) about the multidimensional shape of clusters to be obtained. The ability of new clustering procedures to handle concepts and semantics in classification (in addition to numerical measurements) will be important for certain applications [52, 51]; we see opportunities here for the adoption of ideas arising in the machine learning literature. With its rich history, our reasonable understanding of the properties and limitations of clustering methodology, and the prospect of increasingly powerful computer systems to use in image analysis systems, we see potential for continued contributions to this fruitful research area.

Figure 9. Segmentation of a vehicle's contour from fused motion, color, and edge information.

6: Acknowledgements and Dedication

We thank Jianchang Mao, Torfinn Taxt, Anne Schistad Solberg, Anne Marie Fenstad, Marie-Pierre Dubuisson Jolly and Yu Zhong for their contributions to this chapter. Our research in clustering methodology and its applications in computer vision has been supported over the past 20 years by the National Science Foundation, NASA, Michigan State University, Washington State University, and a number of industrial organizations. We dedicate this chapter to the memory of Dr. Richard C. Dubes (1934-1993): a pioneering researcher in clustering, collaborator, teacher, mentor, and friend.

References

[1] A. Rosenfeld and A.C. Kak, *Digital Picture Processing* (2nd ed.), Academic Press, 1982.

[2] A. Rosenfeld, M.K. Huang and V.B. Schneider, "An Application of Cluster Detection to Text and Picture Processing," *IEEE Trans. on Information Theory* 15(6):672–681, 1969.

[3] C.A. Sher and A. Rosenfeld, "Pyramid Cluster Detection and Delineation by Consensus," *Pattern Recognition Letters* 12(8):477–482, 1991.

[4] S. Dunn, L. Janos and A. Rosenfeld, "Bimean Clustering," *Pattern Recognition Letters* 1:169–173, 1983.

[5] B.J. Schachter, L.S. Davis and A. Rosenfeld, "Some Experiments in Image Segmentation by Clustering of Local Feature Values," *Pattern Recognition* 11:19–28, 1979.

[6] J.-M. Jolion and A. Rosenfeld, "Cluster Detection in Background Noise," *Pattern Recognition* 22(5):603–607, 1989.

[7] A. Scher, M. Shneier and A. Rosenfeld, "Clustering of Collinear Line Segments," *Pattern Recognition* 15(2):85–91, 1982.

[8] A. Rosenfeld and Y.H. Lee, "A Clustering Heuristic for Line-Drawing Analysis," *IEEE Trans. on Computers* C-21:904–911, 1972.

[9] F.R.D. Velasco and A. Rosenfeld, "Some Methods for the Analysis of Sharply Bounded Clusters," *IEEE Trans. on Systems, Man, and Cybernetics* SMC-10(8):511–518, 1980.

[10] T. Pavlidis, *Algorithms for Graphics and Image Processing*, Computer Science Press, 1982.

[11] R. Jain, R. Kasturi and B.G. Schunck, *Machine Vision*, McGraw-Hill, 1995.

[12] A.K. Jain and R.C. Dubes, *Algorithms for Clustering Data*, Prentice-Hall, 1988.

[13] A.K. Jain and P.J. Flynn (eds.), *Three Dimensional Object Recognition Systems*, Elsevier, 1993.

[14] R.O. Duda and P.E. Hart, *Pattern Classification and Scene Analysis*, Wiley-Interscience, 1973.

[15] E. Backer, *Computer-Assisted Reasoning in Cluster Analysis*, Prentice–Hell, 1995.

[16] P. Brodatz, *Textures: A Photographic Album for Artists and Designers*, Dover, 1966.

[17] G.J. McLachlan and K.E. Basford, *Mixture Models: Inference and Applications to Clustering*, Marcel Dekker, 1988.

[18] F. Farrokhnia and A.K. Jain, "A Multi-channel Filtering Approach to Texture Segmentation," *Proc. IEEE Computer Society Conf. on Computer Vision and Pattern Recognition (CVPR '91)*, Maui, 364–370, 1991.

[19] S. Nadabar and A.K. Jain, "MRF Model-Based Segmentation of Range Images," *Proc. Third International Conf. on Computer Vision*, Osaka, 667–671, 1990.

[20] R. Ohlander, K. Price and D.R. Reddy, "'Picture Segmentation Using a Recursive Region Splitting Method," *Computer Graphics and Image Processing* 8(3):313–333, 1979.

[21] K.S. Fu and J.K. Mui, "A Survey on Image Segmentation," *Pattern Recognition* 13:3–16, 1981.

[22] J. Weszka, "A Survey of Threshold Selection Techniques," *Pattern Recognition* 7:259–265, 1978.

[23] O.D. Trier and A.K. Jain, "Goal-Directed Evaluation of Binarization Methods," *IEEE Trans. on Pattern Analysis and Machine Intelligence* 17(12):1191–1201, 1995.

[24] R.C. Dubes and A.K. Jain, "Clustering Techniques: The User's Dilemma," *Pattern Recognition* 8:247–260, 1976.

[25] M. Amadasun and R.A. King, "Low-Level Segmentation of Multispectral Images via Agglomerative Clustering of Uniform Neighborhoods," *Pattern Recognition* 21(3):261–268, 1988.

[26] R.M. Haralick and G.L. Kelly, "Pattern Recognition with Measurement Space and Spatial Clustering for Multiple Images," *Proc. IEEE* 57(4):654–665, 1969.

[27] J.F. Silverman and D.B. Cooper, "Bayesian Clustering for Unsupervised Estimation of Surface and Texture Models," *IEEE Trans. on Pattern Analysis and Machine Intelligence* 10(4):482–495, 1988.

[28] G.B. Coleman and H.C. Andrews, "Image Segmentation by Clustering," *Proc. IEEE* 67(5):773–785, 1979.

[29] Y. Fukuda, "Spatial Clustering Procedures for Region Analysis," *Pattern Recognition* 12:395–403, 1980.

[30] K.C. Gowda, "A Feature Reduction and Unsupervised Classification Algorithm for Multispectral Data," *Pattern Recognition* 17(6):667–676, 1984.

[31] H.H. Nguyen and P. Cohen, "Gibbs Random Fields, Fuzzy Clustering, and the Unsupervised Segmentation of Textured Images," *CVGIP: Graphical Models and Image Processing* 55(1):1–19, 1993.

[32] Z. Wu and R. Leahy, "An Optimal Graph Theoretic Approach to Data Clustering: Theory and Its Application to Image Segmentation," *IEEE Trans. on Pattern Analysis and Machine Intelligence* 15(11):1101–1113, 1993.

[33] V.V. Vinod, S. Chaudhury, J. Mukherjee and S. Ghose, "A Connectionist Approach for Clustering with Applications in Image Analysis," *IEEE Trans. on Systems, Man, and Cybernetics* 24(3):365–384, 1994.

[34] J. Zhang and J.W. Modestino, "A Model-Fitting Approach to Cluster Validation with Application to Stochastic Model-Based Image Segmentation," *IEEE Trans. on Pattern Analysis and Machine Intelligence* 12(10):1009–1017, 1990.

[35] J.-M. Jolion, P. Meer, and S. Bataouche, "Robust Clustering with Applications in Computer Vision," *IEEE Trans. on Pattern Analysis and Machine Intelligence* 13(8):791–802, 1991.

[36] Y.W. Lim and S.U. Lee, "On the Color Image Segmentation Based on the Thresholding and the Fuzzy c-Means Techniques," *Pattern Recognition* 23(9):935–952, 1990.

[37] T. Uchiyama and M. Arbib, "Color Image Segmentation Using Competitive Learning," *IEEE Trans. on Pattern Analysis and Machine Intelligence* 16(12):1197–1206, 1994.

[38] R.L. Hoffman and A.K. Jain, "Segmentation and Classification of Range Images," *IEEE Trans. on Pattern Analysis and Machine Intelligence* PAMI-9(5):608–620, 1987.

[39] A.K. Jain, N.K. Ratha and S. Lakshmanan, "Object Detection using Gabor Filters," *Pattern Recognition*, to appear in 1996.

[40] A.K. Jain and S. Bhattacharjee, "Text Segmentation using Gabor Filters for Automatic Document Processing," *Machine Vision and Applications* 5:169–184, 1992.

[41] P.J. Flynn and A.K. Jain, "BONSAI: 3D Object Recognition Using Constrained Search," *IEEE Trans. on Pattern Analysis and Machine Intelligence* 13(10):1066–1075, 1991.

[42] A. Hoover, G. Jean-Baptiste, X. Jiang, P.J. Flynn, H. Bunke, D. Goldgof, K. Bowyer, D. Eggert, A. Fitzgibbon and R. Fisher, "An Experimental Comparison of Range Image Segmentation Algorithms," *IEEE Trans. on Pattern Analysis and Machine Intelligence*, to appear.

[43] A.K. Jain and F. Farrokhnia, "Unsupervised Texture Segmentation Using Gabor Filters," *Pattern Recognition* 24(12):1167–1186, 1991.

[44] R.C. Dubes, "How Many Clusters are Best? – An Experiment," *Pattern Recognition* 20:645–663, 1987.

[45] D. Judd, P. McKinley and A.K. Jain, "Large-Scale Parallel Data Clustering," *Int. Conf. on Pattern Recognition*, under review.

[46] M. Tuceryan and A.K. Jain, "Texture Analysis," in *Handbook of Pattern Recognition and Computer Vision* (C. Chen, L. Pau and P. Wang (eds.)), World Scientific, 1993, pp. 235–276.

[47] J. Bryant, "On the Clustering of Multidimensional Pictorial Data," *Pattern Recognition* 11:115–125, 1979.

[48] J. Mao and A.K. Jain, "A Self-Organizing Network for Hyperellipsoidal Clustering (HEC)," *IEEE Trans. on Neural Networks* 7(1):16–29, 1996.

[49] C.T. Zahn, "Graph-theoretical Methods for Detecting and Describing Gestalt Clusters," *IEEE Trans. on Computers* C-20:68–86, 1971.

[50] N. Ahuja and M. Tuceryan, "Extraction of Early Perceptual Structure in Dot Patterns: Integrating Region, Boundary, and Component Gestalt," *Computer Vision, Graphics, and Image Processing* 48:304–356, 1989.

[51] M.N. Murty and A.K. Jain, "Knowledge-Based Clustering Scheme for Collection Management and Retrieval of Library Books," *Pattern Recognition* 28(7):949–964, 1995.

[52] R.S. Michalski and R. Stepp, "Automatic Construction of Classifications: Conceptual Clustering *versus* Numerical Taxonomy," *IEEE Trans. on Pattern Analysis and Machine Intelligence* 5:396–409, 1983.

[53] T. Taxt and A. Lundervold, "Multispectral Analysis of the Brain Using Magnetic Resonance Imaging," *IEEE Trans. on Medical Imaging* 13(3):470–481, 1994.

[54] A. Lundervold, A.M. Fenstad, L. Ersland and T. Taxt, "Brain Tissue Volumes from Multispectral 3D MRI–A Comparative Study of Four Classifiers," *Proc. Society of Magnetic Resonance* (fourth meeting), New York, p. 33, 1996.

[55] R.A. Redner and H.F. Walker, "Mixture Densities, Maximum Likelihood and the EM Algorithm," *SIAM Review* 26(4):195–239, 1984.

[56] J. Mao and A.K. Jain, "Texture Classification and Segmentation Using Multiresolution Simultaneous Autoregressive Models," *Pattern Recognition* 25(2):173–188, 1992.

[57] H. Sawhney, S. Ayer and M. Gorkani, "Dominant and Multiple Motion Estimation for Video Representation," *Proc. International Conf. on Image Processing (ICIP '95)*, Washington DC, vol. 1, 322–325, 1995.

[58] J.Y.A. Wang and E.H. Adelson, "Layered Representation for Motion Analysis," *Proc. IEEE Computer Society Conf. on Computer Vision and Pattern Recognition (CVPR '93)*, New York, 361–366, 1993.

[59] A. Jepson and M.J. Black, "Mixture Models for Optical Flow Computation," *Proc. IEEE Computer Society Conf. on Computer Vision and Pattern Recognition (CVPR '93)*, New York, 760–761, 1993.

[60] J. Zhang, J.W. Modestino and D.A. Langan, "Maximum-Likelihood Parameter Estimation for Unsupervised Stochastic Model-Based Image Segmentation," *IEEE Trans. on Image Processing* 3(4):404–420, 1994.

[61] M.P. Dubuisson and A.K. Jain, "Contour Extraction of Moving Objects in Complex Outdoor Scenes," *International J. of Computer Vision* 14:83–105, 1995.

[62] A.H. Schistad Solberg, "Data Fusion and Texture in Classification of Synthetic Aperture Radar Imagery," Ph.D. thesis, Department of Informatics, University of Oslo, June 1995.

Boundary Encoding Revisited

Herbert Freeman
Department of Electrical and Computer Engineering
Rutgers University, Piscataway, NJ 08855-0909

Abstract

In 1969, at a workshop organized by Azriel Rosenfeld, a paper was presented dealing with boundary encoding and processing. In the intervening years, many advances and extensions of the techniques presented then have been made. This paper reviews some of these developments, with particular emphasis on various generalizations of the chain coding scheme.

1: Introduction

The processing of two-dimensional shape is a topic of abiding interest to those concerned with image processing, computer vision, and pattern recognition. A two-dimensional shape, familiarly often referred to as a "blob," is fully defined by its bounding contour. This allows us to convert a problem dealing with the shape of a region into a more tractable one-dimensional one dealing with a closed contour. Why do we want to process a blob? There are a variety of reasons. We may want to analyze it, that is, determine its area, its perimeter, its maximum diameter, whether or not it is convex, the location of its centroid, whether or not it has one or more axes of symmetry, etc. Or we may want to establish the degree of similarity it has with another blob, either with or without regard to size or orientation. Finally we may want to classify it on the basis of one or more of its shape characteristics.

In 1970 the author published a paper in a book co-edited by Bernice Lipkin and Azriel Rosenfeld under the title of *Boundary Encoding and Processing* [1]. The paper addressed the problem of describing a blob's boundary lines, noted the distinctions among the geometric boundary, "black" or interior boundary, and "white" or exterior boundary, and pointed out how sets of directed contours can be used to describe graytone blobs. The chain-coding technique was demonstrated in the context of multi-level graytone blobs and extended to the direct quantization and encoding of boundary contours, whether open or closed. A variety of processing and manipulation algorithms for chain-encoded curves (or boundary contours) were described, and the now well-known three properties of a digital straight line were postulated. It appears fitting that today, 26 years later, as we honor Professor Rosenfeld's many years of contributions to the field of image processing and its allied sciences, the author should

 0-8186-7644-2 $5.00

re-visit the subject of boundary encoding and describe what has happened in the interval, where we have progressed, and where difficult problems still challenge us.

2: A Blob and Its Boundary

When we speak of a blob, we think of a closed, homogeneous region, simply- or multiply-connected, which models some physical entity, e.g., an area feature on a map (state, county, lake), a blood cell viewed through a microscope, or a defect in an otherwise clear sheet of glass. If the blob corresponds to something in the physical world, it must be digitized before it can be processed by computer. With shape as our only concern, we have the choice of either digitizing an image of the blob and then extracting the boundary from the resulting pixel array, or digitizing the boundary directly. Depending on which approach we follow, somewhat different problem situations are encountered.

2.1: Extracting a contour from a digitized blob

If we first digitize the blob and then look for its boundary, we find that there are at least three different ways of defining the boundary. We can use the contour that literally bounds the connected set of pixels forming the blob, and obtain what in Fig. 1 is marked as the *geometric boundary* (also sometimes referred to as the *crack boundary* [2]). Or we can connect the centers of bounding pixels - either those inside the blob or those outside, obtaining, as shown in Fig. 1, the *interior boundary* in the former case and the *exterior boundary* in the latter. As was proved in [1], the interior boundary will always be shorter than the geometric boundary, and the exterior boundary will be shorter if the geometric boundary consists of 10 or more distinct straight-line segments. All three are functions of the digitization process, as well as of any smoothing and thresholding that was applied, and none can be said a priori to be a more faithful rendition of the original analog contour than another.

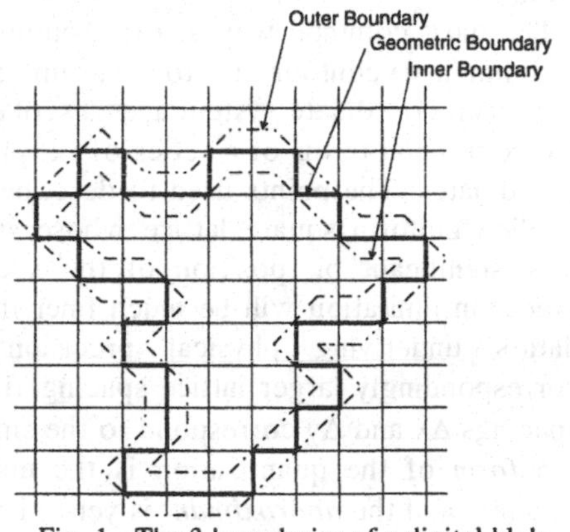

Fig. 1. The 3 boundaries of a digital blob.

During the last few years some researchers have introduced additional ways of describing a digital blob's contour. One interesting variant is the so-called mid-crack scheme [3, 4], which uses straight-line segments of length 1 or $\sqrt{2}/2$ (times the size of a pixel edge) to connect the mid-points of bounding edges. It leads to a bounding contour that more closely "hugs" the geometric boundary but involves more segments than either the interior or exterior boundaries, as illustrated in Fig. 2. Whether or not it has advantages over the more straight-forward interior or exterior boundaries, which use segments of length 1 and $\sqrt{2}$ has not been established. Also, as we shall show later, since precision is solely a function of the underlying quantization lattice, but processing performance is a function of the number of segments, it may be preferable to look for boundary representations that use the fewest possible segments.

2.2: Direct digitization of a contour

The alternate way to obtain the digital representation of a blob's bounding contour is to trace it in the analog domain and digitize it directly. For a multiply-connected blob, we would trace the exterior boundary in the clockwise sense, and any "holes" in a clockwise sense, thus assuring that the interior will always be to the right. The process of digitization is one of description, and description implies quantization [5]. Quantization is defined in term of *form*, the *size* of the quanta, and the *approximant* used to represent the quanta. Many tradeoffs are possible among these, and the choices made can affect the performance of the processing to be carried out on the resulting digital data.

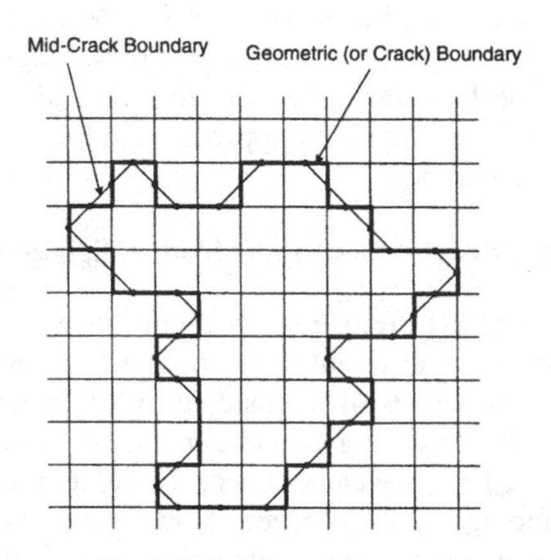

Fig. 2. The mid-crack boundary.

The most common way of representing a boundary contour is to set up a Cartesian coordinate system and describe the contour in terms of a series of (x, y)-coordinates. The points used to describe the contour will necessarily be nodes of an implied uniform square lattice whose spacing corresponds to a unit change in the least-significant bit position of the x and y coordinates. In general, this built-in precision limitation will be much finer than what can be justified on the basis of the data's underlying physical precision limitations, suggesting the use of a correspondingly larger lattice spacing. This is illustrated in Fig. 3, where the lattice spacings Δx and Δy correspond to the smallest significant change in coordinates. Here the *form* of the quantization is the uniform square lattice, the *size* is the lattice spacing, and the *approximant* is yet to be specified. A node becomes a curve point (a selected node for the digital approximation) when the curve crosses a lattice line within half a lattice spacing of the node [6].

If the lattice spacing is chosen as indicated, no useful (i.e., "no significant") information about the curve exists between two adjacent nodes. It follows that to obtain the digital representation of the contour, we should join successive nodes with only the most primitive approximant, namely, a short straight-line segment. This then yields the digital approximation or *chain* representation of a curve, where for each selected curve point the next in sequence is one of the eight lattice points surrounding it, connected with a segment of length 1 or $\sqrt{2}$ (times the lattice spacing) [6, 7].

It has long been known that if a contour is to be approximated by a sequence of straight-line segments, the segments should be short where the curvature is great, and should be correspondingly longer where the curvature is slight. This suggests that we represent a contour by means of line segments of varying length, selected so as to keep some error measure within a specified bound. The error measure could be the maximum distance between the curve and the approximating line segment, the average distance, the mean-square distance, or the in-between area. The result will be a polygonal approximation of the contour. As long as we insist that wherever a line segment crosses one of the lattice lines, the distance from the curve to the nearest node be no greater than half a lattice spacing, we shall retain the full precision set by the lattice.

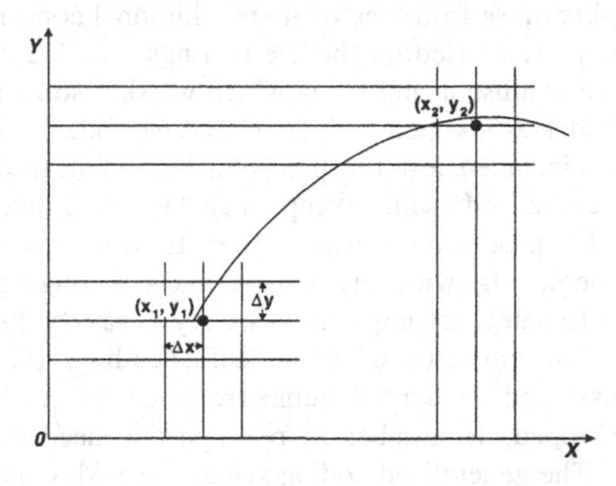

Fig. 3. Quantization of a contour

In Fig. 4 we show a point ***A*** on a square lattice, surrounded by concentric "rings" of nodes. The nodes in the inner-most ring are labeled counterclockwise from 0 to 7; those in the next ring, from 8 to 23, etc. Each ring contains *8n* nodes, where *n* is the ring number. The sum of all nodes for rings 1 through *n* is equal to $4n(n+1)$. Clearly, if we can identify a longer segment, i.e., one that would connect point ***A*** to a node of ring $n > 1$, while satisfying the half lattice-spacing precision requirement, then our approximation will consist of fewer segments. For $n = 1$, we have the basic chain-code, and for rings of all values from 1 to *N*, where *N* is such that the contour lies entirely within this ring, we have unrestricted polynomial approximation.

A contour-approximation scheme that provides for many line segments that rarely occur will tend to be inefficient to encode. This suggests that we curtail the number of permissible line segments. In the familiar chain-coding scheme, only the nodes in ring 1 of Fig. 4 are permissible "next" nodes. Contour representation based on this scheme has been widely used for more than 30 years; it is simple, efficient, compact, and facilitates the design of simple processing and manipulation algorithms [1, 7]. A detailed analysis of the performance of chain codes can be found in [8]

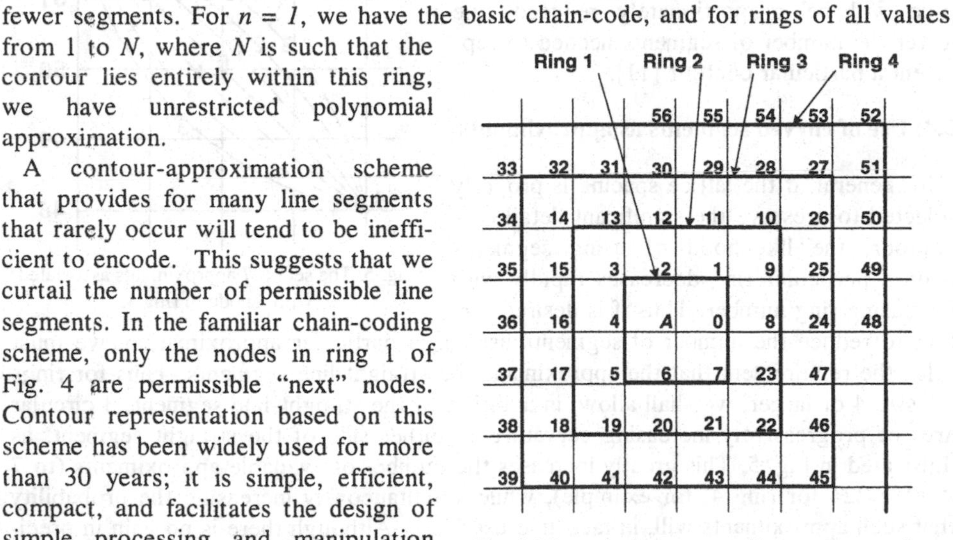

Fig. 4. The node rings surrounding a current node.

2.3 The generalized chain representation.

In the late 1970s the *generalized chain coding scheme* was introduced [9,10]. It allows the building of a contour's digital approximation by using nodes from ring 1 plus those from one or more additional concentric rings. Generalized codes have been explored based on the use of rings 1-2, 1-2-3, 1-2-3-4, and larger. And, of course, we can use a ring set in which we skip some intermediate rings, such as ring sets 1-3 and 1-2-5, which contain 32 and 64 nodes, respectively. Note that ring 1 must always be included lest the precision be compromised. Digitization of a contour using the generalized chain concept is analogous to that used for the basic 8-direction chain [7], The procedure always attempts first to use the largest possible line segments (beginning with the largest allowed ring) and then steps down to smaller rings (ultimately to ring 1) if necessary to satisfy the precision requirement [10].

For purposes of numerical encoding, the segments in a ring set are normally assigned sequential numbers, in a counterclockwise spiraling manner. If a ring is skipped, the number are reassigned as needed, e.g., from 0 to 31 for ring set 1-3.

The generalized coding scheme provides an expanded "vocabulary" of line segments for representing a contour, thereby offering the possibility of a smoother representation as well as of fewer approximating segments. Generally, algorithms for analyzing or manipulating such contour representations process them one segment at a time. With the use of appropriate table-lookup schemes, as long as the ring size is kept to a modest value (e.g., 5 or 7), the complexity of processing a segment is virtually independent of the segment's length. Hence the processing time will be proportionately reduced, the fewer the number of segments needed to represent a particular contour [11].

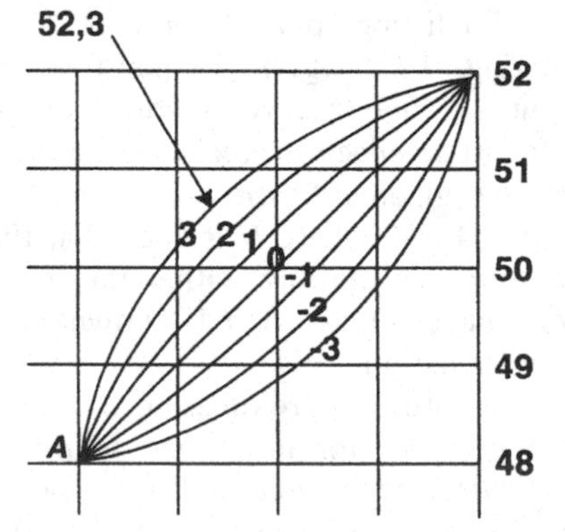

Fig. 5. The set of 7 approximants associated with a node in ring 4.

2.4: Use of curved segments as approximants

In general, if the lattice spacing is properly selected to preserve the significant detail of a contour, the likelihood of using segments from a particular ring decreases rapidly with increasing ring number. Thus if is desired further to reduce the number of segments used in a particular approximation, we must relax the requirement that the approximants be straight-line segments. Thus for rings of size 4 or larger, we shall allow, in addition to the straight-line segment, 3 circular arcs of progressively increasing curvature to either side of the straight segment, as illustrated in Fig. 5. This greatly increases the number of available approximants (to 7 x 32 = 224 for ring 4, for example), while simultaneously increasing the probability that such approximants will, in fact, find use [12]. Although there is no gain in precision (as long as we use the same lattice spacing), this generalized "polycurve" representation can yield faster processing and smoother displays.

The effects of the various generalizations on the chain code are illustrated in Fig. 6. The reduction in the number of approximating segments as well as the increasing smoothness are clearly evident, as we go from the basic 8-direction chain representation to the ring 1-3 generalized chain and the ring 1-4 polycurve representation. The number of segments required to represent this particular blob was found to be 87 in (b), 48 in (c), 42 in (d), and 27 in (e).

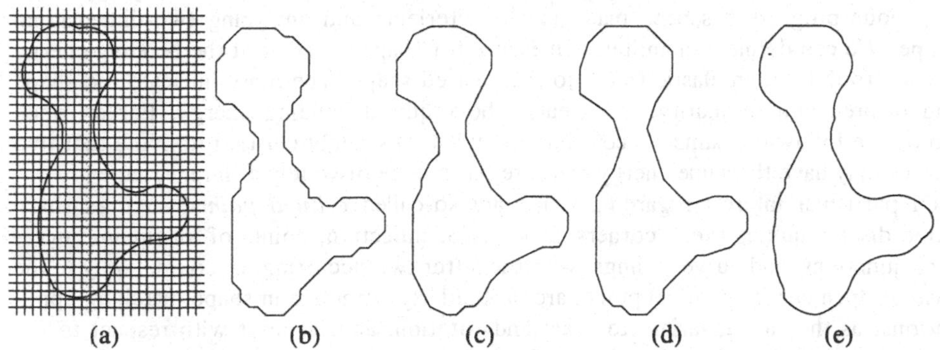

Fig, 6. (a) A given blob with the overlaid digitizing lattice. (b) Representation using ring 1 only (basic chain code), (c) using rings 1 and 2, (d) using rings 1 and 3, and (e) using a 1-4 ring polycurve.

2.5 Open curves

The techniques described for directly representing the boundary of a blob apply equally when the contour is not closed. However, if we want to apply the techniques to general line drawings, we must allow for junctions. One approach is to represent each segment separately, and then make provision for specifying the starting-point coordinates of each segment, say, by embedding special control codes in each chain-code string [7]. An elegant alternative is provided by the *primitives chain code* introduced by O'Gorman, which directly retains branching and junction topology information [13].

3: Hierarchical Representations

Paralleling the development of techniques for serially processing contours in full description, have been efforts at parallel processing and processing at different levels of resolution. The main motivation for considering parallel processing is the obvious one of reduced processing time. Parallel processing has long been considered for image arrays, but its application to curves (open or closed) and general line drawings had not been given much attention until fairly recently, mainly because of the inherently serial nature of many curve representation schemes.

An effective framework for parallel processing of contours is provided by resorting to multi-resolution extensions of the chain coding scheme [14]. Lattices of different resolution are used, with the finest level corresponding to the precision limit, and the coarsest to a single cell. Developments of this approach have led to the RULI chain code [15] and the chain pyramid [16]. Hierarchical multi-resolution approaches additionally are of interest for transmitting line-drawing information, as such schemes allow useful tradeoffs between resolution and transmission time, an important current research topic.

4: Shape Characterization

During the 27 years since the paper on boundary encoding and processing appeared, enormous progress has been made in characterizing and analyzing two-dimensional shape. We can define a quantitative measure for shape in terms of the bending energy needed to deform an elastic ring into the desired shape. The more intricate the shape and desired non-circularity, the greater the required bending energy. For an open curve, we follow the same concept but apply it to a straight elastic rod. Note that two curves may have the same energy measure but still be of widely different shape [17].

Of particular interest regarding shape are so-called *critical points*, such as sharp slope discontinuities (i.e., "corners"), points of inflection, points of maximum curvature, junctions, and curve endings, with the latter two occurring, of course, only in the case of open curves. Critical points are particularly attractive in shape matching applications, as they are invariant to scale and rotation, and, at least with respect to the corners, tend to be relatively robust [18-20].

5: Conclusion

A brief overview has been provided of some selected advances in boundary representation made during the past 25 years. The total effort in this area has been vast, and any attempt to provide a comprehensive review would have been far beyond the scope of this paper. All that was possible was to highlight some of the key developments relating to chain codes, in particular, their generalization to straight and curved segments of varying length, as well as to some variations of these, including those based on the use of hierarchical approaches.

References

1) H. Freeman, "Boundary encoding and processing," in *Picture Processing and Psychopictorics*, ed. by B. Lipkin and A. Rosenfeld, Academic Press, New York, NY, 1970, 241-266.
2) A. Rosenfeld and A.C. Kak, *Digital Picture Processing,* v. 2, Academic Press, New York, 1982, p.196.
3) K. Dunkelberger and O. Mitchell, "Contour tracing for precision measurement, *Proc. IEEE Int'l. Conf. on Robotics and Automation,* St. Louis, MO, 1985, 22-7.
4) F.Y. Shih and W-T Wong, "A new single-pass algorithm for extracting the mid-crack codes of multiple regions," *Jour Visual Commun. and Image Repr., 3,* 3, Sept. 1992, 217-24.
5) C. Cherry. *On Human Communication*, John Wiley & Sons, New York 1957, pp. 46-7.
6) H. Freeman, "On the Encoding of Arbitrary Geometric Configurations", *IRE Trans., EC-10,* (2), 260-268, June 1961.
7) H. Freeman, "Computer Processing of Line-Drawing Images," *Computing Surveys, 6,* (1), March 1974, 57-97.
8) J. Koplowitz,, "On the performance of chain codes for quantization of line drawings," IEEE *Trans. Pattern Analysis and Machine Intell., Vol. PAMI-3,* (2), 1981, 180-185.

9) H. Freeman, "Application of the generalized chain coding scheme to map data processing," *Proc. IEEE Computer Soc. Conf. Pattern Recog, and Image Proc.*, Chicago, IL, 1978, 220-6.
10) H. Freeman and J. Saghri, "Analysis of the Precision of Generalized Chain Codes for the Representation of Planar Curves," *IEEE Trans. Pattern Analysis and Machine Intell., Vol. PAMI-3*, (5), Sept. 1981, 533-539
11) D.L. Neuhoff and K.G. Castor, "A rate and distortion analysis of chain codes for line drawings," *IEEE Trans. Information Theory, IT-31*, (1),. 1985, 53-67.
12) C..Y Choo and H. Freeman, "Computation and features of 2-d polycurve-encoded boundaries," *Proc. IEEE Int'l. Conf. on Systems , Man, and Cybernetics,* Cambridge, MA, 1989.
13) L. O'Gorman, "Primitives chain code," in *Progress in Computer Vision and Image Processing*, ed. by A. Rosenfeld, and L. Shapiro, Academic Press, San Diego, 1992, 167-183.
14) H. Samet and R.E. Webber, "On encoding boundaries with quadtrees," *IEEE Trans. Pattern Analysis and Machine Intell., Vol. PAMI-6*, (3),. 1984, 365-369
15) W.G. Kropatch, "Curve representations in multiple resolutions', *Pattern Recog. Letters, 6,* (3), 1987, 179-184.
16) P. Meer, C.A. Sher, and A. Rosenfeld, "The chain pyramid: hierarchical contour processing," *IEEE Trans. Pattern Analysis and Machine Intell., Vol. PAMI-12*, (4), 1990, 363-376.
17) H. Freeman and J. Glass, "On the quantization of line-drawing data, " *IEEE Trans. System Science and Cybernetics, vol. SSC-5*, (1), January 1969, 70-79.
18) H. Freeman, "Shape Description via the Use of Critical Points," *Pattern Recognition 10*, (3), 1978, 159-66.
19) C-H.Teh and R.T. Chin, "On the detection of dominant points on digital curves," *IEEE Trans. Pattern Analysis and Machine Intell., Vol. PAMI-11*, (8),. 1989, 859-872.
20) P. Zhu and P.M. Chirlian, "On critical point detection of digital shapes," *IEEE Trans. Pattern Analysis and Machine Intell., Vol. PAMI-17*, (8), Sept. 1995 737-748.

Section 2

Feature Extraction

Optimal Edge Detection in Images

Richard J. Qian and Thomas S. Huang
Beckman Institute and Coordinate Science Laboratory
University of Illinois at Urbana-Champaign
405 North Mathews Avenue, IL 61801

Abstract

In this paper, we present a new two-dimensional (2D) edge detection algorithm. The algorithm detects edges in 2D images by a curve segment based edge detection functional that uses the zero-crossing contours of the Laplacian of Gaussian (LOG) as initial conditions to approach the true edge locations. We prove that the proposed edge detection functional is optimal in terms of signal-to-noise ratio and edge localization accuracy for detecting general 2D edges. In addition, the detected edge candidates preserve the nice scaling behavior that is held uniquely by the LOG zero-crossing contours in scale space. The algorithm also provides: 1) an edge regularization procedure that enhances the continuity and smoothness of the detected edges; 2) an adaptive edge thresholding procedure that is based on a robust global noise estimation approach and two physiologically originated criteria to help generate edge detection results similar to those perceived by human visual systems; and 3) a scale space combination procedure that reliably combines edge candidates detected from different scales.

1. Introduction

Although researchers in computer vision have experienced great difficulties in finding a robust edge detection algorithm for visual scene analysis, most of them still believe that edges have an important role in visual perception and therefore the long journey for pursuing a good edge detection algorithm has to continue. Such a belief may stem mainly from the evidence found in a series of psychophysical experiments on biological visual systems [17]. In summary, those experiments found: 1) some patterns are less easily seen if the sharpness of contours associated with those patterns is reduced; 2) apparent brightness is greatly influenced by contour-contour interaction; and 3) the appearance of a single edge is relatively independent of the luminance profile that assembles it. Based on these results and some other emerging evidence found in physiological experiments, especially the findings by Hubel and Wiesel [5], it is often believed that there exist special mechanisms in biological visual systems that perform the edge detection process [17]. However, despite decades of intensive research on biological visual systems, there is still a high degree of disagreement on the basic mechanisms underlying visual signal

0-8186-7644-2 $5.00 © 1996 IEEE

processing. In light of this, most researchers in computer vision have developed their edge detection algorithms from the point of view of computational information processing. One of the most notable pioneering work on computational approaches to edge detection is due to Rosenfeld and Thurston [16].

The difficulties in designing a robust and accurate edge detection algorithm for two-dimensional (2D) images mainly come from two aspects. Firstly, there are tradeoffs in choosing an operator to pursue the best overall edge detection performance. Based on Yuille and Poggio's result [22], the 2D Laplacian-of-Gaussian (LOG) operator [9] should be used because it is the only operator whose zero-crossings have a constrained behavior in 2D scale space, which in turn lays a necessary foundation for scale space manipulations. However, an isotropic operator like the LOG is not optimal in terms of signal-to-noise ratio (SNR) and edge localization accuracy (ELA). The Canny edge detector [3] has better SNR and ELA than the LOG. However, the local extrema of its output may have unconstrained behaviors in 2D scale space [22]. Moreover, the 2D version of the Canny edge detector is obtained by simply extending its 1D version based on a linear constant cross-section edge model. As a result, except in the special case where the detected edge is a straight line and has a constant intensity along the line, the 2D Canny edge detector is not optimal in terms of SNR and ELA. This will be analyzed in details later in this paper. Recent efforts on finding an optimal edge detector can be found, e.g., in the work by Tagare and deFigueiredo [18] and the work by Boyer and Sarkar [2]. Unfortunately the issue of establishing a more accurate 2D edge model has still been overlooked. Secondly, it is very difficult to find a reliable edge combining approach in 2D scale space since the zero-crossings of the second derivatives of the filtered 2D signals generally behave in a much more complex way in scale space than the 1D zero-crossings. General discussion of scale space theory can be found in Witkin's paper [21] and in the book by Lindeberg [7]. Recent effort on detecting edges in scale space can be found, e.g., in the work by Perona and Malik [12]. For a more thorough discussion on the general issues related to edge detection, see our technical report [14].

In this paper, we introduce a new 2D parametric edge model and develop a new 2D edge detection functional based on such a model. We prove that the new edge detection functional not only achieves the optimality in terms of SNR and ELA for detecting edges in 2D images but also preserves the nice scaling property of the LOG zero-crossings in scale space. In addition, we present: 1) an edge regularization procedure that enhances the continuity and smoothness of the detected edges; 2) an adaptive edge thresholding procedure that is based on a robust global noise estimation approach and two physiologically originated criteria; and 3) a scale space edge combination procedure that reliably combines edge detection results from different scale channels. In Section 2, we will present the 2D parametric edge model and develop the optimal edge detection functional. The edge regularization procedure will also be described there. In Section 3, we will discuss issues of how to evaluate edge salience and how to combine edge detection results in scale space. In Section 4, we will describe the implementation details of our algorithm and present some experimental results. Finally in Section 5, we will summarize our algorithm and discuss some future work.

2. Optimal 2D edge detection and edge regularization

Because in general the linear constant cross-section edge model can represent only a short piece of a 2D edge in reasonable precision, edge detection algorithms assuming such a model usually detect edges by extracting edge pixels in a point by point manner. The drawbacks of this type of methods are two folds. Firstly, it limits the best possible edge detection performance in terms of SNR and ELA as we will see in Section 2.3. Secondly, it directly causes the well-known streaking problem in edge detection results. We address these problems in this section by introducing a new 2D parametric edge model and developing an optimal 2D edge detection functional based on such a model.

2.1. A 2D edge and its models

A 2D edge model is a surface model that represents the intensity surface around a 2D edge in an intensity image. Among 2D edge models, the linear constant cross-section model has been widely adopted by previous edge detection investigators for its simplicity. The model is built upon the assumption that to some extent a 2D edge segment remains straight and at a constant intensity along the edge direction, see Figure 1(b). The illumination is usually not uniform in real world, however, therefore intensity can vary in any direction including the edge direction in real images. In many cases, the intensity changes along edge directions are so substantial and rapid that the constant cross-section assumption cannot be taken as a good approximation. Figure 1(a) shows the intensity surface around a short edge segment taken from the picture 'Lena'. We can see that the edge is curved and the intensity changes rapidly along the edge direction. Namely, in order to make the linear constant cross-section model more plausible for any intensity varying or curved edges in real images, one has to confine the extent of the model to a very small local neighborhood. As a result, an edge detection approach built upon such a model tends to detect edges locally, very often point by point.

Here we introduce a more accurate parametric model for general 2D edges. Assume that a 2D edge has a 2D trajectory $\vec{\alpha}(u) = (x(u), y(u))^{\mathrm{T}}$ in the image plane, where u is the parameter of the trajectory. Then the intensity surface around the 2D edge can be modeled as a parameterization [4] of the local coordinates (u, v)

$$E(u,v) = A(u)P(u,v), \quad -L \le u \le L, -D \le v \le D, \tag{1}$$

where v is the parametric coordinate in the gradient direction of intensity, $A(u)$ the amplitude function, $P(u,v)$ the profile function of the edge, and $[-L,L]\times[-D,D]$ the region of support of the edge surface. Note that if we let $\vec{\alpha}(u)$ be a straight line in the image plane and $A(u)$ be constant, the above model degenerates to the linear constant cross-section model. For a step edge in real image, $P(u,v)$ is the convolution of a Gaussian function and an ideal step function, i.e.,

$$P(u,v) = \frac{1}{\sqrt{2\pi}\sigma(u)}\int_0^{\infty}[\exp(-\frac{(v-\xi)^2}{2\sigma^2(u)}) - \exp(-\frac{(v+\xi)^2}{2\sigma^2(u)})]d\xi,$$
$$-L \le u \le L, \quad -D \le v \le D. \tag{2}$$

where the variance $\sigma(u)$ of the Gaussian determines the scale of the edge profile. Note that we have subtracted the DC component from the step function before the convolution. The profile functions for other types of edges can be found in a similar way as for the step edge. Figure 1(c) illustrates a 2D step edge model based on Equation (1) and Equation (2) for the edge segment in Figure 1(a).

2.2. An optimal 2D edge detection functional

Let $E(u,v)$ be an edge with a trajectory $\vec{\alpha}(u)$ to be detected in an image. We assume that $\vec{\alpha}(u)$ is a regular curve, i.e., the tangent of the curve is defined at every point on the curve. We construct a 2D edge detection functional Q_{edge} with two functions $f_t(u)$ and $f_g(u,v)$ to be optimized as follows

$$Q_{edge}(u_0, v_0(u)) = \int_{-L}^{L} f_t(u)\left(\int_{-D}^{D} f_g(u_0-u, v)E(u_0-u, v_0(u_0-u)-v)dv\right)du, \tag{3}$$

where $v_0(u)$ is a trajectory of parameter v with respect to parameter u.

Assuming an additive white Gaussian noise with variance n_0^2 in the image and using the 2D edge model defined by Equation (1), we show in Appendix that the output SNR of the edge detection functional Q_{edge} is

$$\mathrm{SNR} = \frac{\left|\int_{-L}^{L} f_t(u)A(-u)\left(\int_{-D}^{D} f_g(-u,v)P(-u,-v)dv\right)du\right|}{n_0\sqrt{\int_{-L}^{L} f_t^2(u)\left(\int_{-D}^{D} f_g^2(-u,v)dv\right)du}}, \tag{4}$$

and the ELA is

$$\mathrm{ELA} = \frac{n_0\sqrt{\int_{-L}^{L} f_t^2(u)\left(\int_{-D}^{D} \left(\frac{\partial}{\partial v} f_g(-u,v)\right)^2 dv\right)du}}{\left|\int_{-L}^{L} f_t(u)A(-u)\left(\int_{-D}^{D} \frac{\partial}{\partial v} f_g(-u,v)\frac{\partial}{\partial v}P(-u,-v)dv\right)du\right|}. \tag{5}$$

If we assume that the profile functions $P(u,v)$ of a 2D edge have uniform scale factor $\sigma(u) = \sigma(0)$, $-L \le u \le L$, by recalling Equation (2), the above output SNR and ELA can then be simplified as follows.

$$\mathrm{SNR} = \frac{\left|\int_{-L}^{L} f_t(u)A(-u)du\int_{-D}^{D} f_g(0,v)P(0,-v)dv\right|}{n_0\sqrt{\int_{-L}^{L} f_t^2(u)du\int_{-D}^{D} f_g^2(0,v)dv}}, \tag{6}$$

and

$$\mathrm{ELA} = \frac{n_0 \sqrt{\int_{-L}^{L} f_t^2(u)du \int_{-D}^{D} (\frac{\partial}{\partial v} f_g(0,v))^2 dv}}{\left| \int_{-L}^{L} f_t(u) A(-u)du \int_{-D}^{D} \frac{\partial}{\partial v} f_g(0,v) \frac{\partial}{\partial v} P(0,-v)dv \right|} \quad (7)$$

The edge detection functional Q_{edge} can be optimized by maximizing the SNR and minimizing the ELA. If no additional constraint is selected, the optimal solution can be obtained by using the Cauchy-Schwartz inequality. Namely, the optimal detection function along the edge trajectory $f_t^*(u)$ is

$$f_t^*(u) = A(-u), \quad -L \leq u \leq L, \quad (8)$$

and the optimal detection function in the gradient direction $f_g^*(u,v)$ for a step edge is

$$f_g^*(u,v) = P(u,-v) = \frac{1}{\sqrt{2\pi}\sigma(u)} \int_0^{\infty} [\exp(-\frac{(v+\xi)^2}{2\sigma^2(u)}) - \exp(-\frac{(v-\xi)^2}{2\sigma^2(u)})]d\xi,$$

$$-L \leq u \leq L, \quad -D \leq v \leq D. \quad (9)$$

The above results are also recognized as the matched filters in information theory.

In practice, the exact locations of edge trajectories in an image cannot be known beforehand. That is indeed the purpose of edge detection. However, the edges may be approximately located by first applying some non-optimal edge detecting operators, in our case the LOG, to the image. In order to determine the true location of an edge, we then search for a trajectory $v_0^*(u)$ that maximizes the response of the detection functional Q_{edge} in some neighborhood around the corresponding LOG zero-crossing contour. We choose the LOG as our preliminary edge detector because it posses the best scaling property in scale space. Using the LOG zero-crossing contours as its initial conditions, the optimal edge detection functional Q_{edge} therefore inherits the nice scaling property from the LOG. We will discuss more about the implementation issues including how to obtain the amplitude function $A(u)$ of an edge in real images in Section 4.

2.3. Output SNR and ELA improvement over the Canny edge detector

Let $E(u,v)$ be an edge with a trajectory $\vec{\alpha}(u)$ to be detected in an image. Assuming $\vec{\alpha}(u)$ is curved or $E(u,v)$ has nonuniform amplitude, then we have

$$\frac{\mathrm{SNR}^{Q_{\text{edge}}}}{\mathrm{SNR}^{\text{Canny}}} = \frac{\mathrm{ELA}^{\text{Canny}}}{\mathrm{ELA}^{Q_{\text{edge}}}} = \frac{\sqrt{\int_{-L}^{L} A^2(u)du}}{A(u)}, \quad -L \leq u \leq L. \quad (10)$$

Namely, the edge detection functional Q_{edge} has improved the SNR and ELA simultaneously over the Canny edge detector by a factor of the number defined in Equation (10). In the case where $E(u,v)$ has a curved trajectory $\vec{\alpha}(u)$ but uniform amplitude $A(u)$, the above factor is simply the square root of the total length of the edge, i.e.,

$$\frac{\mathrm{SNR}^{Q_{edge}}}{\mathrm{SNR}^{Canny}} = \frac{\mathrm{ELA}^{Canny}}{\mathrm{ELA}^{Q_{edge}}} = \sqrt{2L}, \quad -L \le u \le L. \tag{11}$$

The reason for that the edge detection functional Q_{edge} has better SNR and ELA than the Canny edge detector is mainly due to the fact that Q_{edge} always detects edges piecewisely while the Canny detector is pointwise in the case where edges are curved and/or the amplitude along edges is nonuniform. The increased detecting range along edge direction improves the output SNR and lower the probability of declaring spurious response as edges therefore improves the ELA.

2.4. Edge regularization

In Section 2.2, we introduced an optimal 2D edge detection functional Q_{edge}. Due to noise and discretization error, however, a numerical optimal solution of the trajectory $v_0^*(u)$ that maximizes the edge detection functional Q_{edge} defined in Equation (3) may not be continuous or smooth. One way to improve the continuity and/or smoothness is to regularize the problem [20, 13, 6]. More specifically, we may define energy functionals associated with an edge and then minimize the total energy functional of the edge to obtain an optimal trajectory that has some desired characteristics for that edge.

In the Cartesian coordinate system, a trajectory $v(u)$ can be expressed as $\mathbf{v}(u)$ where $\mathbf{v}(u) = (x(u), y(u))^{\mathrm{T}}$. We define the energy functionals of an edge as

$$\boldsymbol{E}^*_{edge} = \int_{-L}^{L} [\boldsymbol{E}_{int}(\mathbf{v}(u)) + \boldsymbol{E}_{con}(\mathbf{v}(u))] du \tag{12}$$

where

$$\boldsymbol{E}_{int}(\mathbf{v}(u)) = (\alpha(u)|\dot{\mathbf{v}}(u)|^2 + \beta(u)|\ddot{\mathbf{v}}(u)|^2)/2 \tag{13}$$

and

$$\boldsymbol{E}_{con}(\mathbf{v}(u)) = \gamma(u)|\mathbf{v}(u) - \mathbf{v}_0^*(u)|^2/2. \tag{14}$$

The total energy $\boldsymbol{E}^*_{edge}$ of an edge can be minimized over an optimal trajectory $\mathbf{v}^{opt}(u)$ by solving the following Euler-Lagrange equation

$$\gamma(u)(\mathbf{v}^{opt}(u) - \mathbf{v}_0^*(u)) - \frac{d}{du}(\alpha(u)\dot{\mathbf{v}}^{opt}(u)) + \frac{d^2}{du^2}(\beta(u)\ddot{\mathbf{v}}^{opt}(u)) = 0. \tag{15}$$

The details of the numerical method to solve the above equation will be discussed in Section 4.

3. Detecting salient edges at different scales

Edges can occur over a wide range of strengths as well as scales in real images. In order to generate a complete yet clean edge map, some thresholds are needed and they have to be selected adaptively. And detection results from different scales are also needed to be combined. We discuss all these issues in this section.

3.1. Evaluating the salience of an edge

In order to derive an adaptive thresholding mechanism for edge detection, we have to measure certain characteristics of the detected images. The first and perhaps also the most important measure is to estimate the noise level in the images. Once the noise level can be estimated reliably, some thresholds may be selected to separate the responses of real edges and those of noise. An edge is considered salient in an image if, with respect to noise, it has a distinguishable response to the edge detector.

Wiener filtering is used by Canny [3] to estimate the noise level in an image. Since the procedure eventually needs to estimate the mean-squared noise from local values, the results can be easily biased by some regular patterns, e.g., step edges. In fact, any type of noise estimation based on local values is sensitive to some regular patterns, like step edges, and its performance will be poor if the density and strengths of edges are high [3]. As a result, a local type of noise estimation is usually not robust in practice.

Here we introduce a global noise estimation approach. In contrast to the method based on local values, our approach is carried out entirely on the candidate edge segments in all the selected channels. The approach consists of three steps for each channel. In step 1, it computes the magnitude of gradient in the Gaussian blurred image for every point on the candidate edge segments and assign the magnitude to that point as its strength. After computing the strengths for all those points, the strengths are equalized into the full scale of intensity by a standard histogram equalization procedure. In step 2, the approach computes the normalized strength for each candidate edge segment. The normalized strength is defined as dividing the sum of the strengths of the points on the segment by the length of the segment. We compute the normalized strengths for all the candidate edge segments in all the channels. A histogram of the computed edge segment strengths can then be constructed for each channel. Figure 2 shows the histograms of the normalized strengths of the candidate edge segments of the 'Lena' picture in two different channels. Due to random noise in an image, a Gaussian distribution can always be found at the low intensity end in the histogram constructed as above. Finally in step 3, the approach fits the low intensity part of the histogram into a Gaussian distribution. The mean and the variance of the Gaussian reflect the average global noise level and the spreading range of the noise. Based on these values, a thresholding mechanism can be derived. In Section 4, we will describe such a mechanism that has also been employed in our current implementation.

Although the noise estimation serves as the basis for determining edge salience, additional criteria are often needed to help in generating edge detection results similar to those perceived by human visual systems. Here we introduce two criteria based on the physiological evidence found from biological visual systems. Our first criterion is based on the physiological evidence that unbalanced Difference of Gaussian (DOG) operators are employed in biological visual system [8]. The unbalanced DOG operators imply that if two edges have an equal value of intensity change then the biological visual systems will have stronger response to the edge with higher absolute intensity. We implement this criterion into our edge detection algorithm by applying a weighting function to the evaluation of the normalized strengths of candidate edge segments. The weighting function is constructed such that the strength of a candidate edge segment with low absolute intensity is suppressed. Our second criterion is based on the visual behavior of lateral inhibition found in many biological visual systems [15]. We can also observe the similar phenomenon from the LOG zero-crossing contours in different channels. In Figure 3(a)-(d), we notice that in the neighborhood of the zero-crossing contours that correspond to salient edges in the image, there are much fewer random zero-crossing contours caused by noise. Therefore, a zero-crossing contour that has a cleaner neighborhood may imply a higher probability to be a salient edge in the image.

3.2. Combining edges from different scales

In our edge detection algorithm, a final edge map is obtained by combining the detected edge candidates from a wide range of scale channels. The details of this procedure will become more clear after we give the complete diagram of the implementation of our algorithm in Section 4. Briefly, the detected edge candidates in each individual scale channel are the salient optimized trajectories obtained by previous processes at the corresponding scale. Since smaller scale of the detection functional always gives better ELA and only salient edges with high SNR survive through the salience test introduced in Section 3.1, the final edge map should always incorporate all the detected salient edge candidates from the smallest scale. The combination procedure can then continue to check if there are new salient edges in the detection results from larger scales.

Our scale space approach overcomes the difficulties faced by Marr and Hildreth's method. Firstly, a physically significant edge does not have to exist as matched edge candidates in more than one scale channel to appear in the final edge map as long as it can pass the edge salience test. Secondly, the detected edge candidates in all scale channels have locally continuous and smooth regular shape due to the edge regularization procedure. Thirdly, the edge candidates in large scale channels do not move far away from the true edge position since our edge detection functional has the optimal SNR and ELA performance. Finally, the number of the detected edge candidates in a small scale channel is much less than that of the LOG zero-crossing contours in the corresponding scale channel because very few pure noise caused candidates survive the edge salience test.

Note that the strategy of our scale space approach is fine-to-coarse but there are two important differences that differentiate our method from Canny's approach. Firstly, our method is based on the edge candidates selected from the optimized trajectories from the

LOG zero-crossing contours. Therefore, in scale space these new contours will keep the most important scaling property of the LOG zero-crossing contours, i.e., the edge candidates can only merge into fewer candidates but never split into more as the scale increases[18]. This nice scaling property of our edge candidates in scale space lays the necessary mathematical foundation for ensuring our scale space approach will perform reliably in real situations. Secondly, our edge candidates in each scale channel are curve segments not points. Therefore our method is much less sensitive to noise than Canny's method.

4. Implementation details and experimental results

The implementation of the proposed 2D edge detection algorithm consists of a series of channels. Each channel detects edge candidates at a preferred scale. The final edge detection result is generated by a scale space procedure that combines the detected edge candidates from all the channels. Figure 4 gives a diagram of our implementation for the proposed edge detection algorithm.

There are five stages in each channel. In stage 1, we first convolve the input images with the LOG of appropriate scale for the channel. Then we extract the zero-crossing contours from the convolution result. In stage 2, we segment each zero-crossing contour at points with large curvatures. The purpose of this presegmentation is to prepare contour segments of regular shape, i.e., without sharp corners, for the detection stage. In stage 3, we find the true edge location for each presegmented contour by searching for a trajectory in the neighborhood of the contour that gives the maximum response to our 2D edge detection functional described in Section 2.2. To construct the functional, we select the channel scale as the scale of the function $f_g(u,v)$. The gradient direction of intensity is computed from the original intensity image at each point on the contour. The value of the amplitude function $A(u)$ at each point on the contour is estimated by computing the average intensity difference between the neighbor pixels on the two sides of the contour point in the gradient direction. In stage 4, we construct a histogram of the computed normalized edge strengths for all the candidate edge segments, as introduced in Section 3.1. Then we estimate the mean and the variance of the fitted Gaussian distribution of the noise. Based on the mean and the variance, a low, a medium and a high threshold can be selected. If a candidate edge segment has a normalized strength stronger than the medium threshold, we check if the segment has subsegments of at least a minimum length that have normalized strengths weaker than the low threshold. If so, then only the rest of the segment will pass the strength test. Otherwise the whole segment will pass. If a candidate edge segment has a normalized strength weaker than the medium threshold, we then check if the segment has subsegments of at least a minimum length that have normalized strengths stronger than the high threshold. If so, then those subsegments will pass the strength test. Otherwise, the whole segment fails the test. As mentioned in Section 3.1, two additional criteria have also been employed in our algorithm. The first one has been implemented as a weighting function applied to the evaluation of the edge strength. The weighting function penalizes a candidate edge segment with low absolute intensity. The second criterion has been implemented as an index of the number of edge segments in the neighborhood of the tested candidate edge segment. The larger the number, the less likely the segment is to pass the test. In stage 5, we optimize the trajectories obtained in stage 4 by the edge regularization procedure introduced in Section 2.4. To solve Equation (15), we adopt the numerical method used in Snakes [6], i.e., we use the first-order explicit Euler method with respect to only the externally constrained forces. Since the contours in our case are not necessarily closed, the quindiagonal coefficient matrix in the Euler iteration equation will not be symmetric in general. To compute the inverse of the quindiagonal matrix, we use the algorithm developed by Benson and Evans [1].

As described in Section 3.2, our scale space procedure employs a fine-to-coarse strategy. To accomplish this, we generate two maps. One map will be the final edge map in which an edge in the image will have a single edge trajectory detected from the corresponding channel. In the following we will refer this map as the min map. The other map will contain all the edge candidates detected from all the channels from small to large. This map will be referred as the max map. The procedure first takes all the detected edge candidates from the smallest scale channel and puts them into both the min and max maps. Then it checks if the edge candidates contained in the max map appear in a parallel region of each edge candidates detected by the second smallest scale channel. The width of the region is selected equal to the second smallest scale. If an edge candidate from the second channel or a segment of at least a minimum length of it does not have near parallel trajectories from the max map, then that candidate or segment will be added into the min map for final output. In any case, the complete edge candidates from the second channel will be added into the max map. Such a process is applied repeatedly to the third channel, the fourth, etc. until the largest scale channel is reached. The scales of our various channels have been selected to span the range of filters that operate in the human fovea [9]. The smallest scale in our current implementation has been selected as $\sigma_{min} = 2.5$ and the largest scale as $\sigma_{max} = 6.7$. We decided not to change the scale range or any other preset parameters in our algorithm and applied it directly to many different real images. The essential purpose of this was to test how robust our algorithm is in reality and how well it matches the performance of human visual systems in detecting edges in general cases.

In Section 2.3, we gave the closed-form performance comparison in terms of SNR and ELA between the proposed 2D edge detection functional Q_{edge} and the Canny edge detector. Figure 5 shows the actual edge detection results on the 'Lena' picture by Q_{edge}, the Canny edge detector and the Marr-Hildreth edge detector, all with $\sigma = 2.5$. From Figure 5, it can be seen that the edge detection result by Q_{edge} is clean while the edges remain continuous and smooth. On the other hand, the result by the Marr-Hildreth edge detector appears to be noisy at the selected threshold while the edges have begun to lose their continuity. The result by the Canny edge detector appears to be better than that by the Marr-Hildreth edge detector. However while its edge map still contains more spurious edges than that by Q_{edge} shown in Figure 5(b), its detected edges are not as continuous and smooth as the detected edges by Q_{edge} either. Figure 6(a) shows the overlapping edge candidates on the 'Lena' picture detected by Q_{edge} in all seven channels with $\sigma_{min} = 2.5$ and $\sigma_{max} = 6.7$. Figure 6(b) gives the final edge map of the 'Lena' picture generated by the scale space combination procedure implemented in our 2D edge detection scheme.

Notice that most edges in the final edge map are detected by the smallest channel but some fuzzy edges are effectively detected only by the large scale channels. More experimental results can be found in our technical report [14].

5. Discussion

We have proposed a new 2D edge detection functional derived from an adaptive 2D edge model. The detection functional is optimal in terms of SNR and ELA for detecting edges in 2D images and preserves the nice scaling behavior of the LOG operator in scale space. It detects 2D edges based on edge segments rather than edge points. This guarantees the continuity of the detected edges and greatly reduces the impact of random noise on the detection results. The proposed edge detection algorithm also employs an edge regularization procedure to enhance desired smoothness and stiffness on the detected edges. A global noise estimation procedure and the other two physiologically based criteria have also been introduced to provide a robust estimation for the global noise level in an image and fine tune the edge detection results to make them similar to what is perceived by human visual systems. Finally, a reliable scale space combining procedure has been established based on the continuity in image and the scaling property in scale space of the detected edges.

Although the edge detection approach proposed in this paper has succeeded on many real images, there are still many possibilities to add improvement to this algorithm. Among the future work, one interesting direction is to study the further theoretical basis for evaluating the salience of an edge in an image. It is our belief that the salience of an edge in an image does not solely depends on the stochastic characteristics, e.g., noise level, of the image. It also depends on the perceptual context in the image. Therefore, it seems very relevant to us to study the topics on connecting the edge detection process with other visual signal processes, e.g., region detection, texture analysis under a unified framework of image segmentation and object recognition.

Acknowledgment

This work was supported by ARPA/ONR Grant N00014-93-1-1167.

Appendix SNR and ELA of a 2D edge detection functional

Let $E(u,v)$ be an edge with a trajectory $\vec{\alpha}(u)$ to be detected in an image. Using Equation (1) and Equation (3), we have the response of the edge detection functional Q_{edge} to the edge (signal) at the point $u_0 = 0$ and at the trajectory $\vec{\alpha}(u)$ or equivalently $v_0(u) = 0$ as

$$Q_{\text{edge}}^{\text{signal}}(0,0) = \int_{-L}^{L} f_t(u) A(-u) \left(\int_{-D}^{D} f_g(-u,v) P(-u,-v) dv \right) du . \qquad \text{(A-1)}$$

Assume an additive white Gaussian noise with variance n_0^2 in the image. Then the root-mean-squared response of the edge detection functional Q_{edge} to the noise at the same positions as above can be computed as

$$\sqrt{\mathbf{E}\left[\int_{-L}^{L} f_t(u)(\int_{-D}^{D} f_g(-u,v)N(-u,-v)dv)du \int_{-L}^{L} f_t(w)(\int_{-D}^{D} f_g(-w,z)N(-w,-z)dz)dw\right]} \quad \text{(A-2)}$$

$$= \sqrt{\int_{-L}^{L}\int_{-D}^{D}\int_{-L}^{L}\int_{-D}^{D} f_t(u)f_t(w)\mathbf{E}[N(-u,-v)N(-w,-z)]f_g(-u,v)f_g(-w,z)dvdudzdw} \quad \text{(A-3)}$$

$$= \sqrt{\int_{-L}^{L}\int_{-D}^{D}\int_{-L}^{L}\int_{-D}^{D} f_t(u)f_t(w)n_0^2\delta(u-w,v-z)f_g(-u,v)f_g(-w,z)dvdudzdw} \quad \text{(A-4)}$$

$$= n_0\sqrt{\int_{-L}^{L}\int_{-D}^{D} f_t^2(u)f_g^2(-u,v)dvdu}\,. \quad \text{(A-5)}$$

Therefore, the output SNR of the edge detection functional Q_{edge} is

$$\text{SNR} = \frac{\left|\int_{-L}^{L} f_t(u)A(-u)(\int_{-D}^{D} f_g(-u,v)P(-u,-v)dv)du\right|}{n_0\sqrt{\int_{-L}^{L} f_t^2(u)(\int_{-D}^{D} f_g^2(-u,v)dv)du}}. \quad \text{(A-6)}$$

Next we derive the ELA of the edge detection functional Q_{edge}. Suppose there is a local maximum in the total response Q_{edge} at point $u_0 = 0$ and at trajectory $v_0(u)$. Then we have

$$\frac{\partial}{\partial v_0(u)} Q_{\text{edge}}^{\text{signal}}(0, v_0(u)) + \frac{\partial}{\partial v_0(u)} Q_{\text{edge}}^{\text{noise}}(0, v_0(u)) = 0 \quad \text{(A-7)}$$

where $\frac{\partial}{\partial v_0(u)} Q_{\text{edge}}^{\text{signal}}(0, v_0(u))$ can be expanded around $\vec{\alpha}(u)$ or equivalently around $v_0(u) = 0$ as

$$\frac{\partial}{\partial v_0(u)} Q_{\text{edge}}^{\text{signal}}(0, v_0(u)) = \frac{\partial}{\partial v_0(u)} Q_{\text{edge}}^{\text{signal}}(0,0) + \frac{\partial^2}{\partial v_0^2(u)} Q_{\text{edge}}^{\text{signal}}(0,0)v_0(u) + O(v_0^2(u)). \quad \text{(A-8)}$$

Notice that $\frac{\partial}{\partial v_0(u)} Q_{\text{edge}}^{\text{signal}}(0,0) = 0$ for any unbiased optimal edge detection functional Q_{edge}. Assuming $|v_0(u)|$ is small, then Equation (A-7) becomes

$$\frac{\partial}{\partial v_0(u)} Q_{\text{edge}}^{\text{noise}}(0, v_0(u)) + \frac{\partial^2}{\partial v_0^2(u)} Q_{\text{edge}}^{\text{signal}}(0, v_0(u)) v_0(u) \approx 0 . \quad \text{(A-9)}$$

Since the noise is a white Gaussian process and Q_{edge} is a linear operator, $\frac{\partial}{\partial v_0(u)} Q_{\text{edge}}^{\text{noise}}(0, v_0(u))$ is a zero-mean Gaussian process and differentiable. From the Level-crossing Theorem of normal processes in probability theory [11], we have the probability density of the maxima in the total response Q_{edge} at $u_0 = 0$ with respect to $v_0(u)$ as

$$\lambda(v_0(u)) = \frac{1}{2\pi} \sqrt{\frac{-R''(0)}{R(0)}} \mu(v_0(u)) \quad \text{(A-10)}$$

where

$$R(\tau) = \mathbf{E}\left[\frac{\partial}{\partial v_0(u)} Q_{\text{edge}}^{\text{noise}}(0, v_0(u)) \frac{\partial}{\partial v_0(u)} Q_{\text{edge}}^{\text{noise}}(0, v_0(u) + \tau) \right] \quad \text{(A-11)}$$

and

$$\mu(v_0(u)) = \exp\left(- \frac{\left(\frac{\partial^2}{\partial v_0^2(u)} Q_{\text{edge}}^{\text{signal}}(0, v_0(u)) v_0(u)\right)^2}{2R(0)}\right). \quad \text{(A-12)}$$

In general, the width of the main-lobe of $\mu(v_0(u))$ around $v_0(u) = 0$ reflects the goodness of edge localization of Q_{edge}. For simplicity, one may choose the square-root of the inverse of the curvature of $\mu(v_0(u))$ at $v_0(u) = 0$ as a measure of the ELA [2]. Namely, we have the curvature [4] of $\mu(v_0(u))$ at $v_0(u) = 0$ as

$$\kappa(0) = \frac{\left| \frac{\partial^2}{\partial v_0^2(u)} \mu(0) \right|}{\left(1 + \left(\frac{\partial}{\partial v_0(u)} \mu(0)\right)^2\right)^{3/2}} . \quad \text{(A-13)}$$

It is easy to verify that $\frac{\partial}{\partial v_0(u)} \mu(0) = 0$ and $\left| \frac{\partial^2}{\partial v_0^2(u)} \mu(0) \right| = \frac{\left(\frac{\partial^2}{\partial v_0^2(u)} Q_{\text{edge}}^{\text{signal}}(0,0)\right)^2}{R(0)}$ where $R(0) = n_0^2 \int_{-L}^{L} f_t^2(u) \left(\int_{-D}^{D} \left(\frac{\partial}{\partial v} f_g(-u, v)\right)^2 dv \right) du$. Therefore, the ELA of the edge detection functional Q_{edge} is

$$\mathrm{ELA} = \sqrt{\frac{1}{\kappa(0)}} = \frac{n_0 \sqrt{\int_{-L}^{L} f_t^2(u) \left(\int_{-D}^{D} \left(\frac{\partial}{\partial v} f_g(-u,v) \right)^2 dv \right) du}}{\left| \int_{-L}^{L} f_t(u) A(-u) \left(\int_{-D}^{D} \frac{\partial}{\partial v} f_g(-u,v) \frac{\partial}{\partial v} P(-u,-v) dv \right) du \right|}. \quad \text{(A-14)}$$

References

[1] A. Benson and D. J. Evans, "Algorithm 80: an algorithm for the solution of periodic quindiagonal systems of linear equations," *Computer J.*, vol. 16, no. 3, pp. 278-279, 1973.

[2] K. L. Boyer and S. Sarkar, "Comments on 'On the localization performance measure and optimal edge detection'," *IEEE Trans. Pattern Anal. Machine Intell.*, vol. 16, no. 1, pp. 106-108, 1994.

[3] J. Canny, "A computational approach to edge detection," *IEEE Trans. Pattern Anal. Machine Intell.*, vol. PAMI-8, no. 6, pp. 679-698, 1986.

[4] M. P. d. Carmo, *Differential geometry of curves and surfaces*, Prentice-Hall Inc., 1976.

[5] D. H. Hubel and T. N. Wiesel, "Receptive fields, binocular interaction and functional architecture in the cat's visual cortex," *J. Physiol. (Lond.)*, vol. 160, pp. 106-154, 1962.

[6] M. Kass, A. Witkin and D. Terzopoulos, "Snakes: active contour models," *Int. J. Comput. Vision*, pp. 321-331, 1988.

[7] T. Lindeberg, *Scale space theory in computer vision*. The Kluwer International Series in Engineering and Computer Science, Kluwer Academic Publishers, Dordrecht, The Netherlands, 1994.

[8] R. A. Linsenmeier, L. J. Frishman, H. G. Jakiela and C. Enroth-Cugell, "Receptive field properties of X and Y cells in the cat retina derived from contrast sensitivity measurements," *Vision Res.*, vol. 22, pp. 1173-1183, 1982.

[9] D. Marr and E. Hildreth, "Theory of edge detection," *Proc. R. Soc. (Lond.)*, vol. B 207, pp. 187-217, 1980.

[10] V. S. Nalwa and T. O. Binford, "On edge detection," *IEEE Trans. Pattern Anal. Machine Intell.*, vol. PAMI-8, no. 6, pp. 699-714, 1986.

[11] A. Papoulis, *Probability, Random Variables, And Stochastic Processes*, McGraw-Hill, Inc., 1984.

[12] P. Perona and J. Malik, "Scale space and edge detection using anisotropic diffusion," *IEEE Trans. Pattern Anal. Machine Intell.*, vol. 12, no. 7, pp. 629-639, 1990.

[13] T. Poggio and V. Toree, "Ill-posed problems and regularization analysis in early vision," *Proc. DARPA Image Understanding Workshop,* New Orleans, pp. 257-263, 1984.

[14] R. J. Qian and T. S. Huang, "A 2D edge detection scheme for general visual processing," *Tech. Rep. Beckman Inst., UIUC,* no. UIUC-BI-AI-RCV-94-04, 1994.

[15] K. Ratliff, and F. Hartline, *Studies on Excitation and Inhibition in the Retina*, The Rockfeller University Press, 1974.

[16] A. Rosenfeld and M. Thurston, "Edge and curve detection for visual scene analysis," *IEEE Trans. Comput.*, vol. C-20, no. 5, pp. 562-569, 1971.

[17] R. M. Shapley and D. J. Tolhurst, "Edge detection in human vision," *J. Physiol. (Lond.),* vol. 229, pp. 165-183, 1973.

[18] H. D. Tagare and. R. J. P. deFigueiredo, "On the localization performance measure and optimal edge detection," *IEEE Trans. Pattern Anal. Machine Intell.*, vol. 12, no. 12, pp. 1186-1190, 1990.

[19] D. Terzopoulos, "Regularization of inverse visual problems involving discontinuities," *IEEE Trans. Pattern Anal. Machine Intell.*, vol. 8, no. 4, pp. 413-423, 1986.

[20] A. N. Tihonov, "Regularization of incorrectly posed problems," *Sov. Math. Dokl.*, vol. 4, pp. 1624-1627, 1963.

[21] A. Witkin, "Scale space filtering," *Proc. IJCAI,* Karlsruhe, pp. 1019-1021, 1983.

[22] A. L. Yuille and T. A. Poggio, "Scaling theorems for zero crossings," *IEEE Trans. Pattern Anal. Machine Intell.*, vol. PAMI-8, no. 1, pp. 16-25, 1986.

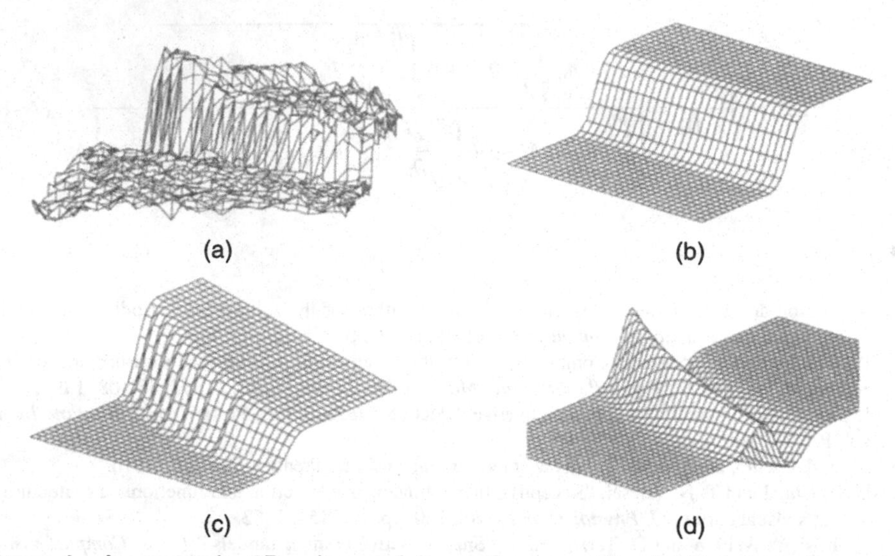

Figure 1. An example 2D edge and its surface models. (a) The intensity surface around a short edge segment in the 'Lena' picture; (b) The constant cross-section edge model; (c) The parametric 2D edge model. See Section 2.1; (d) The constructed optimal 2D edge detection functional based on the edge model in (c). See Section 2.2.

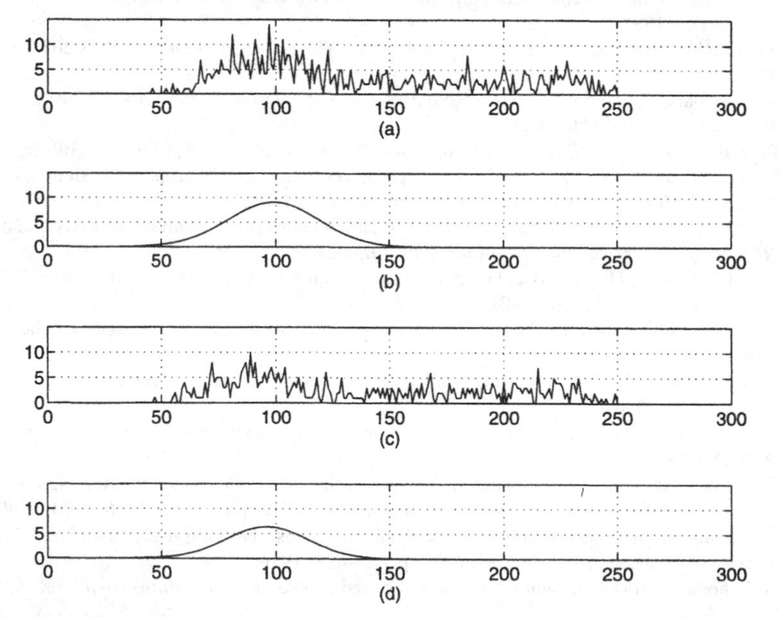

Figure 2. The histograms of the normalized strengths of the candidate edge segments of the 'Lena' picture in two different channels. (a) The histogram in Channel 1 with $\sigma = 2.5$; (b) The fitted Gaussian distribution corresponding to (a); (c) The histogram in Channel 2 with $\sigma = 3.2$; (d) The fitted Gaussian distribution corresponding to (c).

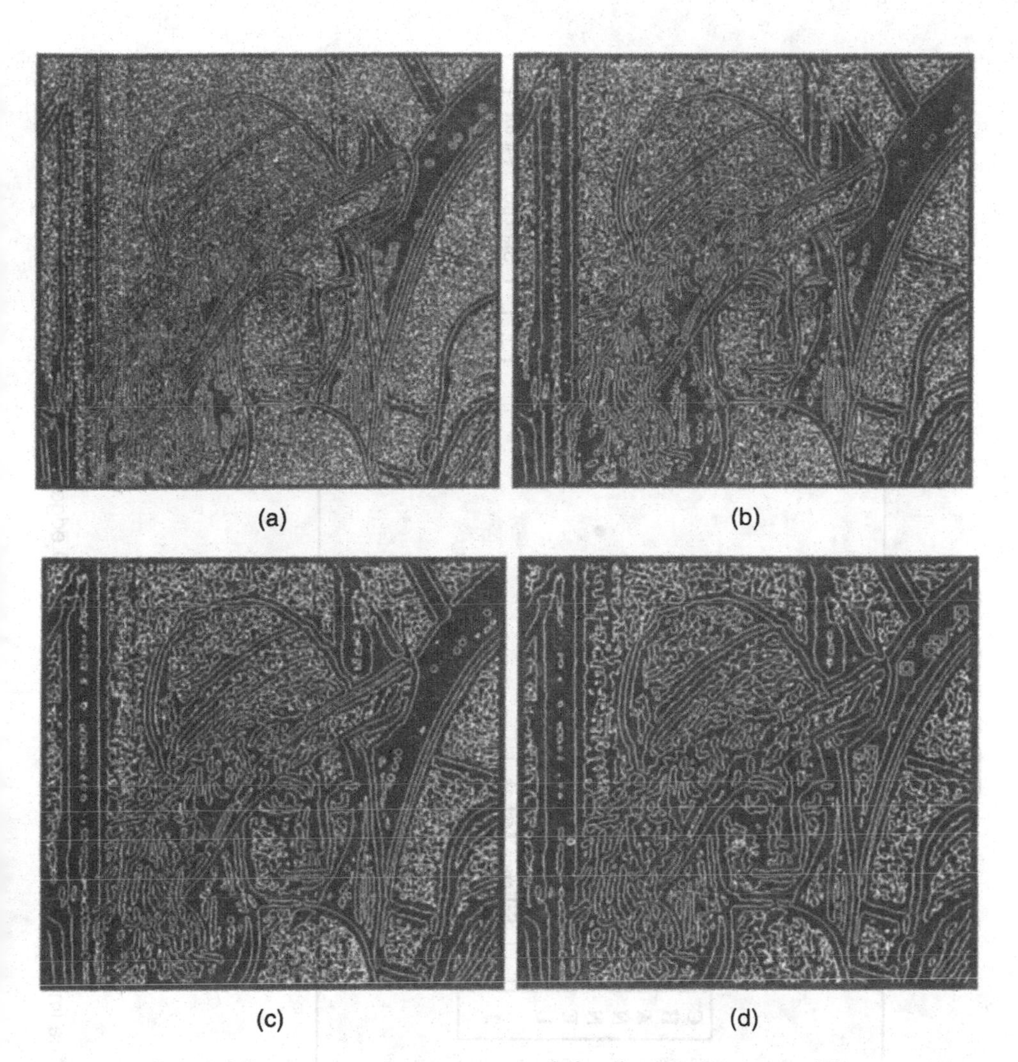

(a) (b) (c) (d)

Figure 3. The LOG zero-crossing contours of the 'Lena' picture at different scales. (a) LOG zero-crossing contours at $\sigma = 2.5$; (b) LOG zero-crossing contours at $\sigma = 3.2$; (c) LOG zero-crossing contours at $\sigma = 3.9$; (d) LOG zero-crossing contours at $\sigma = 4.6$. See Figure 5(a) for the input intensity image. From the above images, it can be seen that a zero-crossing contour that has a cleaner neighborhood usually corresponds to a more salient edge in the picture.

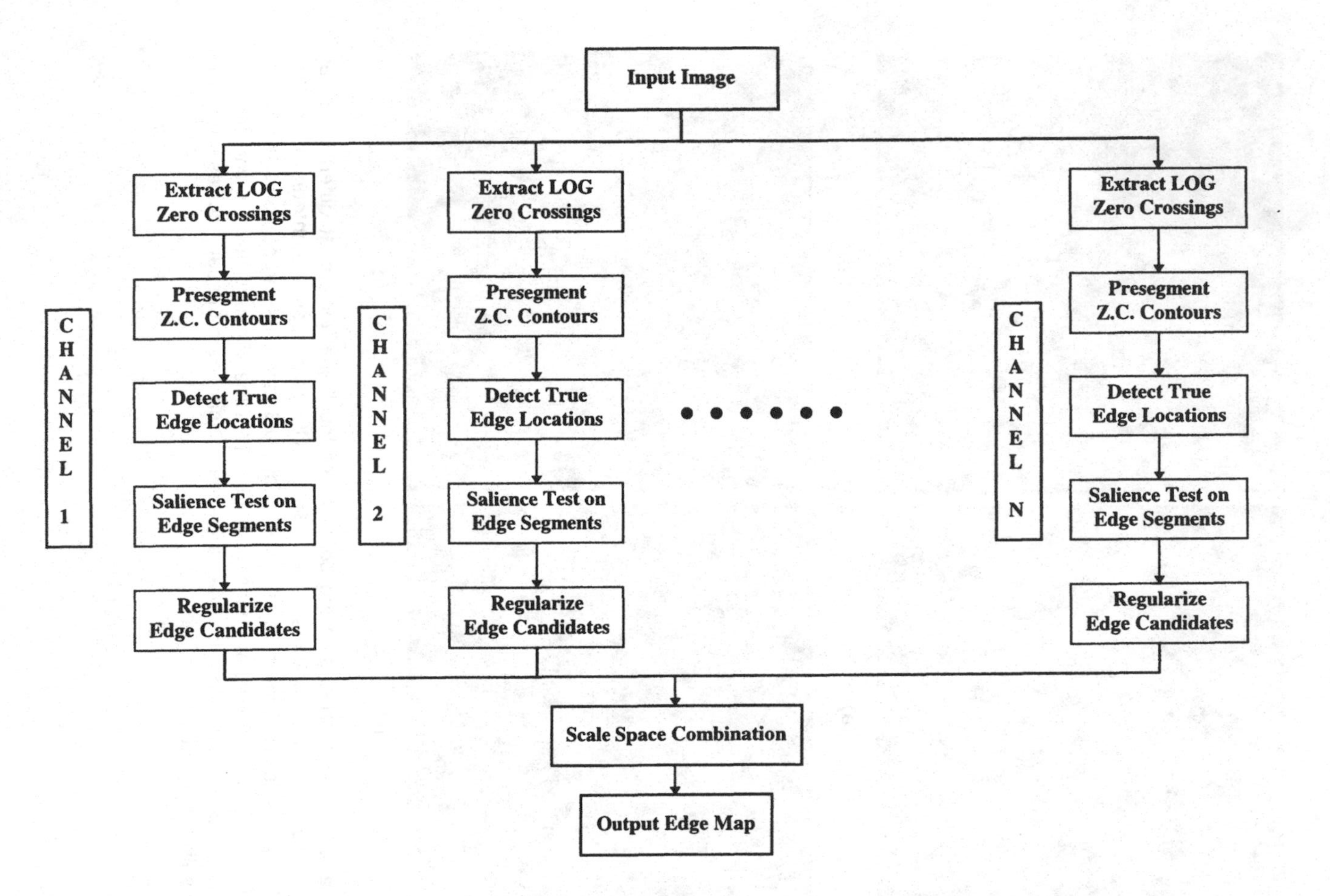

Figure 4. The implementation diagram of the proposed 2D edge detection algorithm.

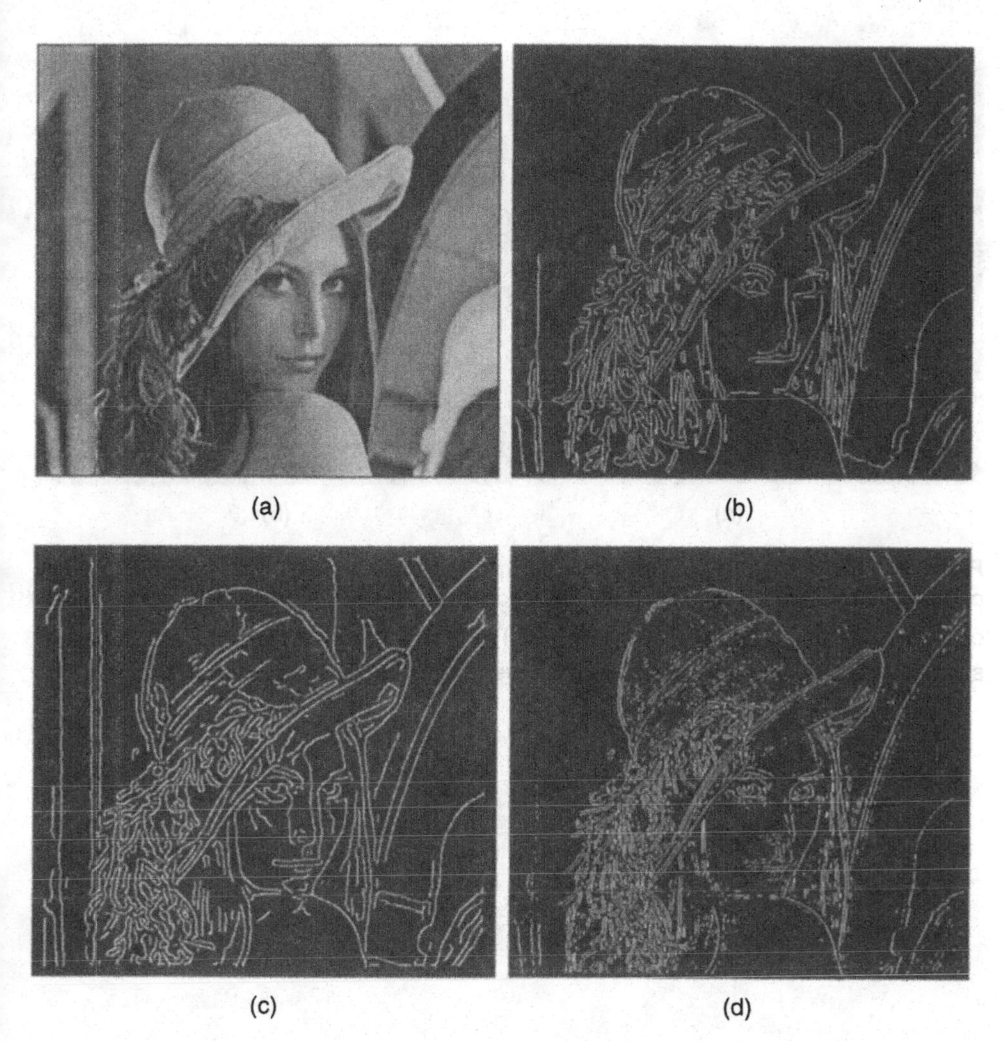

Figure 5. The comparison of the edge detection results on the 'Lena' picture. (a) The input intensity image; (b) The detected edges by the proposed 2D edge detection functional Q_{edge} with $\sigma = 2.5$; (c) The detected edges by the 2D Canny edge detector with $\sigma = 2.5$; (d) The detected edges by the Marr-Hildreth edge detector with $\sigma = 2.5$.

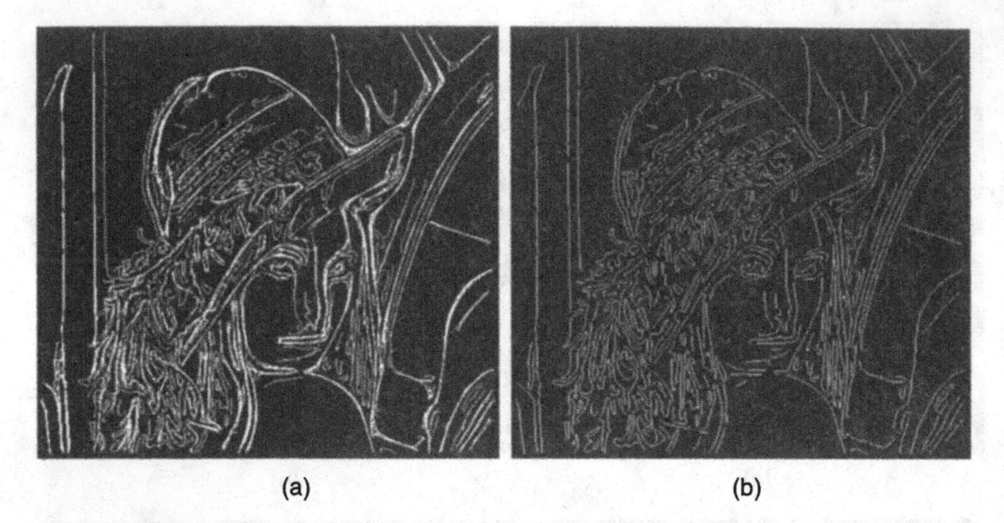

(a) (b)

Figure 6. More edge detection results on the 'Lena' picture. (a) The overlapping edge candidates detected by the proposed 2D edge detection functional Q_{edge} in all seven channels with $\sigma_{\min} = 2.5$ and $\sigma_{\max} = 6.7$; (b) The final edge map generated by the scale space combination procedure employed in the proposed 2D edge detection scheme.

Metamerism in Complete Sets of Image Operators

Jan J.Koenderink and A.J.van Doorn
Utrecht University, the Netherlands

April 7, 1996

Abstract

Consider the *local* description of image structure. It will have to be based upon a necessarily finite number of observations of the image through image operators of limited support. The understanding of the discriminative power of the local representation is a *fundamental* problem in vision. The number of categorically different descriptions will depend on the collection of operators applied to the image, and is a measure of the discriminative power of that set. General practice is to select the operators in an essentially *ad hoc* fashion, and consequently little of general relevance can be said. If the collection is constructed on a principled basis we can formulate a theory of discriminative power of such "local toolkits". One particularly apt choice is the set of differential operators up to (we always mean "and including") the nth–order for some fiducial scale. Such a set is "the nth–order jet $\mathcal{J}^n$ at resolution level s" (say), which is a geometrical object, *i.e.*, all implementations ("representations" of the jet) are equivalent. The jets are essentially *truncated Taylor expansions* of the image at the fiducial resolution, excepting the constant term. For the jets we can exactly state the discriminative power. The possible categories are the local features, or "textons", that the set can distinguish. The set will enforce its categories on any image structure, thus the features are not so much "image properties" as properties of the toolkit. Images that differ in ways not distinguished by the set are bluntly treated as equivalent.

1 Scale–space

By now the structure of "scale–space" has been well established. Azriel Rosenfeld[1] was among the pioneers to forge it into an indispensable tool. If one entertains no prior assumptions that are specific to the scene, then there turns out to exist a unique way in which one may implement "blurring", that

0-8186-7644-2 $5.00 © 1996 IEEE

is lowering of resolution. That method is uniform diffusion of luminance over the visual field, or—equivalently—convolution with Gaußian kernels. Only this operation is translation and rotation invariant, allows both step-wise and one-shot blurring and yields "causal" resolution degradation[3]. By "causal" we mean that luminance maxima invariably are lowered by blurring whereas luminance minima are invariably raised. Intuitively that means that lowering resolution will always lead to a *loss of detail*, never a gain.

1.1 Local jets

In image processing tasks one needs to *differentiate* the optical structure. This is inherently a dangerous and ill defined operation: In effect no natural image will be "differentiable" in the 1st place. One may differentiate blurred images though because these are smooth and differentiation poses no problems at all. Thus differential operations can only be sensibly defined in the context of scale-space.

One has to be very careful though. Let $L(x, y)$ denote the luminance as a function of the Cartesian coordinates (x, y) of an image. Let $G(x, y; s) = (\exp -(x^2 + y^2)/4s)/4\pi s$ denote a Gaußian kernel with resolution parameter s. Then the blurred image L_s is defined as the convolution $L_s = G \otimes L$. Linearity allows you to write ∂L_s in any one of the various forms

$$\partial(G \otimes L) = (\partial L) \otimes G = L \otimes \partial G, \tag{1}$$

where ∂ denotes any differential operator. Although these expressions are indeed formally equivalent, they are pragmatically as different as can be. The 1st *expression* is the derivative of the blurred image. It is numerically unstable, because the blurred image has to be represented in some way, thus will be corrupted by noise (*e.g.*, truncation or rounding errors). The 2nd *expression* is the blurred differential of the image. This expression makes no sense because the differential of the image is not defined to begin with. The 3rd *expression* is the image convolved with the derivative of the Gaußian kernel. Only this expression makes perfect sense and is numerically stable. This is the only sane way to implement differential operators. Notice that the result is equal to the *exact derivative of the blurred image.* Thus we may set up operators with finite support that implement exact local derivative operators! (Here "finite support" means simply that the kernels yield finite observations for images of at most polynomial growth.)

The simplest non-trivial example is that of order 2 in dimension 1. We explore this paradigmatic case in some detail. We consider $\mathcal{J}^2$ at the origin.

2 Metamerism

"Metamerism" is a term borrowed from the science of *colorimetry*[7, 8]. The setting—in a suitably generalized way—is as follows: We consider a space of "images" $\mathcal{I}$ (that are scalar fields on some parameter space) and a space of "observations" $\mathcal{O}$. We cannot assume that an observation suffices to identify an image. In colorimetry one considers the space of spectra (the "images") and the space of colorimetric coordinates (the "observations", a typical observation would be a set of three numbers, say the red, green and blue intensities). A colorimetric specification does not identify a spectrum, but a "metamer", that is a $^{\infty}\mathcal{D}$ family of spectra with the same colorimetric specification (these "look the same color", though they correspond to distinct physical beams). Exactly the same formalism applies to images: If we observe the image via a battery of local image operators we obtain a specification that is not sufficient to identify the image, but instead specifies a large class of images that may be called the "metamer" corresponding to that observation.

Consider the simple case where the observations are obtained via a set of linear operators acting on the image. (The typical "edge finders"[2], "Laplacean operators", *etc.*, fall in this scheme.) Then we may set up a basis for image space such that the 1st $\dim \mathcal{O}$ basis vectors span observation space, whereas the remaining $(\dim \mathcal{I} - \dim \mathcal{O})$ dimensions lead to null observations. This factors the image space into two components, the "fundamental component space" and the "metameric black space". The observation scheme is (by construction) completely blind for the metameric black images (hence the name). An observation is sufficient to uniquely identify a member of the fundamental component space though. Thus the fundamental components may be used as canonical representatives. In practice this is often unattractive though because the fundamental components may well turn out to possess undesirable features. (For instance, the fundamental components typically have pixel values that are not non–negative throughout, thus they fail to correspond to physically possible images.)

We obtain a particularly clear description in the scalespace setting. Here an obvious choice for the set of linear operators is apparent: The nth–order jet $\mathcal{J}^n$ of truncated Taylor expansions[5]. Is is comparatively easy to analyze this important case in detail. It is not merely a nice example though: In actual image analysis the jet representation has many practical advantages quite apart from being the obvious choice from 1st–principles.

2.1 The $^1\mathcal{D}$ case

Here we treat the simplest case, that is $^1\mathcal{D}$. In the $^1\mathcal{D}$ case the 2nd–order jet $\mathcal{J}^2$ has 3–degrees of freedom. (This is exactly the case of colorimetry.) Thus we can map the sample obtained by a local $\mathcal{J}^2$ as a point in $^3\mathcal{D}$ "pattern space". A "pattern" represents an ∞ class of images, namely those images that all yield the given sample, thus agree in their local $\mathcal{J}^2$'s. Such a class is called a "metamer" in colorimetry, and the fact that *many* images correspond to a given sample is known as "metamerism". The case of colorimetry is indeed formally identical to $\mathcal{J}^2$ in $^1\mathcal{D}$. We may employ the powerful methods developed in colorimetry by Schrödinger[7, 8] in the 1920's.

2.1.1 Metamerism in the local 2nd order jet

For the $\mathcal{J}^2$ we have the kernels (from here on we set $s = 1/4$): $\psi_0(x) = G(x; 1/4)$, $\psi_1(x) = \partial G(x; 1/4)/\partial x$ and $\psi_2(x) = \partial^2 G(x; 1/4)/\partial x^2$.

Suppose we let the operators (ψ_0, ψ_1, ψ_2) act on an image. We obtain the responses $(\varrho_0, \varrho_1, \varrho_2)$. In an operational sense all we may know of an image locally are such triples of samples. If you move an impulse function over the full axis you describe a closed curve in "response space", that is $(\varrho_0, \varrho_1, \varrho_2)$–space. The curve starts and ends at the origin. For various amplitudes of the impulse you obtain scaled versions of this loop. Together these describe a conical surface. (See figure 1).

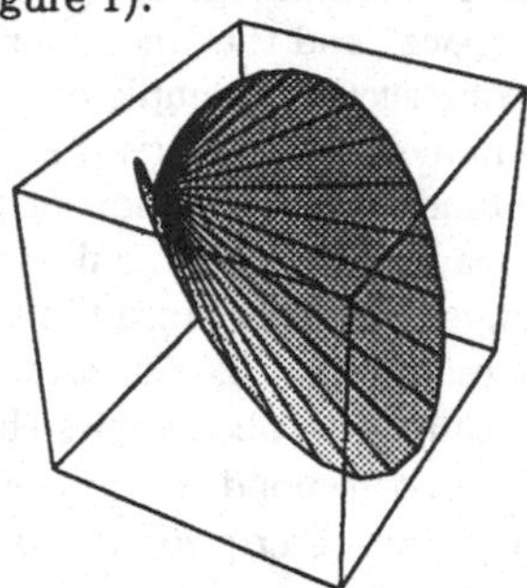

Figure 1: *The cone of impulse patterns. The length of the generators is the intensity for the corresponding impulse.*

We limit the discussion to functions that are nowhere negative, since we are only interested in *images*. In this case the region *inside* this cone corresponds to possible responses, whereas the region outside the cone corresponds to responses that can never occur. The reason is simple enough: Since *every* function can be decomposed into impulses, with positive weight, all responses must lie in the *convex hull* of the locus of impulse responses. Thus you see that the responses of Gaußian derivatives are not independent: Only certain combinations of responses can occur at all.

In practice it is often useful to consider a lower dimensional representation. Because images that differ only in contrast (overall intensity scaling) are essentially similar, we often use the representation $(\zeta_1, \zeta_2) = (\varrho_1, \varrho_2)/\varrho_0$. This is a central projection from the origin of $(\varrho_0, \varrho_1, \varrho_2)$–space to the plane $\varrho_0 = 1$. In this plane the cone appears as a parabola (conveniently parameterized by the coordinate x), that is to say $(\zeta_1, \zeta_2) = (x\sqrt{2}, (2x^2-1)/\sqrt{2})$. Instead of using the coordinate x to indicate the impulse functions we also use the phase angle $\lambda = \arctan(-\zeta_2, \zeta_1)$ (the phase angle corresponds roughly to the "hue" in color science). This is convenient because we deal with a finite parameter, moreover, the implied periodicity turns out to make much sense later. We can parameterize any response in terms of its *intensity* ζ_0, its *phase angle* λ or "equivalent position" $\overline{x} = \tan\lambda$ and its *contrast* $\chi = \sqrt{(\zeta_1^2 + \zeta_2^2)}/\sqrt{(4\overline{x}^4 + 1)/2}$. Notice that the contrast vanishes for the uniform image and reaches the maximum value of unity for the impulse functions. This effectively interprets the (unknown) local image in terms of the superposition of an impulse function (at the equivalent position) and a constant overall intensity. Notice that *any* local image can be represented in this manner. The representation picks out a definite canonical metamer of the image. It is not a particularly fortunate choice in most cases though. This is because the amplitude of the impulse may be very high, whereas the image intensities are typically bounded from above (are in the range 0 to 255 for a typical image say).

If you *add two images* you obtain a 3rd one. It is always possible to find an image that will exactly *cancel* the impulse component of any given image: One simply picks an image with the equivalent position such that the line segment defined by the given image and the 2nd one in (ζ_1, ζ_2)–space includes the origin. This is the case because the addition of images simply corresponds to barycentric addition in (ζ_1, ζ_2)–space, which again corresponds to vector addition in $(\varrho_0, \varrho_1, \varrho_2)$–space which is obvious because of linearity. The equivalent position $\alpha^\sharp$ that can be used to cancel a given position α may be called its "complementary equivalent position". One has simply $\alpha\alpha^\sharp = -1/2$. (Thus there exist weights such that an impulse at position α added to an impulse at position $\alpha^\sharp$ may be equivalent to a uniform intensity over the whole axis.) This will turn out to be an extremely important relation because the structure of $\mathcal{J}^2$ is largely due to it. Notice that the origin is the complementary of both $\pm\infty$: This indicates that the parabolic locus has to be considered as *closed* in matters of complementarity. Indeed, in terms of the phase angle one has simply $\lambda^\sharp - \lambda = \pi$ with no exception. This clarifies the periodic character of the phase angle.

Suppose that we know that the image intensities are bounded from above. For the sake of concreteness we assume that image intensities are in the range $[0, 1]$. Then we have quite a few constraints on the possible responses $(\varrho_0, \varrho_1, \varrho_2)$. For instance, the highest intensity is reached for the uniform white image, it amounts to $1/\sqrt{2\sqrt{\pi}} = 0.5311\ldots$. Images of very high contrast $\chi \approx 1$ must be very dim indeed (almost black) because they can only have appreciable pixel intensities in a very narrow region. In fact, it makes sense to ask for the images that have *the highest intensity* for a given phase angle and contrast. One easily demonstrates (following Schrödinger[7, 8] in his classical paper on color space of the twenties) that such images are *bars*: They are either zero or one and have at most two transition loci. We will refer to such patterns as "optimal images". They are optimal in the sense of being the highest intensity ones for a given phase angle and contrast.

Since *any* image that corresponds to a given phase angle and contrast must be darker than the corresponding optimal image, all such images lie on a line segment between the origin (uniform black image) and the optimal image in $(\varrho_0, \varrho_1, \varrho_2)$–space. This implies that all possible responses lie inside a volume that is bounded by the surface described by the optimal images. We call this the "pattern solid" (see figure 2).

The pattern solid has several symmetric features that are worth noticing. First of all the pattern solid is a convex body. This is obvious because if there were a concavity we could immediately fill it through linear combination of images. However, it is not trivial that the surface of optimal patterns describes the boundary of the convex solid. This is due to the nature of the kernels. The uniform (gray) images lie on the linear segment between the white and black point. In the center of this segment we find the uniform medium gray (intensity 0.5 of the maximum) image. (Notice that this "uniform gray image" has ∞ many articulated metameres. However, $\mathcal{J}^2$ can't "see" these.)

The medium gray image is a *symmetry center* of the pattern solid as you easily demonstrate when you consider the "negatives" of images, that is $I'(x) = I_{max} - I(x)$.

The boundary of the pattern solid is smooth except for the white and

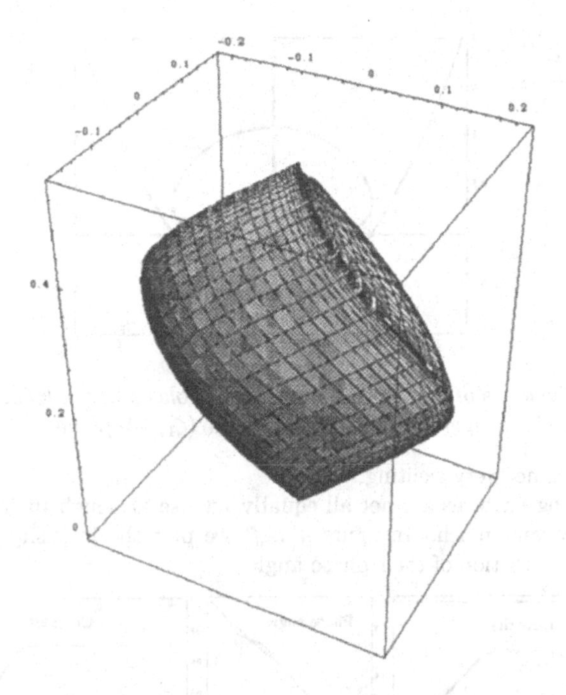

Figure 2: *The pattern solid. Notice the central symmetry and the apices (conical points) at the lowest (black) and highest (white) points. The parameterization is by transition positions of the optimal patterns.*

black points which are conical. At the black point the impulse cone is tangent to the pattern solid, at the white point an inverted copy of the impulse cone is also tangent.

Notice that the central symmetry is closely related to our previous notion of "complementarity". For any phase angle there exists a unique optimal pattern. Such patterns are in a sense "the most articulated" ones. We will refer to them as the "strong patterns". The strong patterns are optimal patterns for which the transition loci are complementary. This follows from the central symmetry if you notice that the strong patterns have the greatest distance from the gray axis: Thus they must lie on the circumscribed cylinder to the pattern solid with generators parallel to the gray axis. We may plot the locus of strong patterns in (ζ_1, ζ_2)–space (see figure 3).

This shows that the region of high intensity images is very limited. Most of the interior of the parabola in (ζ_1, ζ_2)–space corresponds to almost black

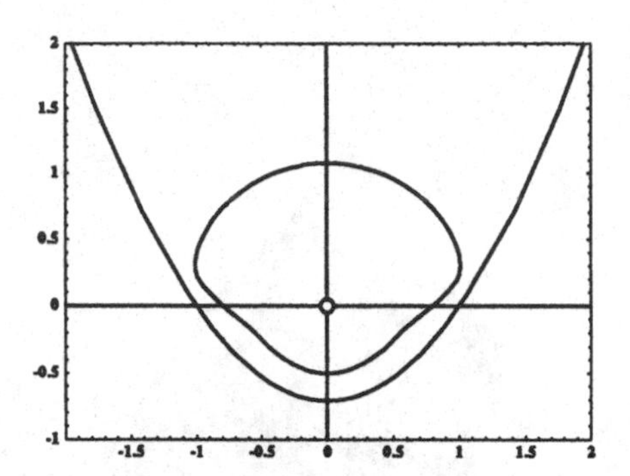

Figure 3: *The locus of impulse functions (parabola) and the locus of strong patterns (the loop encircling the origin) in the* (ζ_1, ζ_2)*–plane.*

images and is not very exciting.

The strong patterns are not all equally intense although their intensity doesn't vary very much. In figure 4 *Left* we plot the intensity of strong patterns as a function of their phase angle.

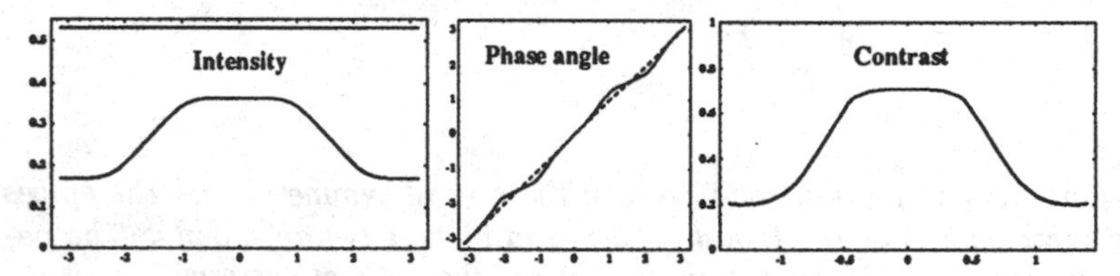

Figure 4: *Left: Intensity of the strong patterns as a function of their phase angle. The intensity of uniform white images is also indicated, it is the upper limit of intensity* per se; *Middle: The phase angle of the strong patterns as a function of their band center; Right: Contrast of the strong patterns as a function of their phase angle.*

Clearly the patterns with near zero phase angles (these are light bars) have the highest intensity (more than half of the maximally attainable intensity) whereas those with phase angles near $\pm\pi$ are darker (about a third of the maximum intensity). Fairly sharp transitions occur at phase angles near $\pm\pi/2$. In this sense $\mathcal{J}^2$ seems to "prefer light bars".

The phase angle of a strong pattern with transitions at $(\lambda-\pi/2, \lambda+\pi/2)$ is very near to λ. We show the actual relation in figure 4 *Middle*. This is often a useful rule of thumb.

The contrast of the strong patterns depends critically on their phase angle. We plot the dependence in figure 4 *Right.* The contrast is almost uniformly low for the bandgap patterns, and uniformly high for the bandpass patterns. Thus $\mathcal{J}^2$ again seems to "prefer light bars": they have both the highest contrast and the highest intensity among all strong patterns.

If we want we can now classify any pattern uniquely as the superposition of an attenuated strong pattern with a uniform gray pattern. Such patterns have two transitions at complementary locations and they switch between two levels that are neatly in the range of allowed intensities ($[0, 1]$). We may parameterize them by phase angle (this defines the transition locations), pattern content p, white content w and black content b, with the relation $p + w + b = 1$. If the intensity of the strong pattern is I_s, of the uniform white pattern I_w, of the uniform black pattern $I_b = 0$, then the levels are simply $wI_w + bI_b$ and $wI_w + bI_b + pI_s$. This classification is analogous to Ostwald's classification of surface colors[8, 4]. It (almost) succeeds in assigning a sensible unique metamer to any observation. The problem is that *very* high contrast images may actually have more contrast than the strong pattern. In that case the pattern, black or white content may become negative. In realistic cases they will turn out to lie in the range $[0, 1]$ though. The problem is easy to fix[4], but we won't concern ourselves with such matters in this report.

Before going on with the discussion we illustrate the structure of the pattern solid into somewhat richer detail. First notice that the optimal patterns occur in 4 distinct kinds: 1^{stly} we have the light bar patterns, 2^{ndly} the dark bar patterns, 3^{rdly} the left edge patterns, and 4^{thly} the right edge patterns. The bar patterns form 2–parameter families, whereas the edge patterns form only 1–parameter families. Thus the bar patterns will fill areas on the boundary of the pattern solid, whereas the edge patterns will only describe certain curves on it. We find that the light and dark bar patterns occupy mutually symmetric connected areas that meet each other at the curves defined by the edge patterns. A useful representation lets you appreciate the distribution of optimal patterns (figure 5). We plot curves of equal bar center and barwidth in the (ζ_1, ζ_2)–diagram. The division of light and dark bars by the left and right edge patterns is also immediately apparent.

We are not yet done with the discussion of metamerism in $\mathcal{J}^2$: We have as yet developed no appreciation for the essential ***finite support*** of the operators. The aperture function basically lets us "see" only the interval $x \in (-1, +1)$. Everything outside that interval will be sampled by other $\mathcal{J}^2$'s, located at more advantageous positions. Outside the interval the

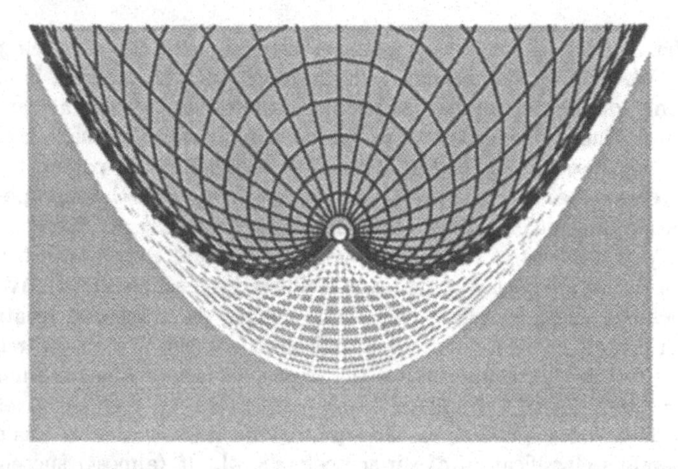

Figure 5: *The optimal patterns parameterized by bar center and width. The curves that fan out from the point of uniform patterns are the loci of constant bar center location. The curves that run transversally to them are the loci of constant bar width. Notice that these are split into two families (depicted in different tone): the light and the dark bar patterns, separated by the loci of left and right edges.*

operators are not oscillatory but just decay, thus they are not fit to capture any useful pattern structure.

We can demonstrate these problems by studying 2 very simple cases: A symmetric Gaußian light bar of variable width ($I(x) = \exp(-x^2/2\sigma^2)$), and an exponential ramp of varying slope ($I(x) = \exp(\alpha x)$). Some remarkable instances are shown in figure 6.

In case of the broad bar the transitions are *outside* of the aperture (that is to say the interval $[-1, +1]$) and it would make more sense to classify this image as *uniform* than as *a bar* purely on the basis of the observation. Likewise the shallow exponential ramp leads to an edge representation with transition near the origin, the steep one to an edge with transition far from the origin. The edge transition position is a monotonic function of the steepness of the exponential ramp. Again, it would make more sense to classify the steep ramp as *uniform* than as *an edge* purely on the basis of the observation. Clearly, we are in need of an extended discussion of metamerism in $\mathcal{J}^2$ that also takes the limited support of $\mathcal{J}^2$ into account.

One simple way to proceed is the following. Given an observation we consider all possible interpretations, *i.e.*, the corresponding metamer. If this

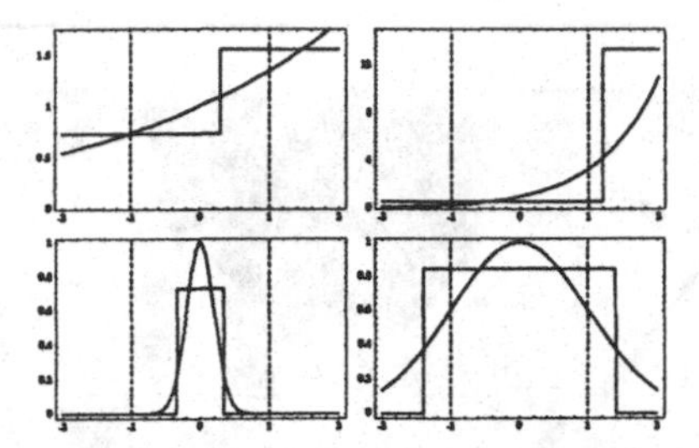

Figure 6: *Upper left: A shallow exponential ramp and its equivalent optimal pattern;Upper right: A steep exponential ramp and its equivalent optimal pattern. Different from the previous case the transition occurs outside of the aperture;Lower left: A narrow Gaußian hill and its equivalent optimal pattern;Lower right: A broad Gaußian hill and its equivalent optimal pattern. Different from the previous case the transitions occur outside of the aperture.*

equivalence class contains a member that happens to be uniform over the interval $[-1,+1]$ we consider the observation to indicate a *featureless image*. In any case we try to pick the least committing interpretation in some reasonable sense. This will typically mean an interpretation with very small variation over the interval $[-1,+1]$. We need strong evidence to commit ourselves to a really articulate interpretation. The optimal patterns are thus excellent candidates. For any given observation we can find a 1–parameter family of optimal patterns that are valid interpretations. It is easy to construct this family: We construct the halfplane on the gray axis that contains our observation. Then we draw the lines through the observation and the white and black point. These lines meet the optimal pattern locus in two points. Thus we define a stretch on the optimal pattern locus: This is the required locus for the observation can be explained by the mixture of an optimal pattern on this stretch and a uniform gray image. Next we consider these possibilities and note whether there exist instances for which one or two of the transition locations fall outside the stretch $[-1,+1]$. If both transitions may fall outside we declare the image to be *locally uniform*. If one transition may fall outside we declare it to be an edge. If both transitions fall inside the interval $[-1,1]$ we declare the image to be a *bar*.

It is fairly easy to find the segmentation of (ζ_1,ζ_2)–space into regions that admit of uniform, edge or bar interpretations (see figure 7).

Thus we find that the observations obtained via the ${}^1\mathcal{D}\,\mathcal{J}^2$ are of one of the following categories:

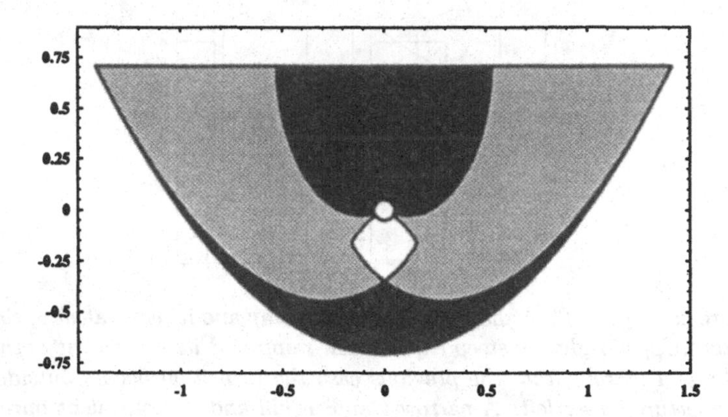

Figure 7: *The segmentation of the (ζ_1, ζ_2)–plane into various regions. The area outside the convex region bounded by the parabola consists of points that can never occur as valid observations. The convex area can be subdivided into 2 disjunct areas for which edge interpretations are the simplest interpretation. One of these areas contains the right, the other the left edges. There are also 2 disjunct areas where bars are the simplest interpretation. One area contains light the other dark bars. The remaining area is the diamond shaped region that contains the white point as a boundary element. In this region a uniform interpretation suffices.*

Invalid observations some observations do not correspond to any interpretation with nonnegative pixel intensities. Such observations should probably be treated as "error messages";

Uniform patterns some observations can be explained by patterns that are uniform within the region $[-1, +1]$. Outside this interval they are articulated, but other observations (centered at neighboring points) should take over there;

Edges some observations cannot be explained by uniform patterns but can be explained by assuming the presence of a single transition from one constant value to another occurring in the interval $[-1, +1]$;

Bars again, some valid observations cannot be explained by uniform patterns, nor by edge patterns. Such observations can be interpreted in terms of bars, *i.e.*, the occurrence of 2 transitions within the interval $[-1, +1]$ where transitions between 2 constant levels occur.

For a given observation there will still be some leeway left, *i.e.*, a limited 1–parameter variation of levels and transition locations. Thus one might be induced to augment the categorization scheme with a discretization of transition locations and/or levels.

The consequences of these considerations are important. For one thing, one should not think of ψ_1 as "edge finders", nor of the ψ_2 as "bar detectors". The *meaning* of the observation ϱ_1 or ϱ_2 depends on the total activity of $\mathcal{J}^2$. This meaning can be fixed once the order of the observation has be decided upon. (For instance, in $\mathcal{J}^1$ one can *only* find uniform patterns and edges, but in $\mathcal{J}^2$ one may also find bars. Thus the "meaning" of ϱ_1 depends on the order of the jet.) For a given order the meaning of ϱ_1 still depends on the values of ϱ_0 and ϱ_2, *etc.* We have shown how one may categorize local image structure by way of observations in $\mathcal{J}^2$. The jet $\mathcal{J}^2$ as a whole can recognize the categories "'bar", "edge", "uniform" or yield an "error message" and *only those.* In order to discriminate more (local) patterns one has to increase the order. Another way to discriminate more patterns is of course to consider multilocal observations.

2.2 Two–dimensional example: The visual field

The case of the 2–dimensional visual field is not all that different from the (almost trivial) 1–dimensional case of color vision. Different from the case of color, in the case of the visual field one is also interested in more complicated configurations with jets attached to each vertex of a lattice of sample points. Moreover, the interest is not restricted to any special order of representation. Often one is actually interested in the trade off between multilocal and local methods. In this paper we consider only the local case though.

Because the Gaußian kernel is separable in Cartesian coordinates there exists a close connection between the ${}^1\mathcal{D}$ and the ${}^2\mathcal{D}$ cases. If one has an image that changes only in 1 direction, the ${}^2\mathcal{D}$ case simply degenerates into the ${}^1\mathcal{D}$ case. Different from what one may perhaps expect this is actually an important case because *edges* and *lines* occur so frequently in practically interesting images.

Truly ${}^2\mathcal{D}$ image features are more or less circular blobs, corners and such important entities as "T–junctions". In such cases one is again mainly interested in the effects of variation of a *single* parameter, in these cases the distance from the fiducial point is largely irrelevant. However, this case is quite different from that of a unidirectional variation of course. The main problem (not considered here!) is of course how to apply the coordinate system that makes things simple.

2.2.1 Metamerism in the visual field

One can easily transpose Schrödinger's method to higher dimensions and higher order representations. However, the number of categories will grow fast as either of these are increased. The case of ${}^2\mathcal{D}$, order–2 is still comparatively simple. Notice that $\mathcal{J}^2$ in ${}^2\mathcal{D}$ is a *six* dimensional vector. One can construct a unique metameric image for any given image by considering additive combinations of sextuplets of point sources. Thus Schrödinger's argument works just as well for the lemma that optimal patterns have to be ***binary*** ones. The constraint on the ***transitions*** is now that it should be impossible to perturb it at 6 points in general position. If the transition curve is a general quadric one has exactly this situation: Since a quadric is fully determined by 5 points one can't find a sextuplet in general position. (In retrospect the pair of transition wavelengths in Schrödinger's original case are simply quadrics in one dimension.) One can indeed prove this hunch easily by writing down the condition $\det|\psi_j(x_k, y_k)| = 0$, where $j = 00, 10, 01, 20, 11, 02$, $k = 1, \ldots, 6$ and (x_k, y_k) are points on the transition locus. This constraint is satisfied if (x_6, y_6) satisfies a quadratic equation with the coefficients determined by the other coordinates: Thus the transition is indeed a quadric. That it is indeed a *general* quadric is clear from the fact that one may specify the points $(x_1, y_1), \ldots, (x_5, y_5)$ arbitrarily.

Thus the optimal patterns are binary patterns with quadrics as boundaries. A general quadric in the plane is $1+a_1x+a_2y+a_3x^2+a_4xy+a_5y^2 = 0$, and has 5 degrees of freedom. These degrees of freedom can be used to parameterize the boundary of the volume of possible jet activities. This volume is a convex subset of the space of *a priori* possible jet activities. Activities outside the volume cannot be induced by any optical structure and if they nevertheless would occur can only be interpreted as "error messages".

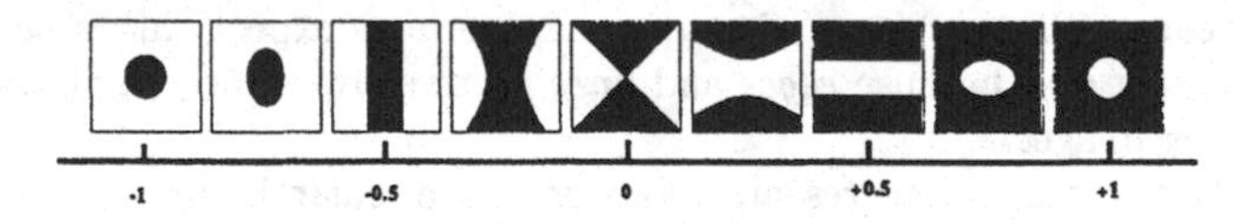

Figure 8: *The shape descriptor.*

By a translation and a rotation of the coordinate system one can bring the quadric in the canonical form $(k_1x^2 + k_2y^2)/2 = 1$. Thus (k_1, k_2) specify the *shape* of the quadric. Notice that one can still factor out the overall size and that interchange of the axes leaves the shape invariant. If one introduces the parameter $\sigma = -(2/\pi)\arctan(k_{max} + k_{min})(k_{max} - k_{min})$,

$k_{max} \geq k_{min}$, one has a convenient pure shape descriptor. (Figure 8.) The parameter σ takes values on the segment $[-1, +1]$, where a sign change indicates a change of orientation (*e.g.*, white blob versus dark blob of the same shape). When $|\sigma| < 0.5$ one has hyperbolæ, for $|\sigma| > 0.5$ ellipses. For $|\sigma| = 1$ the quadrics are circles, whereas $|\sigma| = 0.5$ denotes the degenerated quadric (pair of parallel lines). This parameterization is useful because it gives us a handle on the topology of the image. When you consider maps from the image plane on the σ domain you immediately see that generically you will have *regions* where the quadrics will be either elliptical or hyperbolical throughout, *curves* (that separate these regions) on which we find degenerated quadrics, and *isolated points* at which the quadrics are circular. In analogy with the classical differential geometry of surfaces I call the curves of degenerated quadrics "parabolic curves" and the points of circularity "umbilical points" or "umbilics".

Thus you can generically have light or dark blobs (elliptical boundaries) and light or dark bars (hyperbolical boundaries). If one restricts the attention to a circular disk of radius $\sqrt{4s}$ one finds

Edges, *concave or convex according to some convention,*

Bars, *black and white, curved and tapered and*

Blobs, *either white or black on a black or white ground.*

You see that these categories are merely labels for certain canonical representatives of the metamers. A metamer is an *equivalence class* of local image structure, not any specific structure. Only if the "saturation" (to borrow a convenient term from color science) is very high does the metamer almost shrink to the canonical representative itself. The "meaning" of the activity of a jet is most uncertain for the low saturations. However, the uncertainty is due to not observable power at the higher orders, that is of higher resolution than one should consider in the 1st place: It is actually reasonable to disregard it. The fine detail should be studied at *another level of resolution.* Thus one obtains both a *category* and a measure of *confidence.*

It is easy enough to characterize the local image structures for which the toolkit is completely *blind*: Any linear combination of fuzzy derivative operators of orders higher than that of the jet will fail to evoke any activity. Such patterns may be called "metameric black images" because they are really "black" as far as the toolkit is concerned! Such patterns can be detected only if one moves to another level of resolution.

For the higher order jets the theory is in all respects similar. The $3^{rd}$and 4th order algebraic curves have been classified according to various criteria, some of them perhaps of little interest for the case of vision. One needs a novel classification, based on *genericity* (many of the classical curves are

Figure 9: *Some textons in $\mathcal{J}^4$. These are just a collection of instances, many more are possible. From left to right and top to bottom we have: A uniform pattern, a disk, a triangle, a square, a split disk, a four petal leaf cluster, an annulus, a Landolt C, a blob with indentation; an edge, edge with hump, edge with single saw tooth, edge with island, edge with branch, edge with "fruit", bar, bar with hump, bar with constriction, truncated bar, dash, cross, 2 windmill patterns and an edge with truncated bar.*

singular cases) and up to affinities (many of the classical taxonomies are up to projectivities). The major classes from a vision point of view are (articulated) blobs, bars and edges. The curves with a number of asymptotes exceeding 3 seem to hold little interest from a pragmatic point of view. In figure 9 we show some of the local structures that are differentiated by $\mathcal{J}^4$.

3 Conclusion

"Local image structure" can be described in a principled way that takes the nature of the observation into account. One aspect of the observation is the choice of a *fiducial resolution.* Another aspect is the choice of the size of the "toolkit" and we have shown that this is equivalent to the choice of the *order of representation.* Once the level of resolution and the order of representation have been fixed, the discriminative power of the observation is settled. We can predict from 1st principles what the categorically different image features are ("blobs', "edges", "bars", ...). In a very real sense such "features" are properties of the observation, rather than of the image. Higher order features (*i.e.*, the features in the next higher order) cannot be detected at all and images that differ in these respects are classified as equivalent or "metameric". The only way to break through metamerism is to adjust resolution and/or perform multilocal comparisons (which need geometrical expertise). The canonical features are all that can be computed from the bunch of numbers yielded by the observation without any such additional expertise.

References

[1] Burt, P.J., Hong, Tsai–Hong, Rosenfeld, A., *Segmentation and estimation of image region properties through cooperative hierarchical computation*, IEEE Trans.SMC–**11**, pp.802–825, 1981.

[2] Canny, J.F., *Finding edges and lines in images*, Tech.Rep. 720, MIT, Cambridge Mass., 1983.

[3] Koenderink, J.J. and A.J. van Doorn, *The structure of images*, Biol.Cybern. **50**, pp.363–370, 1984.

[4] Koenderink, J.J., *Color atlas theory*, J.Opt.Soc.Am. **A4**, pp. 1314–1321, 1987.

[5] Koenderink, J.J. and A.J. van Doorn, *Generic neighborhood operators*, IEEE Trans. PAMI **14**, pp.597–605, 1992.

[6] Koenderink, J.J. and A.J. van Doorn, *Receptive field assembly pattern specificity*, J.Visual Comm. and Image Representation **3**, pp.597–605, 1992.

[7] Schrödinger, E., *Grundlinien einer Theorie der Farbenmetrik im Tagessehen, I, II*, Ann.Physik **63**, p.397 and 427, 1920.

[8] Schrödinger, E., *Theorie der Pigmente von größter Leuchtkraft*, Ann.Physik **62**, p.603, 1920.

Acknowledgements This research was done for the Esprit Basic Research Action REALISE.

Structural Scales in Computational Vision

Steven W. Zucker
Center for Intelligent Machines
McGill University
Montreal, Quebec, Canada

Abstract

It is commonly held that scale is an image-based notion. Scale spaces are built as contraction mappings over pixels, with coarse-scale functions implemented via large operators, and fine-scale functions implemented via small ones. This runs counter to structural considerations, however, which dictate that scale is related not to image-based properties but rather to object-based ones. Fingers are fine-scale with respect to the hand, and the hand with respect to the arm; it is only incidental when such relationships hold in images. We therefore propose that scale is an abstract property, and argue in this paper that it should be defined over abstract representations rather than images. Two examples of such abstract scale spaces are illustrated, one over the tangent map, and the other over an object's bounding contour.

1: Introduction

Basic physics dictates that there is an effective length scale for photometric phenomena: it is commonly observed that the edges bounding objects are defined at a fine scale, while shading is defined over much larger scales. This is sometimes modulated by statistical considerations, which introduces scales associated with noise that are related to the the spatial extent required for averaging. This, in turn, implies computational considerations about the scale necessary for interaction between local units, and raises issues of computational efficency. Finally, cartoonists employ a satyrical scale to enhance anatomical characteristics. Richard Nixon's nose has become famous in this respect, especially when, for emphasis, it is drawn larger than would be consistent with the scale of his face.

It is thus not surprising that the concept of scale has secured itself a position in the repertoire of computational vision. This is supported by the physiological observation that receptive fields for visual neurons can be observed to be sensitive to a range of spatial support, an observation normally interpreted to imply that receptive field size defines the spatial scale of an operator. The existence of many such receptive fields covering each retinotopic position further suggests a "pyramid"-like organization of operators, and it is common to build a pyramid by e.g., a Gaussian blurring operation followed by subsampling. Such image pyramids are also well known for their computational efficency. To illustrate, if there were a single dark object in a large field of view, then locating it would require searching a large number of pixels. However, if a pyramid of images were constructed over the original, then search could proceed from coarse-to-fine scale images with a logarithmic speedup in computational efficency. Azriel Rosenfeld was both instrumental in developing such ideas, and in establishing their significance for the field ([22]). To quote:

0-8186-7644-2 $5.00 © 1996 IEEE

> Pyramids, in general, are data structures that provide successively condensed representations of the information in the input image. What is condensed may be simply image intensity, so that the successive levels of the pyramid are reduced-resolution versions of the input image; but it may also be descriptive information about features in the image, so that successive levels represent increasingly coarse approximations to these features. (p 2)

This quote makes it clear that Rosenfeld understood pyramids as contraction mappings over structure, and his chapter in his 1984 collection: "Multiresolution Image Processing and Analysis" develops the notion that pyramids facilitate interaction across different length scales. He searched widely for a structural criterion to define these interactions, and is still working toward this goal.

I share this desire to find the correct mechanism for interaction over distance, and have collaborated with Azriel and with Bob Hummel on developing formal relaxation labeling techniques as one such mechanism ([23]). I further share with Azriel a fascination with biological vision, and was influenced early on in my work on texture analysis by his statement: "However, one would not ordinarily use arbitrary local properties for texture analysis, but only properties having significance to the human visual system." [24].

Since my collaboration with Rosenfeld I have taken up further modeling of the physiology of primate visual systems, and this has forced me to consider whether orientation is *the* fundamental organizing feature for visual systems. This question relates to the scale issue, as orientation is abstract from image properties, and suggests very different scale spaces from the ones normally considered in computer vision. Other researchers, notably Witkin [26], Koenderink [13], Marr [17], and Lindeberg [16]. have defined types of scale spaces in an attempt to follow structure across image size. However, none of these different scale spaces have clarified the differences around orientation-based structure. For example, how can a scale space be composed from image operations over a hair (texture flow) pattern without averaging the individual hairs away? How can the multiple edge responses in the neighborhood of a blurry edge be unified into a single descriptor of edge position ([5])?

In seeking to answer these questions, we have diverged from the classical position to what we can now pose as the central claim in this paper: that *structural scales, not image scales, are the appropriate scales for computational vision.* Image scales are necessary only insofar as image noise is a statistical problem, and a *minimum reliable scale* can be selected to handle it ([5]); beyond this, scale spaces should be built according to structural representations. As this is a somewhat radical hypothesis, we now proceed to defend it by illustrating two such scale spaces. In the Conclusions we return to the opening motivations.

2: Curves, Complexity, and Corners

Edge detection is normally considered the first stage of image analysis, and Rosenfeld was one of the first to consider non-linear operators for this [20]. . We have developed further non-linearities that are embedded in what we call "logical/linear operators" ([10]) that provide more robust local estimates of edge and line structure, by enforcing continuity conditions along the tangential extent and contrast variations along the normal extent of the spatial support of these operators. (Note how a notion of orientation, or tangent, is essential to defining these conditions.) Finally, since these estimates are still local, a relaxation labeling system ([9]) has been developed to "transport" estimates from one position to another nearby, and to establish the "consistency" between these estimates. This transport process again requires a notion of tangent, and also a notion of curvature; we refer to this relationship as *co-circularity*; see [19] and [27]. The result is a map that associates approximate orientations to pixel positions through which an edge or line curve could be passing.

This map defines our first abstraction over the image. It describes the (differential) geometric structure in the image in what we call a *tangent field*. Assuming that we seek to recover descriptions of objects in our visual world, it follows that the projections of these objects into the image are bounded by curves, that these curves are singular at points of intersection, and that, for efficency reasons, we seek to recover these curves through an intermediate local representation. Thus, in a mathematical sense the most natural such representation is a map that associates a (discretely quantized approximation to the) tangent to each image position through which a curve passes. Discontinuous points, corners, or points of intersection, could then be described as those points to which more than one tangent is assigned, reflecting the limiting process to the discontinuous point from either side. Observe that such a representation naturally models orientation hypercolumns in neurobiology.

The curves in an image might arise from the boundary of an object, from the pinstripes on a shirt, or from the hairs in a texture. It is a fundamental segmentation problem then to select which tangents correspond to which of these different classes of structure. Note that these classes are defined by structure in the world, not structure in the image.

2.1: Complexity of the Tangent Map

We illustrate this segmentation problem with a specific example. Imagine standing on an edge element in an unknown image, as in the first Figure. Is this edge part of a curve, or perhaps part of a texture? If the former, which is the next element along the curve? If the pattern is a texture, is it a hair pattern, in which nearby elements are oriented similarly, or a spaghetti pattern, in which they are not. The question is in part one of complexity, since curves are "simpler" than textures and in part one of dimensionality, since curves are 1-D and textures are 2-D.

Benoit Dubuc and I ([4]) have proposed a measure of representational complexity that seeks to answer these questions, and it involves building a scale space over the tangent field. We sketch these ideas in this Section.

The mathematical foundation for our measure of complexity is geometric measure theory. This differs from classical differential geometry in that curves are not given as prespecified maps, but rather we use a parameterization-free characterization of curve-like sets as those with Besicovitch tangents; informally, these are the locations in the image at which contrast organizes into locally dense, oriented arrangements.

Viewing curves as curve-like sets of points, we now consider dilations over it. Minkowski dilations are routinely used in mathematical morphology and for the estimation of fractal dimension. The approach consists in creating a new set which is the Minkowski sum with a dilating (structuring) element. The dilation is done isotropically over the set. More formally, given a set $E \subset \mathbf{R}^2$ and a compact convex set $F \subset \mathbf{R}^2$, the *dilation* of E by F is given by $E \oplus F = \{a + b, a \in E, b \in F\}$, where $+$ here denotes the vector sum. Given $\epsilon > 0$, we can also scale the structuring element F and obtain $\epsilon F = \{\epsilon x : x \in F\}$. An example of the isotropic dilation of a line segment with a ball is shown in the Figure.

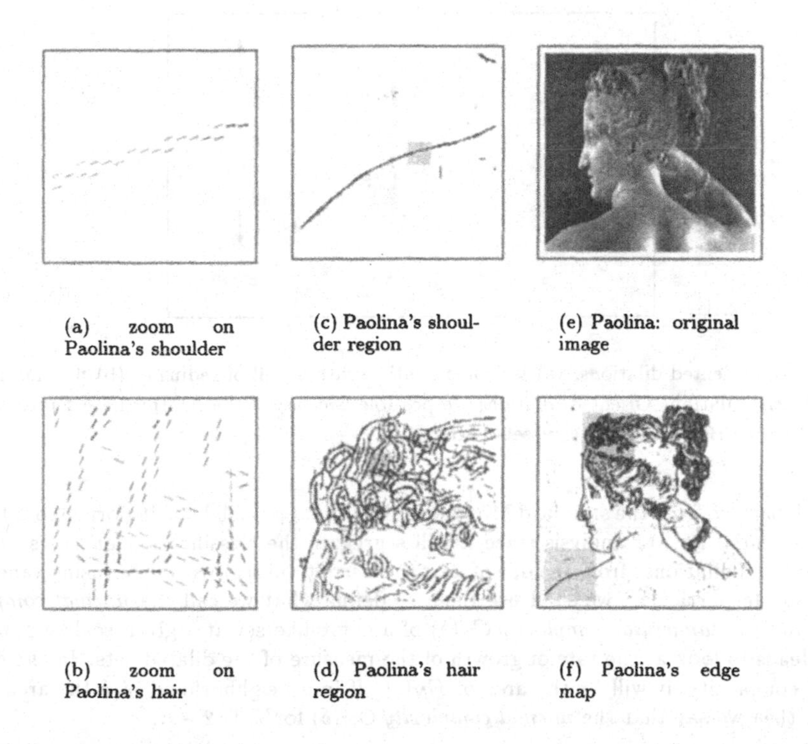
(a) zoom on Paolina's shoulder

(c) Paolina's shoulder region

(e) Paolina: original image

(b) zoom on Paolina's hair

(d) Paolina's hair region

(f) Paolina's edge map

The subtlety of "walking through" a tangent map: (top row) curves, (bottom row) texture. Notice how it is relatively easy to "walk" from one tangent to another when the tangents depict a curve, but that there is substantial confusion for the more complex texture (hair) pattern. For many of the tangents in the hair it is functionally impossible to determine which one is "next" along the same hair.

Our complexity measure will be based on *oriented dilations* that will adapt to the local structure of the set. Reconsidering the line segment example, the same Figure illustrates the concepts of both *normal and tangential dilations*:

Let E be a curve-like set and T its tangent map. The NORMAL DILATION $E_N(\epsilon)$ *of E at a scale ϵ is the dilation of the set E with the segment $(-\epsilon, \epsilon)$ in the direction normal to the tangents $\theta \in \Theta(x)$ at x. The* TANGENTIAL DILATION $E_T(\epsilon)$ *is obtained by dilating with the segment $(-\epsilon, \epsilon)$ in the direction of the tangent (see Figure.)*

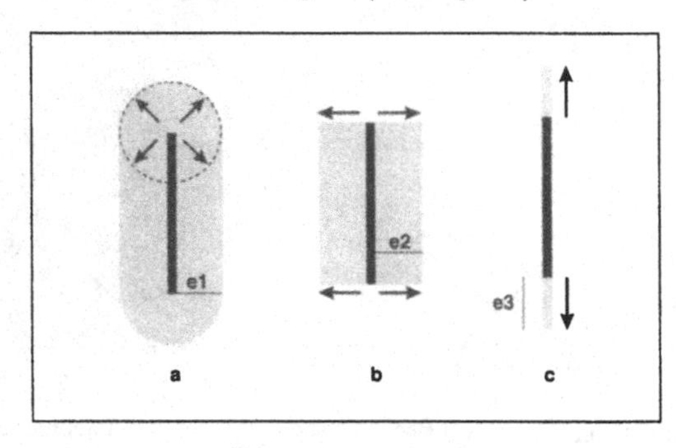

Isotropic and oriented dilations. (a) isotropic dilation with a ball of radius ϵ. (b) normal dilation (c) tangential dilation. Oriented dilations are possible because of the intermediate representation provided by the Besicovitch tangent sets. After [4].

The departure from the standard Minkowski dilation approach by using oriented dilations will be essential for our analysis since it will segregate the classification *curve* vs. *texture* (using normal dilations) from the one of *dust* (or discontinuities) vs. *curve* (using tangential dilations). In particular, we shall use them to define what we call the *normal complexity* $C_N(\delta)$ and the *tangential complexity* $C_T(\delta)$ of a curve-like set at a given scale $\delta > 0$. The main idea is to look at the rate of growth of the measure of the dilated sets. In the case of normal complexity, it will be the area of $E_N(\epsilon)$. If in a neighborhood of δ the area grows like δ^α, then we say that the *normal complexity* $C_N(\delta)$ for E is $2 - \alpha$.

Normal complexity *Let E be a curve-like set. If $E_N(\epsilon)$ is the normal dilation of E at scale ϵ, and if $|\cdot|_2$ denotes its area, then the normal complexity log-log plot is*

$$\left(\log\left(\frac{1}{\epsilon}\right), \log\left(\frac{1}{\epsilon^2}|E_N(\epsilon)|_2\right)\right); \tag{1}$$

moreover the NORMAL COMPLEXITY $C_N(\delta)$ *at scale δ will be the left derivative of the normal complexity log-log plot at δ.*

The tangential complexity is defined analogously.

2.2: Segmenting the Tangent Field into Structural Classes

Returning to the Paolina image, we observe that, in the hair region, the rate of growth α is approximatively 0 at the chosen scale δ, thus the complexity is $2 - \alpha \approx 2$ (texture). For the shoulder, the rate of growth is linear ($\alpha \approx 1$), leading to a normal complexity of $2 - \alpha \approx 1$ (curve). This suggests the following principles for segmenting the tangent field:

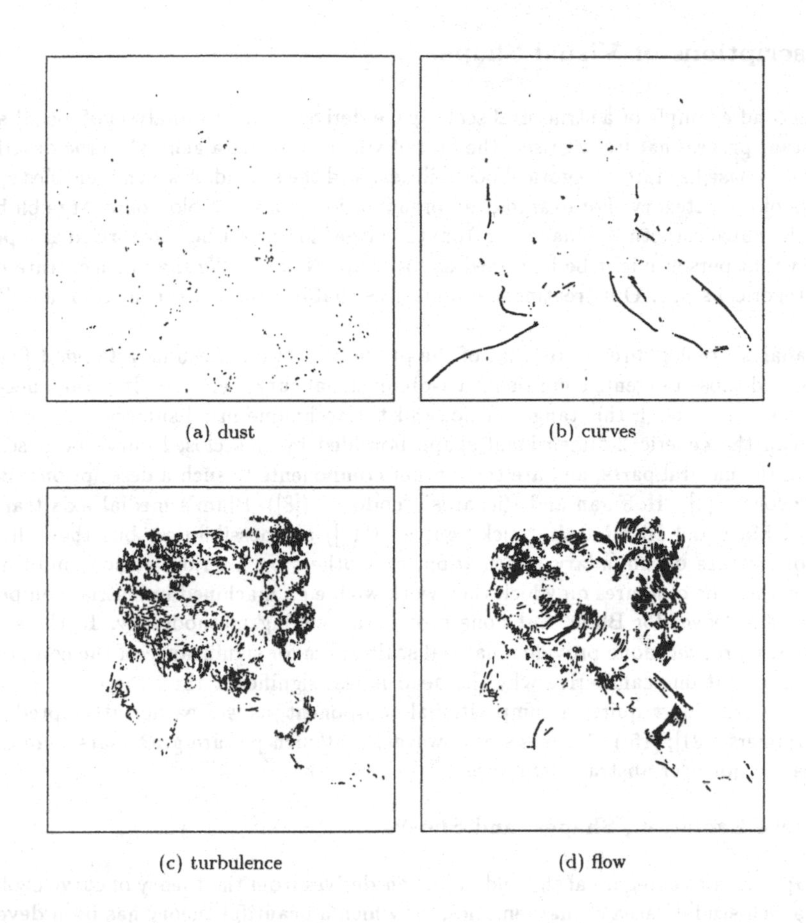

Segmentation can be pulled back to the level of the complexity map. After [4].

dust: sets in which the tangent map is sparse, the object almost nowhere extends along its length locally, and might be thought of as being curve-free (low normal complexity - low tangential complexity);

curves: discrete tangent maps for which a curve representation is completely adequate. Objects extend along their length, like Paolina's shoulder, and the density of other tangents is low almost everywhere along it in a local neighborhood (low normal complexity - high tangential complexity);

turbulence: tangent maps that are characterized by objects that do not extend along their length but are dense in the normal direction; e.g., Paolina's uncombed hair (high normal complexity - low tangential complexity);

flow: tangent maps for which the objects extend along their length and are also dense in the normal direction (high normal complexity - high tangential complexity).

The result of this segmentation applied to the Paolina tangent field is in the Figure.

3: Descriptions of Visual Shape

The second example of a structural scale space derives from the analysis of visual shape. A two-stage process naturally arises, the first of which provides a general shape description to serve as a base key into categorical possibilities, and the second of which identifies specific members of the category. For example, an organization of a small blob on a large blob with several elongated structures emanating from the large blob might be categorized as a person, and then that person might be identified as "Richard Nixon". We shall concentrate on the first, categorical stage. Our treatment summarizes that in Kimia, Tannenbaum, and Zucker ([11]).

The analysis builds directly on that of the previous Section. Assume a tangent field was inferred, and those tangents comprising a 1-dimensional curve selected. It is then necessary to infer a curve through this tangent field, and the technique in [3] suffices.

How can the generic 2-dimensional shape bounded by this closed curve be described? What are the natural parts, and are there other components to such a description? Biederman's "geons" ([1]), Hoffman and Richards' "codons" ([8]), Blum's medial axis transform ([2]), and Marr and Nishihara's "stick figures" ([17]) are possibilities, but they all differ. Some concentrate on boundary information, and others on interior (region) information. Each has a regime of figures on which they work well, e.g., machined industrial components for geons and leaves for Blum, but none provide a unifying methodology. In the spirit of this paper, moreover, none provide a natural scale space for significance of the components, in the sense that one can derive why the head is less significant than the torso, etc. This significance ordering requires a compositional component, as was earlier attempted in picture grammers ([21]; [15]). Note again, however, that such picture grammars were defined over images, not over abstract structure.

3.1: Curve Evolution, Shapes, and Shocks

Our approach to categorical shape description derives from the theory of curve evolution. This is a well-studied area of mathematics, for which a beautiful theory has been developed over the past half century ([7], [25, 18], [14], [6]. What will be of primary significance for us is that such evolutions develop singularities—or *shocks*—and these singularities define the natural shape descriptors. They also imply a natural significance hierarchy over shapes, which is the structural scale space that we seek.

We begin with the priciple that: *slight changes in the boundary of an object cause only slight changes to its shape*, which suggests that we consider slight boundary deformations. Let the shape be represented by the curve $C_0(s) = (x_0(s), y_0(s))$, where s is the parameter along the curve (not necessarily arclength), x_0 and y_0 are the Cartexian coordinats and the subscript) denotes the initial curve prior to deformation. We consider the specific deformation (see Kimia et al. ([11]) for derivation, background, and related references):

$$\begin{aligned} \frac{\partial C}{\partial t} &= (1 + \epsilon\kappa)\vec{N} \\ C(s, 0) &= C_0(s) \end{aligned}$$

An illustration of a curve evolving according to this equation is shown in the Figure. The ϵ parameter runs along the x-axis, and the time of evolution, or scale, along the vertical

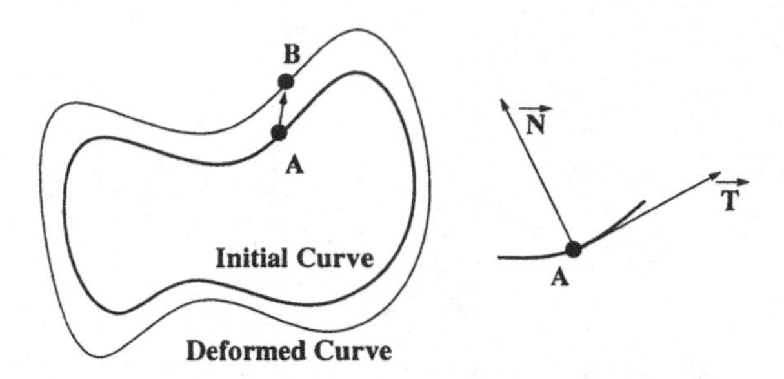

The points on the initial curve A move to B to generate a new curve. The direction and magnitude of this motion is arbitrary in order to capture general deformations. However, with mild restrictions appropriate to shape, one can classify this deformation to be a sum of *constant deformation* and *curvature deformation* along the normal.

axis. The result is a scale space built upon the shape, which we call a *reaction/diffusion space*.

This space is no named because of the parameter ϵ, which controls the amount of diffusion during the evolution. For large amounts of ϵ it dominates, and the result is a *geometric heat equation*:

$$\begin{aligned} \frac{\partial C}{\partial t} &= \kappa \vec{N} \\ C(s,0) &= C_0(s) \end{aligned}$$

An interesting analogy is thus revealed: we earlier remarked that image-based scale spaces are often built by blurring the image with Gaussian operators of increasing scale (e.g., Koenderink ([12]), Witkin ([26])); this equation shows that curve evolution according to the curvature in the normal direction amounts to a similar process but based on the curve itself. Moreover, it is non-linear, since it involves an effective "re-parameterization" at each iteration.

3.2: Shocks and Shape Categories

At the opposite extreme, when $\epsilon = 0$, there is no smoothing and shocks are created. Four such shocks are illustrated in the Figure; they are:

1. *First-Order Shocks* are discontinuities in orientation of the boundary of a shape. They denote *protrusions* and intrusions, such a the nose on a face.
2. *Second-Order Shocks* result when, during the process of deformation, two distinct non-neighboring boundary points join and not all the other neighboring boundary points have collapsed together. Second-order shocks define the *parts* of a shape.
3. *Third-Order Shocks* result when, during the process of deformation, two distinct non-neighboring boundary points join in a manner such that neighboring boundaries of

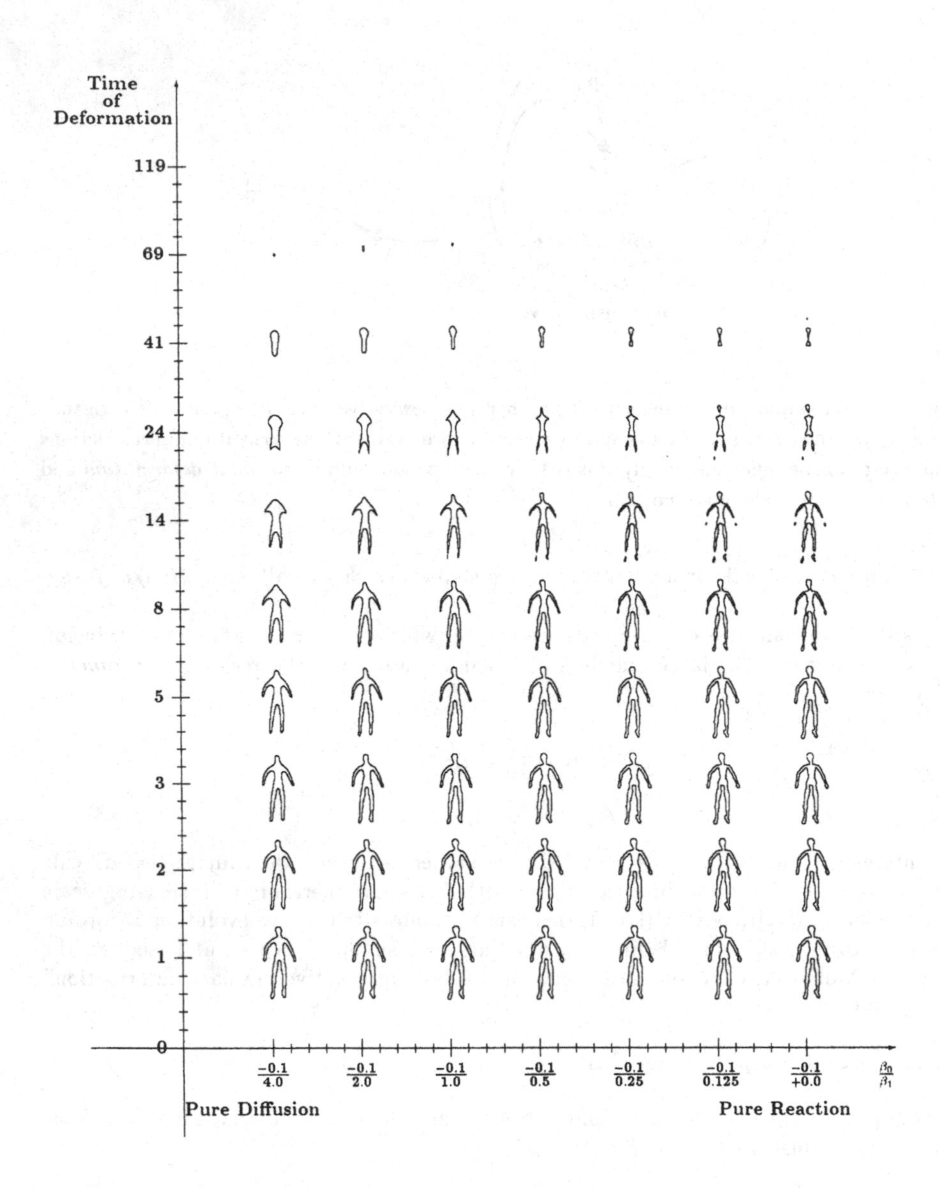

An illustration of a portion of the reaction-diffusion space for a DOLL image (from the National Research Council of Canada's Laser Range Image Library CNRC9077 Cat No 422. The image was thresholded and stored as a 128x128 image. The numbers on the x-axis are indicative of the ratio of reaction to diffusion. Note how the reaction creates shocks, and the diffusion "melts" the boundary of the shape. After [11].

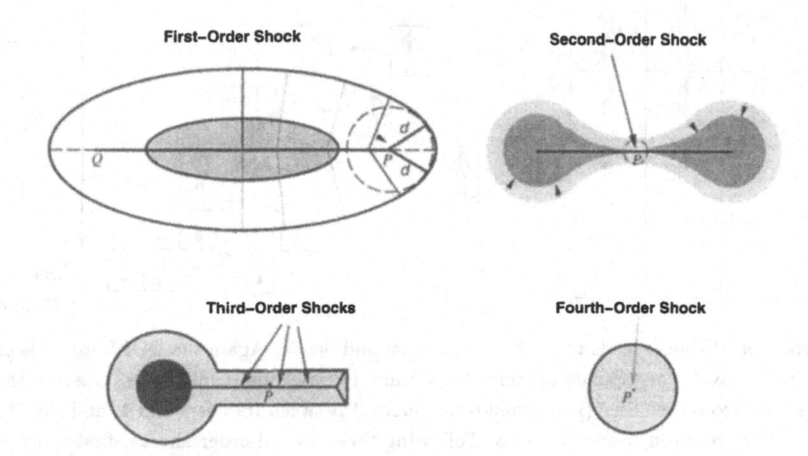

Illustration of the four types of shocks formed in the course of the reactive deformation process; each is correlated with a generic shape category, *i.e.*, *protrusion*, *part*, *bend* and *seed*. For each shock type, the maximal inscribed disc is overlayed. After [11].

each point also collapse together. Third-order shocks indicate an extended axis, as in a tail or a *bend*.

4. *Fourth-Order Shocks* occur when, during the process of deformation, a closed boundary collapses to a point. Thus fourth-order shocks are the *seeds* of a shape, in that they indicate the locations where mass is to be attached.

Second-order shocks persist for only an instant in time, and then give rise to a pair of first-order shocks traversing in opposite directions. The locus of points traced out by the shocks is the Blum skeleton.

3.3: Significance of Shape Components

Because the shocks occur during an evolutionary process, the time of formation is related to significance. In particular, the least significant shocks form first, and the most significant ones last. The result is a hierarchical description that is suitable for categorical object description; see Figure.

4: Conclusions

The notion of scale space is classical in computational vision, but typically they are built over image structure. The assumption behind such scale spaces is that structure in the world projects into equivalent structure in an image. However, image size is a function of imaging configuration, projective mappings, and noise prefiltering. Object structure, on the other hand, is a function of the object. It therefore follows that notions of scale exist that are tied to abstract objects rather than to image size. We have argued that such

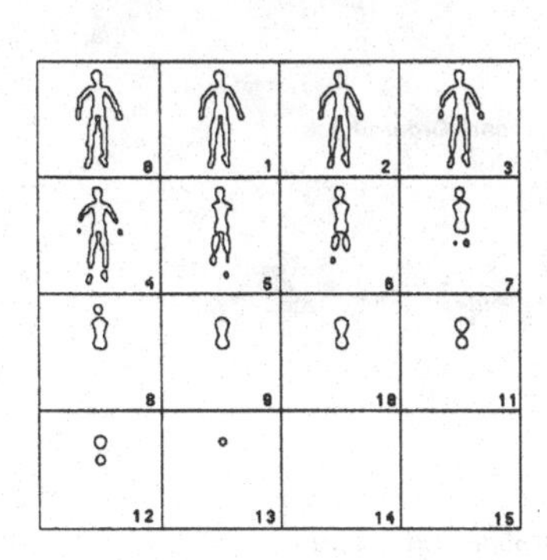

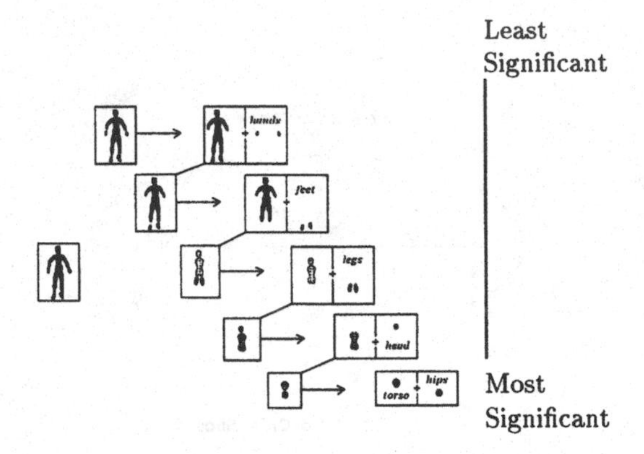

a) The evolution of shocks leads to parts, protrusions, and bends. Again the DOLL image is shown. The contour in box N corresponds to increasing boundary evolution (time) steps. Observe that the "feet" partition from the "legs" (via second-order shocks) between frames 3 and 4, and the "hands" from the "arms" between frames 2 and 3. Following these second-order shocks, first-order shocks develop as the "arms" are "absorbed" into the chest. Running this process in the other direction would illustrate how the arms "protrude" from the chest. b) The hierarchical decomposition of a doll into parts. Selected frames were organized into a hierarchy according to the principle that the significance of a part is directly proportional to its survival duration. After [11]

notions can be developed for computational vision, and have shown two examples. First, a compexity map was developed for the tangent field, a representation of "edge detected" structure. Neighborhood size in the tangent field was used to define an oriented complexity measure, which could then be used to segregate individual curves from textures. Similar ideas could also be used to segregate discontinuous points from regular curves. At a higher level, the reaction/diffusion space provided scale for object description. It was built from the boundary of the object, and based on the mathematics of curve evolution. Smoothing of a boundary was then related to the geometric heat equation, rather than to Gaussian blur of the image. Finally, a natural scale for significance emerged as the time of shock formation.

Both of these constructions are abstract with respect to images, although they are built from image-derived structure. It remains to determine whether additional structural scale spaces are necessary to model generic visual processing.

References

[1] Irving Biederman. Recognition by components. *Psych. Review*, 94:115–147, 1987.

[2] Harry Blum. Biological shape and visual science. *J. Theor. Biol.*, 38:205–287, 1973.

[3] C. David and S. W. Zucker. Potentials, valleys, and dynamic global coverings. *Int. journal of Computer Vision*, 5:219–238, 1990.

[4] Benoit Dubuc and Steven Zucker. Indexing visual representations through the complexity map. 1995.

[5] James Elder and Steven Zucker. Local scale control for edge detection and blur estimation. 1996.

[6] M. Gage and R. S. Hamilton. The heat equation shrinking convex plane curves. *J. Differential Geometry*, 23:69–96, 1986.

[7] M. A. Grayson. The heat equation shrinks embedded plane curves to round points. *J. Differential Geometry*, 26:285–314, 1987.

[8] D. D. Hoffman and Whitman A. Richards. Parts of recognition. *Cognition*, 18:65–96, 1985.

[9] Robert Hummel and Steven Warren Zucker. On the foundations of relaxation labeling processes. 6:267–287, 1983.

[10] Lee A. Iverson and Steven W. Zucker. Logical/linear operators for image curves. 1995.

[11] Benjamin B. Kimia, Allen R. Tannenbaum, and Steven W. Zucker. Shapes, shocks, and deformations, I: The components of shape and the reaction-diffusion space. *International Journal of Computer Vision*, 15:189–224, 1995.

[12] J. J. Koenderink and A. J. van Doorn. Dynamic shape. *Biological Cybernetics*, 53:383–396, 1986.

[13] Jan J. Koenderink. The structure of images. *Biological Cybernetics*, 50:363–370, 1984.

[14] Peter D. Lax. *Shock Waves and Entropy*, pages 603–634. Academic Press, New York, 1971.

[15] Michael Leyton. *Symmetry, Causality, Mind.* MIT press, April 1992.

[16] Tony Lindeberg. *Scale-Space Theory In Computer Vision.* Kluwer Academic Publishers, 1994.

[17] David Marr. *Vision.* W.H. Freeman, San Fransico, 1982.

[18] Stanley Osher and James Sethian. Fronts propagating with curvature dependent speed: Algorithms based on Hamilton-Jacobi formulations. *Journal of Computational Physics*, 79:12–49, 1988.

[19] Pierre Parent and Steven Warren Zucker. Trace inference, curvature consistency and curve detection. 11(8):823–839, August 1989.

[20] Azriel Rosenfeld. A nonlinear edge detection technique. *Proceedings IEEE*, pages 814–816, 1970.

[21] Azriel Rosenfeld. Isotonic grammars, parallel grammars, and picture grammars. *Machine Intelligence*, 6:281–294, 1971.

[22] Azriel Rosenfeld. *Multiresolution Image Processing and Analysis.* Springer-Verlag, 1984.

[23] Azriel Rosenfeld, Robert Hummel, and Steven Zucker. Scene labeling by relaxation operations. *IEEE Trans. Systems Man and Cybernetics*, 6:420–433, 1976.

[24] Azriel Rosenfeld, Y. Lee, and R. Thomas. (edge and curve detection for texture discrimination. *Picture Processing and Psychopictorics, Academic Press, New York*, page 381, 1970.

[25] James A. Sethian. Curvature and the evolution of fronts. *Comm. Math. Physics*, 101:487–499, 1985.

[26] Andrew Witkin, Demetri Terzopoulos, and Michael Kass. Signal matching through scale space. *International Journal of Computer Vision*, 1(2):133–144, 1987.

[27] Steven Warren Zucker, Allan Dobbins, and Lee Iverson. Two stages of curve detection suggest two styles of visual computation. *Neural Computation*, 1:68–81, 1989.

Propagating Covariance In Computer Vision

Robert M. Haralick
Intelligent Systems Laboratory
Department of Electrical Engineering
University of Washington
Seattle, WA 98195

Abstract

This paper describes how to propagate approximately additive random perturbations through any kind of vision algorithm step in which the appropriate random perturbation model for the estimated quantity produced by the vision step is also an additive random perturbation. We assume that the vision algorithm step can be modeled as a calculation (linear or non-linear) that produces an estimate that minimizes an implicit scaler function of the input quantity and the calculated estimate. The only assumption is that the scaler function have finite second partial derivatives and that the random perturbations are small enough so that the relationship between the scaler function evaluated at the ideal but unknown input and output quantities and the observed input quantity and perturbed output quantity can be approximated sufficiently well by a first order Taylor series expansion.

The paper finally discusses the issues of verifying that the derived statistical behavior agrees with the experimentally observed statistical behavior.

1 Introduction

Each real computer vision problem begins with one or more noisy images and has many algorithmic steps. Development of the best algorithm requires understanding how the uncertainty due to the random perturbation affecting the input image(s) propagates through the different algorithmic steps and results in a perturbation on whatever quantities are finally computed. Perhaps a more accurate statement would be that the quantities finally computed must really be considered to be estimated quantities.

Once we have the perspective that what we compute are estimates, then it becomes clear that even though the different ways of estimating the same quantity typically yield the same result if the input quantities are not affected by a random perturbation, it is certainly not the case that the different ways of estimating the same quantities yield an estimate with the same distribution when the input is perturbed by a random perturbation. It is clearly the case that the distribution of the estimate depends on the distribution of the input random perturbation and the method or type of estimate.

With this in mind, it is then important to understand how to propagate a random perturbation through any algorithm step in a vision problem. The difficulty is that the steps are not necessarily linear computations, the random perturbations are not necessarily additive, and the appropriate kinds of perturbations change from algorithm step to algorithm step. Nevertheless, there are many computer vision and image analysis algorithm steps in

 0-8186-7644-2 $5.00

which the appropriate kind of random perturbation is additive or approximately additive. And for these kinds of steps one basic measure of the size of the random perturbation is given by the covariance matrix of the estimate.

In this paper, we describe how to propagate the covariance matrix of an input random perturbation through any kind of a calculation (linear or non-linear) that extremizes an implicit scaler function, with or without constraints, of the perturbed input quantity and the calculated output estimate. The only assumption is that the scaler function to be extremized have finite second partial derivatives and that the random perturbations are small enough so that the relationship between the scaler function evaluated at the ideal but unknown input and output quantities and the observed input quantity and perturbed output quantity can be approximated sufficiently well by a first order Taylor series expansion. The propagation relationships do not depend on what algorithm is used to extremize the given scalar function.

As a related case, the given propagation relationships also show how to propagate the covariance of the coefficients of a function for which we wish to find a zero to the covariance of any zero we can find.

The analysis techniques of propagation of errors is well known in the photogrammetry literature. The Manual of Photogrammetry (Slama, 1980) has a section showing how to determine the variance of Y where $Y = F(X)$ from the variance of X. The generalization of this to find the covariance matrix for Y given the covariance matrix for X is rather straightforward. Just expand F around the mean of X in a first order Taylor expansion and consider that Y is a linear function T of X. Once the coefficients of the linear combination is known, so that the randomness of Y can be approximated by $Y - \mu_Y = T(X - \mu_X)$, then the covariance matrix Σ_Y of Y is easily seen to be given in terms of T and the covariance matrix Σ_X of X by $\Sigma_Y = T\Sigma_X T'$ (Mikhail, 1976; Koch, 1987). This only works well for cases where the function F can be given explicitly. The problem we discuss here is one in which the function F is not given explicitly, but Y is related to X in a specific way. The techniques we employ are well-known in statistical and engineering communities. There is nothing sophisticated in the derivation. However, this technique is perhaps not so well known in the computer vision community. There are many recent vision-related papers that could be cited to illustrate this. See for example Weng, Cohen and Herniou (1992), Wu and Wang (1993), or Williams and Shah (1993).

The paper concludes with a discussion of how to validate that the software which we use to accomplish the calculation we desire actually works. We argue that this validation can be done by comparing the predicted statistical behavior with the experimentally observed statistical behavior in a set of controlled experiments.

2 The Abstract Model

The abstract model has three kinds of objects. The first kind of object relates to the measurable quantities. There is the unobserved $N \times 1$ vector X of the ideal unperturbed measurable quantities. We assume that each component of X is some real number. Added to this unobserved ideal unperturbed vector is an $N \times 1$ unobserved random vector $\triangle X$ of noise. The observed quantity is the randomly perturbed vector $X + \triangle X$.

The second kind of object relates to the unknown parameters. There is the unobserved $K \times 1$ vector Θ. We assume that each component of Θ is some real number. Added to this ideal unperturbed vector is a $K \times 1$ unobserved vector $\triangle\Theta$ that is the random perturbation on Θ induced by the random perturbation ΔX on X. The calculated quantity is the randomly perturbed parameter vector $\hat{\Theta} = \Theta + \Delta\Theta$.

The third kind of object is a continuous scaler valued function F which relates the vectors X and Θ and which relates the vectors $X + \Delta X$ and $\Theta + \Delta\Theta$. The function F has finite first and second partial derivatives with respect to each component of Θ and X, including all second mixed partial derivatives taken with respect to a component of Θ and with respect to a component of X.

The basic problem is: given $\hat{X} = X + \Delta X$, determine a $\hat{\Theta} = \Theta + \Delta\Theta$ to minimize $F(\hat{X}, \hat{\Theta})$ given the fact that Θ minimizes $F(X, \Theta)$.

Of course, if $\hat{\Theta}$ is computed by an explicit function h, so that $\hat{\Theta} = h(\hat{X})$, the function F is just given by $F(X, \Theta) = (\Theta - h(X))'(\Theta - h(x))$. However, our development can handle as well the determining of the covariance of a $\hat{\Theta}$ which is known to minimize $F(\hat{X}, \hat{\Theta})$, without requiring any knowledge of how the minimizing $\hat{\Theta}$ was computed.

3 Example Computer Vision Problems

There is a rich variety of computer vision problems which fit the form of the abstract model. In this section we outline a few of them, specifically: curve fitting, coordinated curve fitting, local feature extraction, exterior orientation, and relative orientation. Other kinds of calculations in computer vision such as calculation of curvature, invariants, vanishing points, or points at which two or more curves intersect, or problems such as motion recovery are all examples of problems which can be put in the abstract form as given above.

3.1 Curve Fitting

In the general curve fitting scenario, there is the unknown free parameter vector, Θ, of the curve and the set of unknown ideal points on the curve $\{x_1, \ldots, x_N\}$. Each of the ideal points is then perturbed. If Δx_n is the random noise perturbation of the nth point, then the observed point n^{th} point is $\hat{x}_n = x_n + \Delta x_n$. The form of the curve is given by a known function f which relates a point on the curve to the parameters of the curve. That is, for each ideal point x_n we have $f(x_n, \Theta) = 0$. We also assume that the parameters of the curve satisfy its own set of constraint equations: $h(\Theta) = 0$. The curve fitting problem is then to find an estimate $\hat{\Theta}$ to minimize $\Sigma_{n=1}^{N} f^2(\hat{x}_n, \hat{\Theta})$ subject to $h(\hat{\Theta}) = 0$. To put this problem in the form of the abstract problem we let

$$\begin{aligned} X &= (x_1, \ldots, x_N) \\ \hat{X} &= (x_1 + \Delta x_1, \ldots, x_n + \Delta x_N) \\ F(X, \Theta, \Lambda) &= \Sigma_{n=1}^{N} f^2(x_n, \psi) + h(\Theta)'\Lambda \end{aligned}$$

Then the curve fitting problem is to find $\hat{\Theta}$ and $\hat{\Lambda}$ to minimize $F(\hat{X}, \hat{\Theta}, \hat{\Lambda})$ where $F(X, \Theta, \Lambda) = 0$.

3.2 Coordinated Curve Fitting

In the coordinated curve fitting problem, multiple curves have to be fit on independent data, but the fitted curves have to satisfy some joint constraint. We illustrate the discussion in this section with a coordinated fitting of two curves and a constraint that the two curves must have some common point at which they are tangent.

Let $(x_1, \ldots, x_I)$ be the ideal points which are associated with the first curve whose parameters are ψ_1 and whose constraint is $h_1(\psi_1) = 0$. Each point x_i satisfies $f_1(x_i, \psi_1) = 0$, $i = 1, \ldots, I$.

Likewise, let $(y_1, \ldots, y_J)$ be the ideal points which are associated with the second curve whose parameters are ψ_2 and whose constraint is $h_2(\psi_2) = 0$. Each point y_j satisfies $f_2(y_j, \psi_2) = 0$, $j = 1, \ldots, J$.

The coordinated constraint is that for some unknown z,

$$\begin{aligned} f_1(z, \psi_1) &= 0 \\ f_2(z, \psi_2) &= 0 \\ \frac{\partial f_1}{\partial z}(z, \psi_1) &= \frac{\partial f_2}{\partial z}(z, \psi_2) \end{aligned}$$

The observed points $\hat{x}_i$ and $\hat{y}_j$ are related to the corresponding ideal points by

$$\hat{x}_i = x_i + \triangle x_i$$
$$\hat{y}_j = y_j + \triangle y_j$$

To put this problem in the framework of the abstract model, we take

$$\begin{aligned} \hat{X} &= (\hat{x}_1, \ldots, \hat{x}_I, \hat{y}_1, \ldots, \hat{y}_J) \\ \hat{\Theta} &= (\hat{\psi}_1, \hat{\psi}_2, \hat{z}) \\ \hat{\Lambda} &= (\hat{\lambda}_1, \hat{\lambda}_2, \hat{\lambda}_3, \hat{\lambda}_4, \hat{\lambda}_5) \end{aligned}$$

and define

$$\begin{aligned} F(\hat{X}, \hat{\Theta}, \hat{\Lambda}) &= \Sigma_{i=1}^{I} f_1^2(\hat{x}_i, \hat{\psi}_1) + \Sigma_{j=1}^{J} f_2^2(y_j, \psi_2) + \hat{\lambda}_1 h_1(\hat{\psi}_1) + \hat{\lambda}_2 h_2(\hat{\psi}_2) \\ &+ \hat{\lambda}_3 f_1(z, \hat{\psi}_1) + \hat{\lambda}_4 f_2(z, \hat{\psi}_2) + \hat{\lambda}_5 [\frac{\partial f_1}{\partial z}(z, \psi_1) - \frac{\partial f_2}{\partial z}(z, \psi_2)] \end{aligned}$$

The coordinated curve fitting problem is then to determine a $\hat{\Theta}$ and $\hat{\Lambda}$ to minimize $F(\hat{X}, \hat{\Theta}, \hat{\Lambda})$, where the perturbed $\hat{\Theta}$ is considered related to the ideal Θ by $\hat{\Theta} = \Theta + \triangle\Theta$.

3.3 Local Feature Extraction

There are a variety of local features that can be extracted from an image. Examples include edges, corners, ridges, valleys, flats, saddles, slopes, hillsides, saddle hillsides, etc. Each local feature involves the calculation of some quantities assuming that the neighborhood has the feature and then a detection is performed based on the calculated quantities. For example, in the simple gradient edge feature, the quantity calculated is the gradient magnitude and the edge feature is detected if the calculated gradient magnitude is high enough. Here we concentrate on the calculation of the quantities associated with the feature and not the detection of the feature itself.

To put this problem in the setting of the abstract problem, we let Θ be the vector of unknown free parameters of the feature and X be the unobserved neighborhood array of noiseless brightness values. We let $\hat{X}$ be the perturbed observed neighborhood array of brightness values, $\hat{X} = X + \triangle X$, and $\hat{\Theta}$ be the calculation of the required quantities from the perturbed brightness values $\hat{X}$. The form the of feature is given by the known function f which satisfies that $f(X, \Theta) = 0$. The feature extraction problem is then to find the estimate $\hat{\Theta}$ to minimize $F(\hat{X}, \Theta) = f^2(\hat{X}, \hat{\Theta})$.

3.4 Exterior Orientation

In the exterior orientation problem, there is a known 3D object model having points $(x_n, y_n, z_n), n = 1, \ldots, N$. The unobserved noiseless perspective projection of the point (x_n, y_n, z_n) is given by (u_n, v_n). The relationship between a 3D model point and its corresponding perspective projection is given by a rotation and translation of the object model point, to put it in the reference frame of the camera, followed by a perspective projection. So if ψ represents the triple of tilt angle, pan angle, and swing angle of the rotation, t represents the x-y-z-translation vector, and k represents the camera constant (the focal length of the camera lens), we can write:

$$\begin{aligned}(u_n, v_n)' &= \frac{k}{r_n}(p_n, q_n)' \text{ where} \\ (p_n, q_n, r_n)' &= R(\psi)(x_n, y_n, z_n)' + t\end{aligned}$$

and where $R(\psi)$ is the 3×3 rotation matrix corresponding to the rotation angle vector ψ.

The function to be minimized can then be written as:

$$f_n(u_n, v_n, \psi, t) = f(u_n, v_n, x_n, y_n, z_n, \psi, t) \text{ where}$$

$$\begin{aligned}f(u_n, v_n, x_n, y_n, z_n, \psi, t) &= [u_n - k\frac{(1,0,0)(R(\psi)(x_n, y_n, z_n)' + t)}{(0,0,1)(R(\psi)(x_n, y_n, z_n)' + t)}]^2 \\ &+ [v_n - k\frac{(0,1,0)(R(\psi)(x_n, y_n, z_n)' + t)}{(0,0,1)(R(\psi)(x_n, y_n, z_n)' + t)}]^2\end{aligned}$$

To put this problem in the form of the abstract description we take

$$\begin{aligned}X &= (u_1, v_1, \ldots, u_n, v_n) \\ \hat{X} &= (\hat{u}_1, \hat{v}_1, \ldots \hat{u}_n, \hat{v}_n) \\ \Theta &= (\psi, t) \\ \hat{\Theta} &= (\hat{\psi}, \hat{t})\end{aligned}$$

and define

$$F(\hat{X}, \hat{\Theta}) = \Sigma_{n=1}^{N} f_n^2(\hat{u}_n, \hat{v}_n, \hat{\Theta})$$

The exterior orientation problem is then to find a $\hat{\Theta}$ to minimize $F(\hat{X}, \hat{\Theta})$, given that $F(X, \Theta) = 0$. And because F is non-negative it must be that Θ minimizes $F(X, \Theta)$.

3.5 Relative Orientation

The relative orientation problem can be put into the form of the abstract problem in a similar way to the exterior orientation problem. We let the perspective projection of the nth point on the left image be (u_{nL}, v_{nL}) and the perspective projection of the nth point on the right image be (u_{nR}, v_{nR}). Then we can write that

$$\begin{aligned}(u_{nL}, v_{nL})' &= \frac{k}{z_n}(x_n, y_n)' \text{ and that} \\ (u_{nR}, v_{nR})' &= \frac{k}{r_n}(p_n, q_n)\end{aligned}$$

where (p_n, q_n, r_n) is the rotated and translated model point as given in the description of the exterior orientation problem.

The observed perspective projection of the nth model point is noisy and represented as $(\hat{u}_n, \hat{v}_n) = (u_n + \Delta u_n, v_n + \Delta v_n)$. Then taking

$$\begin{aligned} X &= (u_{1L}, v_{1L}, u_{1R}, v_{1R}, \ldots, u_{NL}, v_{NL}, u_{NR}, v_{NR}) \\ \hat{X} &= (\hat{u}_{1L}, \hat{v}_{1L}, \hat{u}_{1R}, \hat{v}_{1R}, \ldots, \hat{u}_{NL}, \hat{v}_{NL}, \hat{u}_{NR}, \hat{v}_{NR}) \\ \Theta &= (x_1, y_1, z_1, \ldots, x_N, y_N, z_N, \psi, t) \\ \hat{\Theta} &= (\hat{x}_1, \hat{y}_1, \hat{z}_1, \ldots, \hat{x}_N \hat{y}_N, \hat{z}_N, \hat{\psi}, \hat{t}) \end{aligned}$$

the relative orientation problem is to find $\hat{\Theta}$ to minimize

$$F(\hat{X}, \hat{\Theta}) = \Sigma_{n=1}^{N} f(u_{nR}, v_{nR}, x_n, y_n, z_n, \psi, t) + f(u_{nl}, v_{nL}, x_n, y_n, z_n, 0, 0)$$

4 Zero Finding

Zero finding such as finding the zero of a polynomial in one or more variables occurs in a number of vision problems. Two examples are the three point perspective resection problem and some of the techniques for motion recovery. The zero finding problem is precisely in the form required for computing the covariance matrix $\Sigma_{\Delta\Theta}$ as described in the solution section. Let X be the ideal input vector and $\hat{X}$ be the observed perturbed input vector. Let Θ be a $K \times 1$ vector zeroing the $K \times 1$ function $g(X, \Theta)$; that is, $g(X, \Theta) = 0$. Finally, let $\hat{\Theta}$ be the computed vector zeroing $g(\hat{X}, \hat{\Theta})$; that is, $g(\hat{X}, \hat{\Theta}) = 0$.

5 Solution: Unconstrained Case

For the purpose of covariance determination of the computed $\hat{\Theta} = \Theta + \Delta\Theta$, the technique used to solve the extremization problem is not important, provided that there are no singularities or near singularities in the numerical computation proceedure itself.

To understand how the random perturbation ΔX acting on the unobserved vector X to produce the observed vector $\hat{X} = X + \Delta X$ propagates to the random perturbation $\Delta\Theta$ on the true but known parameter vector Θ to produce the computed parameter vector $\hat{\Theta} = \Theta + \Delta\Theta$, we can take partial derivatives of F with respect to each of the K components of Θ forming the gradient vector g of f. The gradient g is a $K \times 1$ vector function.

$$g(X, \Theta) = \frac{\partial F}{\partial \Theta}(X, \Theta)$$

The solution $\hat{\Theta} = \Theta + \Delta\Theta$ extremizing $F(X + \Delta X, \Theta + \Delta\Theta)$, however it is calculated, must be a zero of $g(X + \Delta X, \Theta + \Delta\Theta)$. Now taking a Taylor series expansion of g around (X, Θ) we obtain to a first order approximation:

$$g^{K\times1}(X + \Delta X, \Theta + \Delta\Theta) = g^{K\times1}(X, \Theta) + \frac{\partial g}{\partial X}'^{K\times N}(X, \Theta)\Delta X^{N\times1} + \frac{\partial g}{\partial \Theta}'^{K\times K}(X, \Theta)\Delta\Theta^{K\times1}$$

But since $\Theta + \Delta\Theta$ extremizes $F(X + \Delta X, \Theta + \Delta\Theta)$, $g(X + \Delta X, \Theta + \Delta\Theta) = 0$. Also, since Θ extremizes $F(X, \Theta)$, $g(X, \Theta) = 0$. Thus to a first order approximation,

$$0 = \frac{\partial g}{\partial X}'(X, \Theta)\Delta X + \frac{\partial g}{\partial \Theta}'(X, \Theta)\Delta\Theta$$

Since the relative extremum of F is a relative minimum, the $K \times K$ matrix

$$\frac{\partial g}{\partial \Theta}(X, \Theta) = \frac{\partial f^2}{\partial^2 \Theta}(X, \Theta)$$

must be positive definite for all (X, Θ). This implies that $\frac{\partial g}{\partial \Theta}(X, \Theta)$ is non-singular. Hence $(\frac{\partial g}{\partial \Theta})^{-1}$ exists and since it is symmetric we can write:

$$\Delta\Theta = -\{\frac{\partial g}{\partial \Theta}(X, \Theta)\}^{-1}\frac{\partial g}{\partial X}'(X, \Theta)\Delta X$$

This relation states how the random perturbation ΔX on X propagates to the random perturbation $\Delta\Theta$ on Θ. If the expected value of ΔX, $E[\Delta X]$, is zero, then from this relation we see the $E[\Delta\Theta]$ will also be zero, to a first order approximation.

This relation also permits us to calculate the covariance of the random perturbation $\Delta\Theta$.

$$\begin{aligned}
\Sigma_{\Delta\Theta} &= E[\Delta\Theta\Delta\Theta'] \\
&= E[-(\frac{\partial g}{\partial \Theta})^{-1}\frac{\partial g}{\partial X}'\Delta X(-(\frac{\partial g}{\partial \Theta})^{-1}\frac{\partial g}{\partial X}'\Delta X)'] \\
&= (\frac{\partial g}{\partial \Theta})^{-1}\frac{\partial g}{\partial X}'E[\Delta X\Delta X']\frac{\partial g}{\partial X}(\frac{\partial g}{\partial \Theta})'^{-1} \\
&= (\frac{\partial g}{\partial \Theta})^{-1}\frac{\partial g}{\partial X}'\Sigma_{\Delta X}\frac{\partial g}{\partial X}(\frac{\partial g}{\partial \Theta})^{-1}
\end{aligned}$$

Thus to the extent that the first order approximation is good, (i.e. $E[\Delta\Theta] = 0$), then

$$\Sigma_{\hat{\Theta}} = \Sigma_{\Delta\Theta}$$

The way in which we have derived the covariance matrix for $\Delta\Theta$ based on the covariance matrix for ΔX requires that the matrices

$$\frac{\partial g}{\partial \Theta}(X, \Theta) \text{ and } \frac{\partial g}{\partial X}(X, \Theta)$$

be known. But X and Θ are not observed. $X + \Delta X$ is observed and by some means $\Theta + \Delta\Theta$ is then calculated. So if we want to determine an estimate $\hat{\Sigma}_{\hat{\Theta}}$ for the covariance matrix $\Sigma_{\hat{\Theta}}$, we can proceed by expanding $g(X, \Theta)$ around $g(X + \Delta X, \Theta + \Delta\Theta)$.

$$g(X, \Theta) = g(X+\Delta X, \Theta+\Delta\Theta) - \frac{\partial g}{\partial X}'(X+\Delta X, \Theta+\Delta\Theta)\Delta X - \frac{\partial g}{\partial \Theta}'(X+\Delta X, \Theta+\Delta\Theta)\Delta\Theta$$

Here we find in a similar manner,

$$\Delta\Theta = -(\frac{\partial g}{\partial \Theta}(X + \Delta X, \Theta + \Delta\Theta))^{-1}\frac{\partial g}{\partial X}(X + \Delta X, \Theta + \Delta\Theta)\Delta X$$

This motivates the estimator $\hat{\Sigma}_{\Delta\Theta}$ for $\Sigma_{\Delta\Theta}$ defined by

$$\hat{\Sigma}_{\Delta\Theta} = (\frac{\partial g}{\partial \Theta}(\hat{X},\hat{\Theta})^{-1} \frac{\partial g}{\partial X}'(\hat{X},\hat{\Theta})\Sigma_{\Delta X} \frac{\partial g}{\partial X}(\hat{X},\hat{\Theta})(\frac{\partial g}{\partial \Theta}(\hat{X},\hat{\Theta})^{-1}$$

So to the extent that the first order approximation is good, $\hat{\Sigma}_{\hat{\Theta}} = \hat{\Sigma}_{\Delta\Theta}$.

The relation giving the estimate $\hat{\Sigma}_{\hat{\Theta}}$ in terms of the computable

$$\frac{\partial g}{\partial \Theta}(\hat{X},\hat{\Theta}) \text{ and } \frac{\partial g}{\partial X}(\hat{X},\hat{\Theta})$$

means that an estimated covariance matrix for the computed $\hat{\Theta} = \Theta + \Delta\Theta$ can also be calculated at the same time that the estimate $\hat{\Theta}$ of Θ is calculated.

As a special and classic case, we consider the regression problem of finding Θ to minimize $F(X,\Theta) = (X - J\Theta)'\Sigma_X^{-1}(X - J\Theta)$. For this F,

$$g(X,\Theta) = \frac{\partial F}{\partial \Theta} = -2J'\Sigma_X^{-1}J\Theta$$

Hence,

$$\frac{\partial g}{\partial \Theta} = 2J'\Sigma_X^{-1}J$$

and

$$\frac{\partial g}{\partial X} = -2\Sigma_X^{-1}J$$

Then,

$$\begin{aligned}\Sigma_\Theta &= (2J'\Sigma_X^{-1}J)^{-1}(-2\Sigma_X^{-1}J)\Sigma_X(-2\Sigma_X^{-1}J)'(2J'\Sigma_X^{-1}J)^{-1} \\ &= (J'\Sigma_X^{-1}J)^{-1}\end{aligned}$$

As another important case, we consider the general line-fitting problem. Assume that the unobserved points unperturbed points (x_n, y_n), $n = 1,\ldots,N$, lie on a line $x_n \cos\theta + y_n \sin\theta - \rho = 0$. In the line-fitting problem, we observe $(\hat{x}_n, \hat{y}_n)$, noisy instances of (x_n, y_n). $(\hat{x}_n, \hat{y}_n)$ are related to (x_n, y_n) by the noise model:

$$\begin{pmatrix} \hat{x}_n \\ \hat{y}_n \end{pmatrix} = \begin{pmatrix} x_n \\ y_n \end{pmatrix} + \xi_n \begin{pmatrix} \cos\theta \\ \sin\theta \end{pmatrix}$$

where ξ_n are independent and identically distributed as $N(0,\sigma^2)$.

To estimate the best fitting line parameters $(\hat{\theta},\hat{\rho})$ using the least squares method, we use the criterion function:

$$F(X,\Theta) = \sum_{n=1}^{N}(x_n \cos\theta + y_n \sin\theta - \rho)^2$$

where $X = (x_1, y_1, \ldots, x_N, y_N)$ and $\Theta = (\theta, \rho)$.

Now,

$$g^{2\times 1}(X,\Theta) = \frac{\partial F}{\partial \Theta} = \begin{pmatrix} \frac{\partial F}{\partial \theta} \\ \frac{\partial F}{\partial \rho} \end{pmatrix}$$

Letting

$$\begin{aligned}
\mu_x &= \frac{1}{N}\sum_{n=1}^{N} x_n \\
\mu_y &= \frac{1}{N}\sum_{n=1}^{N} y_n \\
\sigma_x^2 &= \sum_{n=1}^{N}(x_n - \mu_x)^2 \\
\sigma_y^2 &= \sum_{n=1}^{N}(y_n - \mu_y)^2 \\
\sigma_{xy} &= \sum_{n=1}^{N}(x_n - \mu_x)(y_n - \mu_y);
\end{aligned}$$

we can compute

$$\frac{\partial F}{\partial \theta} = (\sigma_x^2 - \sigma_y^2 + N(\mu_y^2 - \mu_x^2))\sin 2\theta + 2(\sigma_{xy} + N\mu_x\mu_y)\cos 2\theta + 2N\rho(\mu_x \sin\theta - \mu_y cos\theta)$$

$$\frac{\partial F}{\partial \rho} = -2N(\mu_x\cos\theta + \mu_y\sin\theta - \rho)$$

Then,

$$\frac{\partial g}{\partial \Theta}^{2\times 2} = \begin{pmatrix} \frac{\partial g}{\partial \theta} \\ \frac{\partial g}{\partial \rho} \end{pmatrix} = \begin{pmatrix} \frac{\partial^2 F}{\partial \theta^2} & \frac{\partial^2 F}{\partial \theta \partial \rho} \\ \frac{\partial^2 F}{\partial \rho \partial \theta} & \frac{\partial^2 F}{\partial \rho^2} \end{pmatrix}$$

where

$$\frac{\partial^2 F}{\partial \theta^2} = 2[\sigma_x^2 - \sigma_y^2 + N(\mu_y^2 - \mu_x^2)]\cos 2\theta - 4(\sigma_{xy} + N\mu_x\mu_y)\sin 2\theta + 2N\rho(\mu_x\cos\theta + \mu_y\sin\theta)$$

$$\frac{\partial^2 F}{\partial \rho^2} = 2N$$

$$\frac{\partial^2 F}{\partial \theta \partial \rho} = \frac{\partial^2 F}{\partial \rho \partial \theta} = 2N(\mu_x\sin\theta - \mu_y\cos\theta)$$

And,

$$\frac{\partial g}{\partial X}'^{\,2\times 2N} = \underbrace{\begin{pmatrix} \frac{\partial^2 F}{\partial \theta \partial x_1} & \frac{\partial^2 F}{\partial \theta \partial y_1} & \frac{\partial^2 F}{\partial \theta \partial x_2} & \frac{\partial^2 F}{\partial \theta \partial y_2} & \frac{\partial^2 F}{\partial \theta \partial x_n} & \cdots & \frac{\partial^2 F}{\partial \theta \partial y_n} \\ \frac{\partial^2 F}{\partial \rho \partial x_1} & \frac{\partial^2 F}{\partial \rho \partial y_1} & \frac{\partial^2 F}{\partial \rho \partial x_2} & \frac{\partial^2 F}{\partial \rho \partial y_2} & \cdots & \frac{\partial^2 F}{\partial \rho \partial x_N} & \frac{\partial^2 F}{\partial \rho \partial y_N} \end{pmatrix}}_{2\times 2N} \tag{1}$$

where

$$\frac{\partial^2 F}{\partial\theta\partial x_n} = 2[(y_n - \mu_y)\cos 2\theta - (x_n - \mu_x)\sin 2\theta + [\mu_y \cos 2\theta - \mu_x \sin 2\theta + \rho \sin\theta)]$$

$$\frac{\partial^2 F}{\partial\theta\partial y_n} = 2[(x_n - \mu_x)\cos 2\theta + (y_n - \mu_y)\sin 2\theta + (\mu_x \cos 2\theta + \mu_y \sin 2\theta - \rho\cos\theta)]$$

$$\frac{\partial^2 F}{\partial\rho\partial x_n} = -2\cos\theta$$

$$\frac{\partial^2 F}{\partial\rho\partial y_n} = -2\sin\theta$$

For the given noise model, the covariance matrix Σ_X is given by:

$$\Sigma_X = \sigma^2 \begin{pmatrix} \cos^2\theta & \sin\theta\cos\theta & \dots & 0 & 0 & 0 \\ \sin\theta\cos\theta & \sin^2\theta & 0 & \dots & 0 & 0 \\ 0 & 0 & \cos^2\theta & \sin\theta\cos\theta & \dots & 0 \\ 0 & 0 & \sin\theta\cos\theta & \sin^2\theta & \dots & 0 \\ \vdots & & & & & \\ 0 & 0 & \dots & 0 & \cos^2\theta & \sin\theta\cos\theta \\ 0 & 0 & 0 & \dots & \sin\theta\cos\theta & \sin^2\theta \end{pmatrix}$$

Using these expressions, the covariance matrix of Θ, Σ_Θ, can be computed as:

$$\begin{aligned} \Sigma_\Theta^{2\times 2} &= \begin{pmatrix} \sigma_{\theta\theta} & \sigma_{\theta\rho} \\ \sigma_{\rho\theta} & \sigma_{\rho\rho} \end{pmatrix} \\ &= \frac{\partial g}{\partial\Theta}^{-1}(X,\Theta)\frac{\partial g}{\partial X}'(X,\Theta)\Sigma_X\frac{\partial g}{\partial X}(X,\Theta)\frac{\partial g}{\partial\Theta}^{-1}(X,\Theta) \end{aligned}$$

We will find that

$$\begin{aligned} \sigma_{\theta\theta} &= \frac{4\sigma^2(\sigma_y^2\cos^2\theta + \sigma_x^2\sin^2\theta - \sigma_{xy}\sin 2\theta)}{(2NT)^2} \\ \sigma_{\theta\rho} &= \sigma_{\rho\theta} = \frac{-4\sigma^2[(\mu_x\sin\theta - \mu_y\cos\theta)(\sigma_y^2\cos^2\theta + \sigma_x^2\sin^2\theta - \sigma_{xy}\sin 2\theta)]}{(2NT)^2} \\ \sigma_{\rho\rho} &= \frac{4\sigma^2[(\mu_x\sin\theta - \mu_y\cos\theta)^2(\sigma_y^2\cos^2\theta + \sigma_x^2\sin^2\theta - \sigma_{xy}\sin 2\theta) + N((\mu_x\sin\theta - \mu_y\cos\theta)^2 - T)^2}{(2NT)^2} \end{aligned}$$

where

$$T = \frac{\sigma_y^2 - \sigma_x^2}{N}\cos 2\theta - \frac{2\sigma_{xy}}{N}\sin 2\theta - (\mu_y\sin\theta + \mu_x\cos\theta)^2 + \rho(\mu_y\sin\theta + \mu_x\cos\theta)$$

The geometry of this result can be made easier to understand by re-expressing it. If (x, y) is a point on the line $x\cos\theta + y\sin\theta - \rho = 0$ and k is the signed distance of (x, y) to the point on the line closest to the origin, then

$$k = \begin{cases} +\sqrt{x^2+y^2-\rho^2} & \text{if } y\cos\theta \geq y\sin\theta \\ -\sqrt{x^2+y^2-\rho^2} & \text{otherwise} \end{cases}$$

It is not hard to show that

$$\begin{aligned} x &= -k\sin\theta + \rho\cos\theta \\ y &= k\cos\theta + \rho\sin\theta \end{aligned}$$

Let

$$\begin{aligned} \mu_k &= \frac{1}{N}\sum_{n=1}^{N} k_n \\ \sigma_k^2 &= \sum_{n=1}^{N}(k_n - \mu_k)^2 \end{aligned}$$

then it follows that

$$\begin{aligned} \mu_x &= \rho\cos\theta - \mu_k\sin\theta \\ \mu_y &= \rho\sin\theta + \mu_k\cos\theta \\ \sigma_x^2 &= \sigma_k^2\sin^2\theta \\ \sigma_y^2 &= \sigma_k^2\cos^2\theta \\ \sigma_{xy} &= -\sigma_k^2\sin\theta\cos\theta \end{aligned}$$

Substituting the above expressions in the covariance matrix results in

$$\Sigma_\Theta = \sigma^2 \begin{pmatrix} \frac{1}{\sigma_k^2} & \frac{\mu_k}{\sigma_k^2} \\ \frac{\mu_k}{\sigma_k^2} & \frac{1}{N} + \frac{\mu_k^2}{\sigma_k^2} \end{pmatrix}$$

This result has a simple geometric interpretation. In the coordinate system of the line where 0 is the point on the line closest to the origin, μ_k is the mean position of the points and σ_k is the scatter of the points. μ_k acts like a moment arm. If the mean position of the points on the line is a distance of $|\mu_k|$ from the origin on the line, then the variance of the estimated ρ increases by $\mu_k^2\sigma^2/\sigma_k^2$. This says that the estimate ρ is not invariant to the translation of the coordinate system.

6 Solution: Constrained Case

In the case of the constrained optimization, the function to be minimized is $F(X,\Theta) + s(\Theta)'\Lambda$. As before, we define $g(X,\Theta) = \frac{\partial}{\partial\Theta}F(X,\Theta)$. We must have at the minimizing (X,Θ),

$$\frac{\partial}{\partial\Theta}(F(X,\Theta) + s(\Theta)'\Lambda) = 0$$

And in the case of no noise with the squared criterion function as we have been considering, $F(X,\Theta) = 0$. And this is the smallest F can be. Hence it must be that $g(X,\Theta) = \frac{\partial F}{\partial X}(X,\Theta) = 0$. This implies that $\frac{\partial s}{\partial \Theta}(\Theta)\Lambda = 0$, which will only happen when Λ=0 since we expect $\frac{\partial s}{\partial \Theta}$, a $K \times L$ matrix where $K > L$, to be of full rank.

Define

$$S(X,\Theta,\Lambda) = \begin{pmatrix} g(X,\Theta) + \frac{\partial s}{\partial \Theta}\Lambda \\ s(\Theta) \end{pmatrix}$$

Taking a Taylor series expansion of S,

$$S(X,\Theta,\Lambda) = S(X+\Delta X, \Theta+\Delta\Theta, \Lambda+\Delta\Lambda) - \frac{\partial S'}{\partial X}\Delta X - \frac{\partial S'}{\partial \Theta}\Delta\Theta - \frac{\partial S'}{\partial \Lambda}\Delta\Lambda$$

Because $g(X,\Theta) = 0, \Lambda = 0$, and $s(\Theta) = 0$, it follows that $S(X,\Theta,\Lambda) = 0$. Furthermore, at the computed $\hat{\Theta} = \Theta + \Delta\Theta$ and $\hat{\Lambda} = \Lambda + \Delta\Lambda$, $S(X+\Delta X, \Theta+\Delta\Theta, \Lambda+\Delta\Lambda) = 0$. Hence,

$$-\frac{\partial S'}{\partial X}\Delta X = \frac{\partial S'}{\partial \Theta}\Delta\Theta + \frac{\partial S'}{\partial \Lambda}\Delta\Lambda$$

Writing this equation out in terms of g and s, and using the fact that $\Lambda = 0$, there results

$$\begin{pmatrix} \frac{\partial g}{\partial \Theta} & \frac{\partial s}{\partial \Theta} \\ \frac{\partial s'}{\partial \Theta} & 0 \end{pmatrix} \begin{pmatrix} \Delta\Theta \\ \Delta\Lambda \end{pmatrix} = \begin{pmatrix} -\frac{\partial g'}{\partial X} \\ 0 \end{pmatrix} \Delta X$$

From this it follows that

$$\Sigma_{\Delta\Theta,\Delta\Lambda} = A^{-1} B \Sigma_X B' A$$

where

$$A = \begin{pmatrix} \frac{\partial g}{\partial \Theta}, & \frac{\partial s}{\partial \Theta} \\ \frac{\partial s}{\partial \Theta}' & 0 \end{pmatrix}$$

and

$$B = -\begin{pmatrix} \frac{\partial g'}{\partial X} \\ 0 \end{pmatrix}$$

and all functions are evaluated at Θ and X. For the estimated value $\hat{\Sigma}_{\Delta\Theta\Delta\Lambda}$ of $\Sigma_{\Delta\Theta\Delta\Lambda}$, we evaluate all functions at $\hat{\Theta}$ and $\hat{\Lambda}$.

As a special but classic case of this consider the constrained regression problem to find Θ minimizing

$$F(X,\Theta) = (X - J\Theta)'(X - J\Theta)$$

subject to $H'\Theta = 0$. In this case,

$$A = \begin{pmatrix} 2J'J & H \\ H' & 0 \end{pmatrix}$$

and

$$B = -\begin{pmatrix} 2J' \\ 0 \end{pmatrix}$$

Then

$$A^{-1} = \begin{pmatrix} (2J'J)^{-1}[I - H(H'(2J'J)^{-1}H)^{-1}H'(2JJ')^{-1}] & (2J'J)^{-1}H(H'(2J'J)^{-1}H)^{-1} \\ (H'(2J'J)^{-1}H)^{-1}H'(2J'J)^{-1} & -(H'(2J'J)^{-1}H)^{-1} \end{pmatrix}$$

and

$$A^{-1}B = -\begin{pmatrix} (2J'J)^{-1}[I - H(H'(2J'J)^{-1}H)^{-1}H'(2JJ')^{-1}]2J' \\ (H'(2J'J)^{-1}H)^{-1}H'(2J'J)^{-1}2J' \end{pmatrix}$$

From this it directly follows that if $\Sigma_X = \sigma^2 I$, then

$$\Sigma_\Theta = \sigma^2(J'J)^{-1}[I - H(H'(JJ')^{-1}H)^{-1}H'(J'J)^{-1}]$$

7 Validation

There are two levels of validation. One level of validation is for the software. This can be tested by a large set of Monte-Carlo experiments off-line where we know what the correct answers are.

Another level of validation is on-line reliability. Here all that we have is the computed estimate and estimated covariance matrix for the estimate.

7.1 Software and Algorithm Validation

Software for performing the optimization required to compute the estimate $\hat{\Theta}$ is often complicated and it is easy for there to be errors that are not immediately observable (like optimization software that produces correct answers on a few known examples but fails in a significant fraction of more difficult cases). One approach in testing that the software is producing the right answers is to test the statistical properties of the answers. That is, we can statistically test whether the statistical properties of its answers are similar to the statistical properties we expect. These expectations are whether the mean of the computed estimates is sufficiently close to the population mean and whether the estimated covariance matrix of the estimates is sufficiently close to the population covariance matrix. Rephrasing this more precisely the test is whether the computed estimates could have arisen from a population with given mean and covariance matrix.

Consider what happens in a hypothesis test: a significance level, α, is selected. When the test is run, a test statistic, say $\hat{\phi}$, is computed. The test statistic is typically designed so that in the case that the hypothesis is true, the test statistic will tend to have its values. distributed around zero, in accordance with a known distribution. If the test statistic has a value say higher than a given ϕ_0, we reject the hypothesis that the computed estimate is statistically behaved as we expected it to be. If we do not reject, then in effect, we are tentatively accepting the hypothesis. The value of ϕ_0 is chosen so that the probability that we reject the hypothesis, given that is the hypothesis is true is less than the significance level α.

The key in using this kind of testing is that we can set up an experiment in which we know what the correct answer for the no noise ideal case would be. Then we can additively perturb the input data by a normally distributed vector from a population having zero mean and given covariance matrix. Then using the analytic propagation results derived earlier in the paper, we can derive the covariance matrix of the estimates produced by software.

If we repeat this experiment many times just changing the perturbed realizations and leaving everything else the same, the experiment produces estimates $\theta_1, \ldots, \theta_N$ that will come from a normal population having mean θ, the correct answer for the ideal no noise case, and covariance matrix Σ, computed from the propagation equations. Now the hypothesis test is whether the observations $\theta_1, \ldots, \theta_N$ come fron a Normal population with mean θ and covariance matrix Σ. For this hypothesis test, there is a uniformly most powerful test. Let

$$B = \Sigma_{n=1}^{N}(\theta_n - \bar{\theta})(\theta_n - \bar{\theta})'$$

Define

$$\begin{aligned} \lambda &= (e/N)^{pN/2}|B\Sigma^{-1}|^{N/2} \\ &\times exp(-\frac{1}{2}[tr(B\Sigma^{-1}) + N(\bar{\theta} - \theta)'\Sigma^{-1}(\bar{\theta} - \theta)]) \end{aligned}$$

The test statistic is:

$$T = -2log\lambda$$

Under the hypothesis, T is distributed as:

$$\chi^2_{p(p+1)/2+p}$$

where p is the dimension of θ.

So to perform a test that the program's behavior is as expected we repeatedly generate the T statistic and compute its empirical distribution function. Then we test the hypothesis that T is distributed as the χ^2 variate using a Kolmogorov-Smirnov test.

7.2 On-line Reliability

For the on-line reliablity testing, the estimate is computed by minimizing the scalar objective function. Then based on the given covariance matrix of the input data, an estimated covariance matrix of the estimate is computed using the linearization around the estimate itself. Here a test can be done by testing whether the each of the diagonal entries of the estimated covariance matrix is sufficiently small.

8 Conclusion

Making a successful vision system for any particular application typically requires many steps, the optimal choice of which is not always apparent. To understand how to do the optimal design, a synthesis problem, requires that we first understand how to solve the analysis problem: given the steps of a particular algorithm, determine how to propagate the parameters of the perturbation process from the input to the parameters describing the perturbation process of the computed output. The first basic case of this sort of uncertainty propagation is the propagation of the covariance matrix of the input to the covariance matrix of the output. This is what this paper has described.

This work does not come near to solving what is required for the general problem, because the general problem involves perturbations which are not additive. That is, in mid and high-level vision, the appropriate kinds of perturbations are perturbations of structures. Now, we are in the process of understanding some of the issues with these kinds of perturbations and expect to soon have some results in this area.

References

[1] Karl-Rudolf Koch, *Parameter Estimation and Hypothesis Testing in Linear Models*, Springer-Verlag, Berlin, 1987, p117, p121.

[2] Edward Mikhail, *Observations and Least Squares*, IEP – A Dun-Donnelley Publisher, New York, 1976, p72-90.

[3] Chester Slama (Ed.), *The Manual of Photogrammetry* The American Society of Photogrammetry, Falls Church, VA 22046, 1980, p73-74.

[4] Juyang Weng, Paul Cohen, and Marc Herniou, "Camera Claibration with Distortion MOdels and Accuracy Evaluation," *Pattern Analysis and Machine Intelligence*, Vol 14, No. 10, October 1992, p965-980.

[5] Donna Williams and Mubarak Shah, "Edge Characterization Using Normalized Edge Detector," *CVGIP: Graphical Models and Image Processing*, Vol 55, No. 4, July 1993, p311-318.

[6] Sen-Yen Wu and Mao-Jiun Wang, "Detecting the Dominant Points by the Curvature-Based Polygonal Approximation," *CVGIP: Graphical Models and Image Processing*, Vol 5, No. 2, March 1993, p79-88.

Section 3

3-D Shape from 2-D images

Improving the Vision of Magic Eyes: A Guide to Better Autostereograms

A.M. Bruckstein * , R. Onn ** , and T.J. Richardson *
* Bell Laboratories, Murray Hill, NJ 07974, USA and
** EE Department, Technion, I.I.T., Haifa, 32000, Israel

Abstract

An autostereogram is a single image that has the capability to convey depth information in the same manner as a stereo pair. Given a depth profile, the autostereogram is completely characterized by a two-dimensional basic pattern (a vertical strip.) Some autostereograms are more easily perceived than others depending on the basic pattern chosen to produce them. In this paper we discuss this dependence in terms of the spectrum of the basic pattern. We conclude that samples of 1/f noise yield excellent basic patterns, making it easy for the viewer to lock-into the desired depth profiles and to perceive depth in a stable way.

1: Introduction - On Seeing Depth

The world around us is three dimensional, but eyes and cameras can only see planar projections of spatial scenes. Nevertheless, the third dimension can often be inferred from two dimensional images. Occlusion and prior information on object shapes provide depth information even in single images, and so do shadows and shading. Stereo vision provides a stronger effect enabling us to perceive depth: in stereo, depth is inferred locally from the slight differences in the images of the same scene produced by two horizontally displaced sensors (the eyes). Stereo vision is quantitative, in the sense that the binocular observer is able to evaluate the relative depth of almost all visible objects in a scene, a capability extensively exploited in geodesy. This quantitative depth perception depends on binocular disparity, and disappears when one of our eyes is closed; the visual system always fuses the available images so that a single image is perceived, with or without a sense of depth.

Depth evaluation in stereo vision is independent of the other visual cues that are usually present when viewing the world around us. This fact was shown by Julesz a long time ago, [Julesz,64], via a series of landmark experiments with random dot stereograms, i.e., pairs of similar images consisting of randomly placed dots in the plane, one of them having part of the dots displaced to encode depth. Depth is perceived when such image pairs are (simultaneously) presented to the two eyes of an observer. Independence of the stereo depth perception from other common depth cues was indeed to be expected from the observation that camouflaged objects, invisible in single images often become readily apparent in stereo pairs of images. It is the aim of camouflage to cover objects with a pattern that makes them appear to fuse with the background, their outlines or edges being completely obscured in monocular, or distant views. As in the case of camouflage, a random dot stereogram corresponds to a special (conceptual) coloring of the (imaginary) height profile/object surface

0-8186-7644-2 $5.00 © 1996 IEEE

with randomly spaced spots or dots, so that no information about the depth profile exists in a single image.

Let x, y, z be coordinates for 3D space. To simplify the discussion we will assume that the depth of the surface to be viewed can be represented as a positive function $z = \varphi(x, y)$ and that no occlusions occur. This function will be referred to as the *depth profile*. From the depth profile we can easily construct a stereogram that allows the viewer to perceive it when his eyes are placed at, say $(-x_0, 0, -z_0)$ and $(x_1, 0, -z_0)$. We simply color the surface with some pattern $A(x, y)$ and project this pattern onto the plane $z = 0$ in two ways to produce two images $I_L(x, y)$ and $I_R(x, y)$ corresponding to the two different views of the surface as seen from $(-x_0, 0, -z_0)$ and $(x_1, 0, -z_0)$. In the sequel, we shall artificially assume that the surface is viewed by eyes located far way, with viewing directions of 90 and 45 degrees, i.e., $z_0 \rightarrow \infty$, $x_0 = 0$, $x_1/z_0 = 1$. This assumption lets us deal with parallel projection rather than perspective, simplifying the geometry considerably without sacrificing the essential features of the stereo imaging process. In fact, there is another depth profile φ' which, when projected onto the two eyes, produces the same images as the extreme projection of φ described above. It is this φ' which will actually be perceived.

Stereo matching occurs in the horizontal direction, corresponding to the line through the two eyes. Therefore, images will be viewed as collections of rows, one corresponding to each y -coordinate. For any particular y_0 we have to consider a pair of 1D functions of x , $I_L(x, y_0)$ and $I_R(x, y_0)$. For notational convenience, we shall cease to write the y coordinates for the bivariate functions involved.

Under these simplifying assumptions, we have (see Figure 1):

$$I_L(x) = A(x) \quad \text{and} \quad I_R(x + \varphi(x)) = A(x) \tag{1}$$

Note here that $I_R(x)$ is specified implicitly, via the depth function. If the function $x \rightarrow x + \varphi(x)$ is invertible, and this clearly happens if $|\frac{d}{dx}\varphi(x)| < 1$, then we can readily generate $I_R(x)$ from $I_L(x)$. (In practice, when the two projections are in directions much closer, say 90 and 80 degrees, the condition for invertibility is much milder, i.e. the depth function can have a much steeper slope).

As we have seen, producing the illusion of depth by showing to our eyes two slightly different images can be easily understood. This understanding led to many interesting practical applications, like 3D photography, visualization techniques in computer graphics, methods of photogrammetry, etc.

But, the illusion of depth can also be induced by showing both eyes the very same image, provided this image is carefully designed! This brilliant idea, that first occurred to C. W. Tyler in the seventies, is based on the freedom to choose a coloring of the surface that will yield two *identical* images as a stereo pair, [TylerChang,77], [Tyler,83] and [TylerClarke,90]. Today, countless books, postcards, and posters are dedicated to those "magical" images that, at a first glance look like an almost regular planar pattern of shapes and colors, but a "deeper" and more "defocused" look at them causes a second, and often amazing, three dimensional image to suddenly appear, see e.g. [MagicEye,93]. Tyler called such images *autostereograms*. In the sequel we shall try to explain how these images are are generated, and to analyze the way they are interpreted by the viewer, an analysis that will lead us to some ideas on how to design autostereograms that are more easily and stably interpreted as a three-dimensional image by the viewer. Although many papers in the literature addressed the topic of explaining and efficiently generating (mostly binary, or dot based) autostereograms, see e.g. [ThiIngWit,94], [TerTer,94], the issue of autostereogram

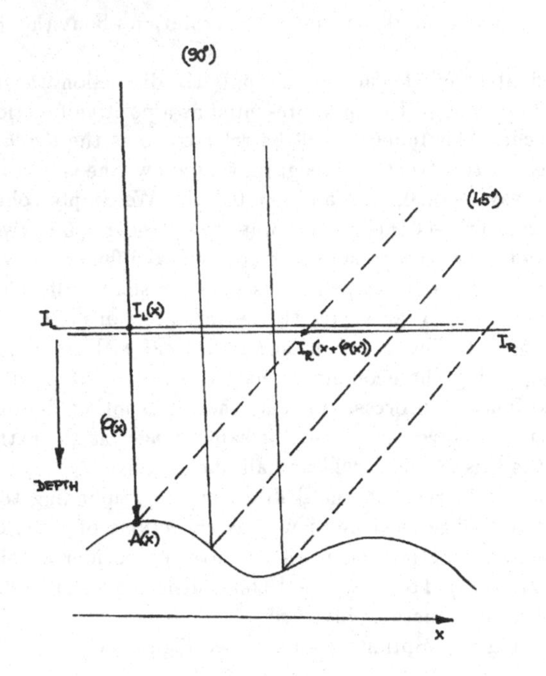

Figure 1. Stereo projection

design for easy interpretation seems to have never been raised and discussed.

2: Interpretation and Design of Autostereograms

As we have seen, an autostereogram is the image that would be obtained as two different projections of a suitably colored depth profile. From Equation (1) it becomes clear that the equality of $I_L(x)$ and $I_R(x)$ forces a coloring that obeys

$$A(x) = A(x + \varphi(x)) \text{ for all } x . \tag{2}$$

Referring to Figure 2, where again the projections are taken to be parallel at 90 and 45 degrees, and a single horizontal line of the image and depth profile are considered, we readily recover geometrically, that the image $I(x) = I_{R/L}(x)$ has to obey the basic functional equation inherited from $A(x)$,

$$I(x) = I(x + \varphi(x)) \text{ for all } x . \tag{3}$$

If the depth profile obeys the slope-limiting condition $|d\varphi(x)/dx| < 1$,[1] then it is easy to see that $I(x)$ is determined everywhere from its values on $x \in [t, t + \varphi(t))$ for any value

[1] It is possible to have perceivable discontinuities in depth in autostereograms but depth information is lost in the portion viewed by only one eye. This complicates the situation in a way which has little bearing on the subject of this paper.

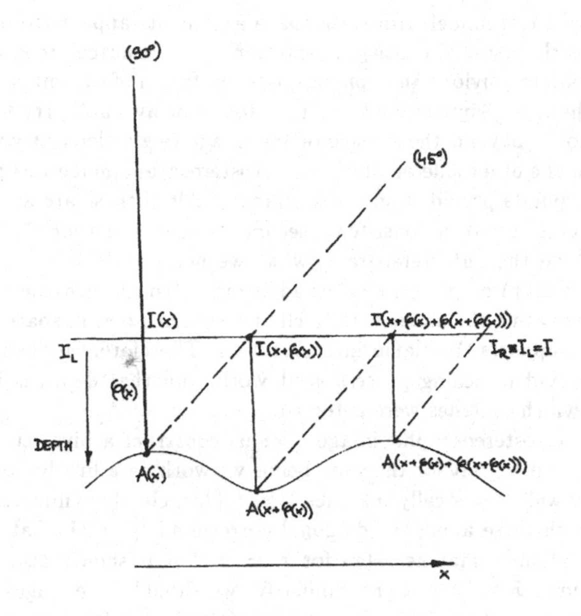

Figure 2. Auto-stereo projection

of t. Indeed, all forward and backward iterations of the transformation $T : x \to x + \varphi(x)$, for any starting point x_0 have exactly one representative inside any interval of the type $[t, t + \varphi(t))$. Thus, we are free to choose $I(x)$ over an arbitrary such interval and the rest of the stereogram (line) is then completely determined. We shall refer to $I(x)$ restricted to this (arbitrary) *basic interval*, as the *basic pattern* of the autostereogram. Note that the full autosterogram image will comprise a set of such parallel profiles (for various y's)-representing horizontal lines in the image.

The freedom to choose the basic pattern can obviously be exploited to get a variety of visual effects and this freedom was indeed exploited, quite amazingly, in commercializing the autostereographic images. Interestingly however, depth is much more easily perceived in some autostereograms than in others. In fact there are autostereograms that satisfy the rules outlined above, and uniquely determine φ, and yet do not produce any depth perception at all. Hence, the following pair of questions arises naturally:

1. What is the underlying mechanism by which we perceive depth in such images?
2. How should we design basic patterns to make it easy for the viewer to perceive the third dimension?

As with almost any question about how biological mechanisms work, we cannot provide definite answers to the first question above. Unfortunately we can not answer Question 2. quantitatively without at least inferring a partial answer to Question 1. In Section 3 we develop a simple model for stereo vision, consistent with the available psychophysical evidence, which can be applied to autostereograms. Researchers studying human vision have proposed various models for stereo vision. The "squared differences" model we consider here was discussed by Sperling [Sperling,81] and Arndt et. al. [ArndtMallotBülthoff,95].

The combined use of this model with scale space arguments appears to be novel.

Recovery of the 3D (usually, depth) information from an autostereogram involves transcending the immediate, obvious and "planar" interpretation of the image seen. The Magic-Eye books, and the many similar books and posters now available, try to help viewers by guiding them to focus beyond the surface of the image (e.g.: "look at your own reflection in the window" on the other side of which an autostereogram image was posted, or "try to merge two feature points provided on the margins"). All of these are attempts to lead the human visual system toward a consistent, second peak of some locally defined matching process. To perceive the autostereogram, what we need to do is to correlate $I(x)$ with $I(x + \Delta(x))$ where $\Delta(x) \simeq \varphi(x)$, rather than being satisfied with the the perfect match obtained when correlating the image with itself at the same point in space, i.e. at $\Delta(x) = 0$, a match that fully supports the planar interpretation. Therefore, we can safely assume that a *local* correlation and matching process is at work, and that depth is inferred from the displacements at which matches were detected.

Let $I(x)$ be an autostereographic image. Let us construct a bivariate function (in fact trivariate, but remember that for the time being we work on a line-by-line basis!), $\Lambda(x, \tilde{x})$ that indicates how well $I(x)$ locally matches $I(\tilde{x})$.[2] This, clearly symmetric, bivariate function will have a high ridge along the diagonal corresponding to the flat image interpretation, since $I(\tilde{x})$ obviously matches $I(x)$ for $\tilde{x} = x$, but it should also have a very high ridge along the curve $\tilde{x} = x + \varphi(x)$. Similarly, we should have ridges along the curves $\tilde{x} = x + \varphi(x) + \varphi(x + \varphi(x))$, etc... The behavior of the 'surface' represented by $\Lambda(x, \tilde{x})$ clearly depends on how we define the local matching of images and on the particular choice of the autostereogram's basic pattern. Suppose that an oracle provided us with the physiologically correct matching function and the process used by the brain to find the best matching. Then, for a given depth profile, it would make sense to ask for the *optimal* autostereogram for a given depth profile. 'Optimal' here shall be interpreted in the sense that the basic pattern chosen yields sharp and high ridges along the curves $\tilde{x} = x$ and $\tilde{x} = x + \varphi(x)$, (and, inevitably, its iterates) and that the "domain of attraction" of the second interpretation should be as large as possible.

As a first exercise, suppose we are told that the depth interpretation is based on a pointwise grey-level match indicator function computed by the brain, "disparity" curves of the type $\tilde{x} = f(x)$ being evaluated by integrating (accumulating) $\Lambda(x, \tilde{x})$ along them. Clearly, with

$$\Lambda(x, \tilde{x}) = \chi\{I(x) = I(\tilde{x})\} = \begin{cases} 1 & if\ I(x) = I(\tilde{x}) \\ 0 & if\ I(x) \neq I(\tilde{x}) \end{cases} \tag{4}$$

the "correct" disparity curves carry a constant distributed weight along them ($\Lambda(x, x) = \Lambda(x, x + \varphi(x)), ...etc$) but, depending on how $I(x)$ was designed, we might have other such ridges too. Furthermore, if the basic pattern of $I(x)$ is one-to-one then no additional ridges will exist, so, from the point of view of this matching function, all stereograms with one-to-one basic patterns will be equally good. However, a glance at the examples of Figure 3 clearly indicate that this matching function fails to capture some important features of the visual system's interpretation of autostereograms. Both basic patterns appearing in Figure 3 are one-to-one: The first is a ramp while the second is obtained by randomly permuting block-wise the ordinate of the ramp function. 3D interpretation turns out to be greatly facilitated by richness of detail and edginess in the basic pattern. One-to-one-ness,

[2] In reality the eye does not process horizontal lines independently: this dependence will be discussed and incorporated into our model in Section 3.

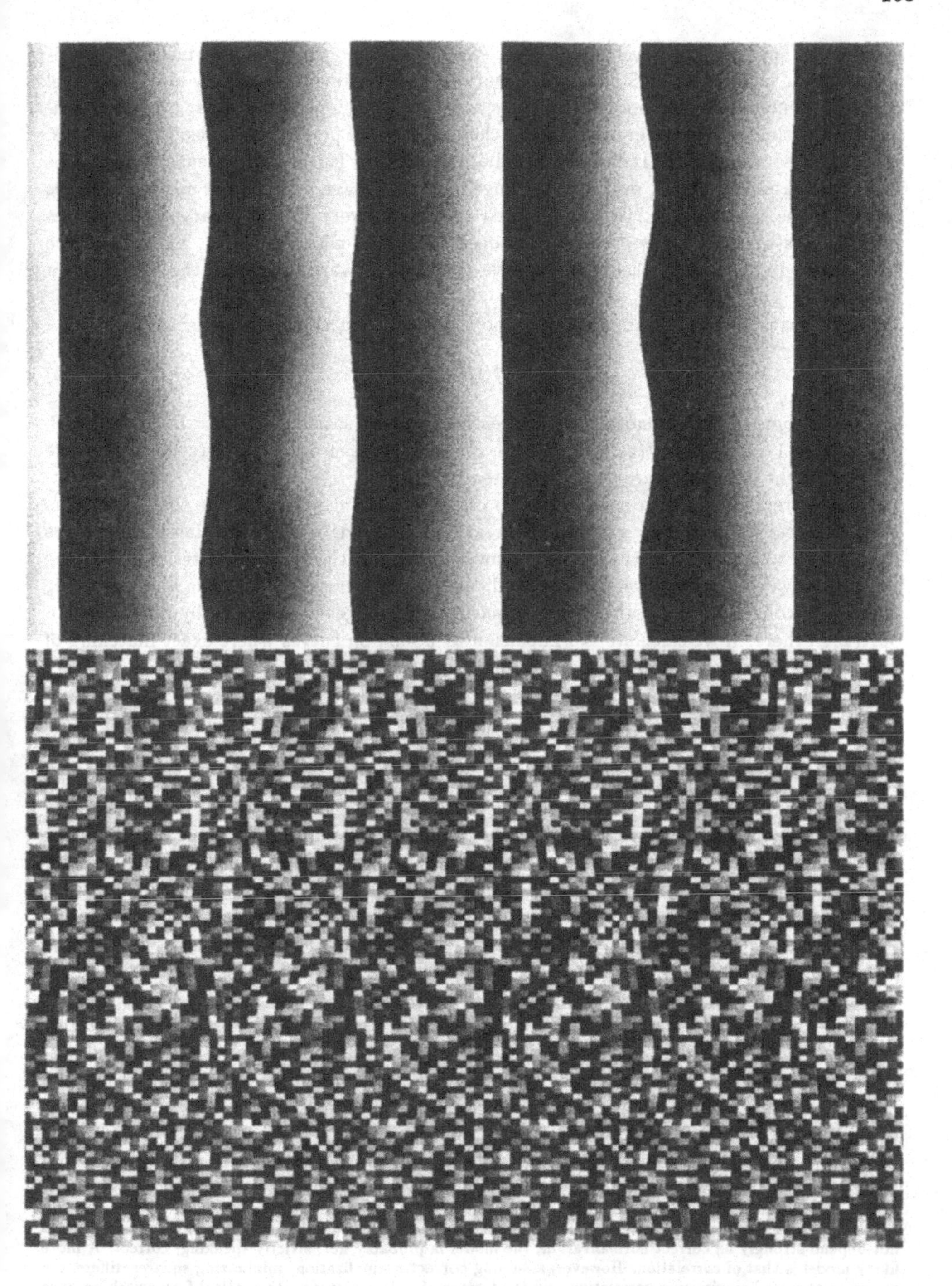

Figure 3. Ramp function and random permutation of ramp function

ensuring no spurious grey-level matches, is clearly less important than some measure of variability in the basic pattern. Stereo interpretation certainly involves not only local gray-level correspondences but also, more importantly, matches of regions with similar gray-level gradients, and similar average grey levels, matches of edges, of points and blobs, and, perhaps, even matches of complex 2D shapes, standing out as local features or tokens. Stereograms and autostereograms work over a wide range of input image types, from complex images requiring the matching of intricate, colored shapes, to very simple binary input images (and Tyler's first autostereograms were indeed black-and-white images.) These facts are also reflected in the many theories that have been put forward to explain the stereo vision process.

Let us next consider a class of slightly more complex matching functions[3] given by

$$\Lambda(x, \tilde{x}) = f([I(x) - I(\tilde{x})]^2) \tag{5}$$

where f is some smooth monotonically decreasing function satisfying $f(0) = 1$ and $\dot{f}(0) < 0$, e.g., $f(z) = 1/(1 + \lambda z)$ or $f(z) = e^{-\lambda z}$. Such functions measure the local grey level distance too, but slight differences are better tolerated by them.

The "correct" disparity curves here also carry a uniform weight of one, but deviations from these curves yield more graceful degradations. The ridges corresponding to the curves $\tilde{x} = x$, $\tilde{x} = x + \varphi(x)$, etc... have downward slopes with steepness determined by the second derivative in the direction perpendicular to the curve. But along the curve the function is constant ($= 1$), therefore sharpness of the ridges is expressed by the Laplacian of $\Lambda(x, \tilde{x})$ there. If we calculate the Laplacian at points on curves where exact matching of grey-levels occurs, e.g. on the curve $\tilde{x} = x + \varphi(x)$, we obtain

$$\nabla^2 \Lambda(x, \tilde{x}) = 2\dot{f}(0) \left\{ [\frac{d}{dx} I(x)]^2 + [\frac{d}{d\tilde{x}} I(\tilde{x})]^2 \right\} \tag{6}$$

Hence, the squared first derivatives of the basic pattern control the shape of the matching function along and around the disparity curves, and in fact at all places where the grey levels match along some curve. So we should (under the assumption that the $\Lambda(x, \tilde{x})$ under consideration is the correct one, and that our goal is to provide the sharpest ridge possible) design basic patterns with high derivatives almost everywhere and as few accidental matches (that do not correspond to desired disparity curves) as possible. Large first derivatives result from edges, hence, a basic pattern with many edges ensures large first derivatives almost everywhere and sharp maxima along the desired disparity curve. This seems to explain part of the results presented earlier (in Figure 3), although we do not yet have an understanding of the behavior of this matching function in between the ridges, for general basic patterns.

This class of matching functions indicates that local maxima are quite sharp if the patterns are rich, but, since the images have finite dynamic range, there will necessarily be many spurious matching points (and perhaps even curves) that will be equally sharp. Therefore, it would be advisable to have several basic patterns encoding each line of the depth pattern so that all of them will have consistent and sharp maxima along the correct disparity curves, but the spurious peaks located at *random* places in the areas in-between.

---

[3]In [Sperling,81] and [ArndtMallotBülthoff,95] the same model is considered. Since human stereo does not depend strongly on correct normalization, the model is probably not, strictly speaking, correct. A more likely model is that of correlation. However, assuming correct normalization, minimizing squares differences approximates well maximizing correlation and the former affords a better mathematical framework because of the uniformity of the matching function along the correct match.

Then, in the actual matching process these spurious peaks would be averaged out. This calls for the use of samples of a noise process for the basic pattern, either in a perhaps far-fetched idea of using time-varying autostereograms (that would have to rely on temporal averaging in the visual system), or in static images, by exploiting the readily-available second dimension in the image plane (the y -direction that we so conveniently disregarded until now!). Indeed, we can safely assume that the depth profile does not vary too fast in the y -dimension, and use several consecutive lines of the autostereogram to encode the same (or slowly changing) depth pattern with a series of samples of a random-process, used to generate the basic patterns. (Physiologically, this amounts to assuming that stereo depth involves a coarsening of resolution.) In this context the question that remains to be answered is: what type of noise processes have the potential to yield good visual results?

Here again, we shall have to postulate the type of matching process that is performed by the visual system. We shall assume that the matching function is, in this case, dependent on averages of squared gray-level differences, over the samples of the process. Therefore, we define

$$V(x,\tilde{x}) := E_\omega(I(x) - I(\tilde{x}))^2 = \mathbf{R}(x,x) + \mathbf{R}(\tilde{x},\tilde{x}) - 2\mathbf{R}(x,\tilde{x}) \tag{7}$$

where $\mathbf{R}(x,\tilde{x})$ is the autocorrelation of the process $I_\omega(x)$, whose samples are the lines of the autostereogram. If the process $I_\omega(x)$ is defined as the extension of a portion of a stationary process sampled over a basic interval, say, $[0,\varphi(0))$, we obtain

$$V(x,\tilde{x}) = 2\mathbf{R}(0) - 2\mathbf{R}(B(x) - B(\tilde{x})) \tag{8}$$

where $B(x)$ and $B(\tilde{x})$ are the $x, \tilde{x}$ "back-projected" into $[0,\varphi(0))$ according to the way $I_\omega(x)$ was extended beyond $\varphi(0)$. If we define the matching function to be

$$\Lambda(x,\tilde{x}) = f(V(x,\tilde{x})) \tag{9}$$

we see that it will have ridges at the correct disparity curves and its behavior around these ridges will be determined by the autocorrelation function of the process used to build $I_\omega(x)$. In particular, if we make sure that the autocorrelation will only attain the maximal value of $\mathbf{R}(0)$ at zero and decay very steeply afterwards, we can shape the matching function into an approximate indicator function. In fact the Laplacian of $\Lambda(x,\tilde{x})$ at the ridges of local maxima, where $B(x) = B(\tilde{x})$, is given by:

$$\nabla^2\Lambda(x,\tilde{x}) = 2\dot{f}(0)\,\mathbf{R}''(0)\left\{[\frac{d}{dx}B(x)]^2 + [\frac{d}{d\tilde{x}}B(\tilde{x})]^2\right\} \tag{10}$$

showing that the shape of the ridge is controlled by the shape of the signal autocorrelation around the origin (we clearly do not have control over the depth-profile-dependent back-projection operator B !). This result is derived under the assumption that $\mathbf{R}(\tau)$ is smooth about the origin and, being symmetric, it has zero first derivative there.

This discussion seems to indicate that we should favor processes with large values of $\mathbf{R}''(0)$. For example, the choice of a completely uncorrelated, white noise for the process generating the basic pattern will lead to very sharp local maxima at the correct disparities, meaning good behavior for autostereogram interpretations, provided we can keep the visual system "locked" into the various possible 3D interpretations. But locking into any particular depth profile, except the trivial one, can be, in this case, extremely hard. The matching function will not direct (via a hill climbing process) the visual interpretation toward any of the secondary maxima. An example of an autostereogram produced with white noise is given in Figure 4. We note that most of the random dot stereograms that were presented in the literature have the appearance of spatial white-noise and were indeed generated using random number generators and thresholds in quite a straightforward manner.

In the next section we present a hypothetical model for the matching process that is based on the common belief that the visual system processes images simultaneously at several scales. The images that are presented to us are "filtered" by several low, or band-pass filters that can be assumed to yield a "pyramid" of coarser and coarser, i.e. more and more blurred, images. It can be argued therefore that, in order for an autostereogram to look good and be easily interpretable, we need to have:

1. images that will look as homogeneous as possible over the entire span of spatial coordinates, in spite of the special way they were generated (which means that the random processes generating the autostereogram lines should be scaling invariant), and
2. images that lead to strong peaks of the image autocorrelation located at the correct disparities, but with with not too narrow "basins of attraction", about them at the coarse scales. In fact, it would help having basins of attraction tuned to the spatial scale, that become narrower as we go from the coarser to the finer scales, in order to direct the interpretation process (via hill-climbing on the corresponding matching functions!) toward the maxima located at the correct disparity curves.

To ensure the appearance of spatial homogeneity (as well as scale-space homogeneity) we would like to have a scale-invariant stochastic process generate the basic pattern, since the depth function locally scales the basic strips (in fact nonlinearly!) to produce the entire image. White noise would again be a reasonable candidate for this, but the requirement of having autocorrelation peaks with basins of attraction widening at a reasonable rate with the scale parameter is not met by a process with $\delta(\tau)$ autocorrelation.

Let $\mathbf{R}_\sigma(\tau)$ be the autocorrelations of the processes obtained when the basic pattern process is low-pass filtered to effective width σ . We can analyze the behavior of $\mathbf{R}''_\sigma(0)$ as a function of σ for various types of processes. It is seen that a white noise basic pattern leads, with decreasing σ, to a very quick narrowing of the peaks of the corresponding scale space of matching functions, $\Lambda_\sigma(x, \tilde{x})$, while a noise whose spectrum that decays like $1/f^2$ in the frequency domain provides too slow a sharpening of the peaks with a decreasing scale parameter. A noise process that has $1/f$ behavior over the frequency range relevant to visual perception seems to be ideally suited for our needs. Indeed, $1/f$ -type noise has the property of selfsimilarity under scalings, needed for spatial homogeneity, and long-range correlation tails that will correctly guide the process of locking into the various depth interpretations!

To substantiate the above claim in a simple case, consider a constant depth profile. Then, $I_\omega(x)$ is a periodic process.[4] Hence the samples of the process can be described by a Fourier series as follows

$$I_\omega(x) = \sum_{i=0}^{\infty} a_i \cos(i w_0 x + \phi_i)$$

where ϕ_i are i.i.d. random phases distributed uniformly over $[0, 2\pi)$, and a_i are positive

[4]The constant depth case is rather trivial and is known in the stereo vision literature as the "wall-paper" effect for periodic patterns (see [Ittelson,60].)

random variables too. The (periodic) auto-correlation of this stationary process is given by

$$\mathbf{R}(\tau) = \frac{1}{2}\sum_{i=0}^{\infty} E(a_i^2)\cos(iw_0\tau)$$

Now assume that we have a scale-space of filtered versions of the process $I_\omega(x)$ so that I_ω^σ is obtained by cutting off frequency components beyond w_0/σ. Then we have

$$\mathbf{R}_\sigma(\tau) = \frac{1}{2}\sum_{i=0}^{\sigma^{-1}} E(a_i^2)\cos(iw_0\tau)$$

and therefore,

$$\frac{\partial^2}{\partial(\tau/\sigma)^2}\mathbf{R}_\sigma(\tau) = -\frac{1}{2}\sum_{i=0}^{\sigma^{-1}} E(a_i^2)(i\sigma w_0)^2\cos(iw_0\tau)\,.$$

Note that here we have normalized τ by σ since σ is the appropriate unit of length on scale σ; we should expand $\mathbf{R}_\sigma(\tau)$ in terms of τ/σ. Let us consider a sequence of matching functions $\Lambda_\sigma(x,\tilde{x})$ corresponding to the filtered versions of $I_\omega(x)$. Since the peaks of the matching function $\Lambda_\sigma(x,\tilde{x})$ is controlled by $\mathbf{R}_\sigma''(0)$, we see that the (σ normalized) peaks of $\Lambda_\sigma(x,\tilde{x})$ get narrower with decreasing σ at a rate described by:

$$F(\sigma) := \sigma^2 R_\sigma''(0) = -\frac{1}{2}\sum_{i=1}^{\sigma^{-1}} E(a_i^2)(i\sigma\omega_0)^2\,.$$

If we choose $E(a_i^2) \propto (i\omega_0)^{-\beta}$ (for $i \geq 1$) and let $\sigma \to 0$ we have

$F(\sigma) \to \sigma^{-1}$, for $\beta = 0$ ("white" noise),
$F(\sigma) \to$ constant, for $\beta = 1$ ("$1/f$" noise),
$F(\sigma) \to \sigma$, for $\beta = 2$ ("$1/f^2$" noise),
$F(\sigma) \to \sigma^2\ln(1/\sigma)$, for $\beta = 3$ ("$1/f^3$" noise),
$F(\sigma) \to \sigma^2$, for $\beta > 3$ ("$1/f^{3+}$" noise).

Hence we see that with $1/f$ noise the peaks (normalized to length scale σ) retain essentially constant normalized width in scale space. More generally, we will have $\mathbf{R}_\sigma(\tau) \simeq \mathbf{R}(\tau/\sigma)$. This is desirable property for many reasons. We hypothesize that, whatever the matching mechanism is, it is invariant across scale. When τ is on the order of σ we expect $\mathbf{R}_\sigma(\tau)$ to be the operative correlation. With $1/f$ noise we see that the width of the peaks of the matching function are directly proportional to scale. Suppose τ is adapted according to some hill climbing process in scale space. Then, in this case, for $\tau \ll \sigma$ we are well within the peak of matching function on scale σ so that scale will contribute little to the correction in τ. Having $\sigma \simeq \tau$ places τ on the steep portion of the peak, hence a strong indication of the appropriate correction to τ is generated. For $\tau \gg \sigma$ we expect to be outside the domain of attraction and small, perhaps random, corrections to τ are indicated. Thus $1/f$ noise appears ideal from the point of view of obtaining stable convergence to the peak over the widest possible range. For $\beta > 1$ we expect convergence to break down on small scales, hence resolution will be lost. For $\beta < 1$ we expect convergence to break down on large scales, so the domain of attraction of matching will be reduced and the autostereogram will be harder to perceive.

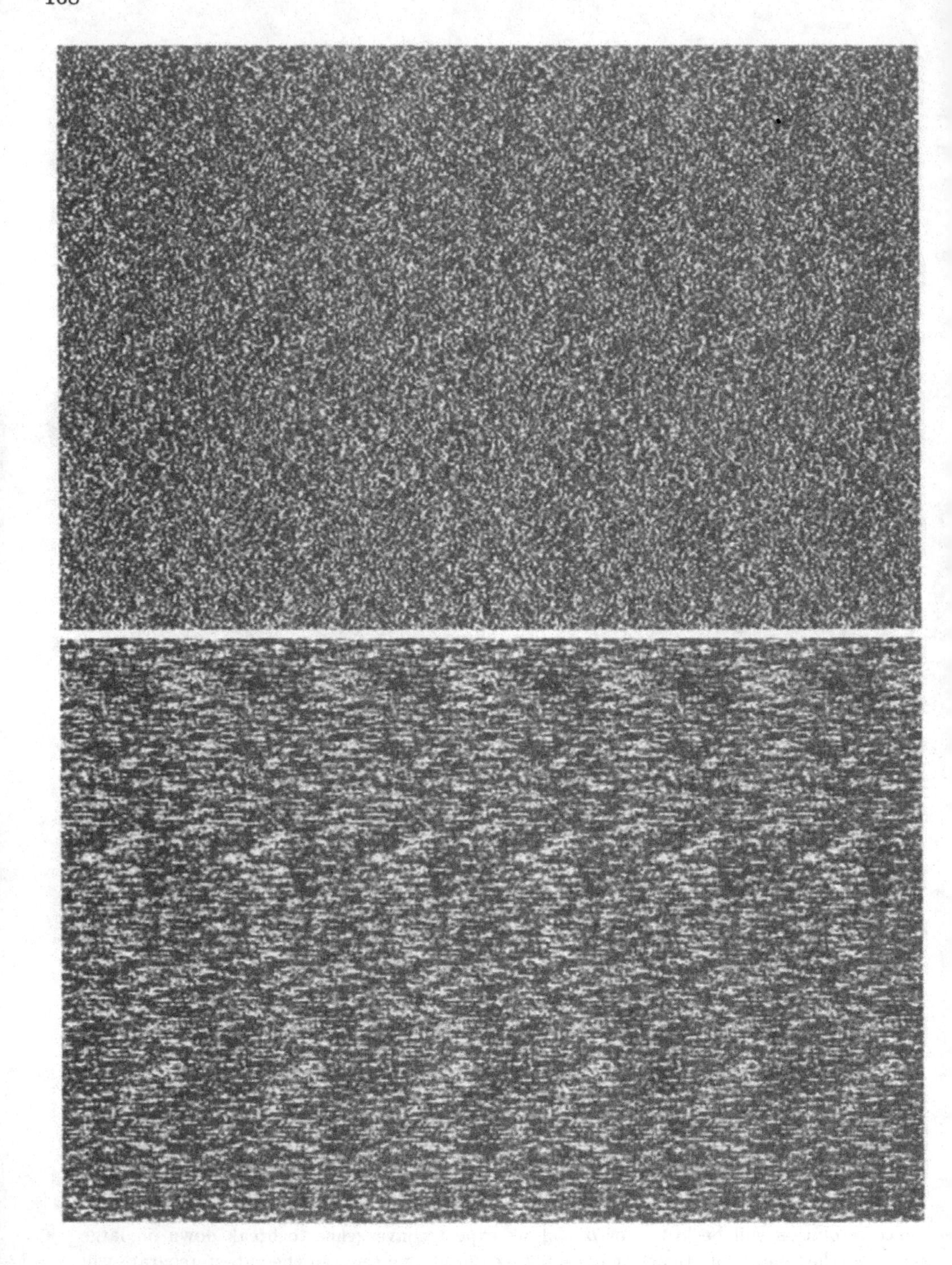

Figure 4. Independent lines of $1/f^0$ (white) noise and $1/f$ noise

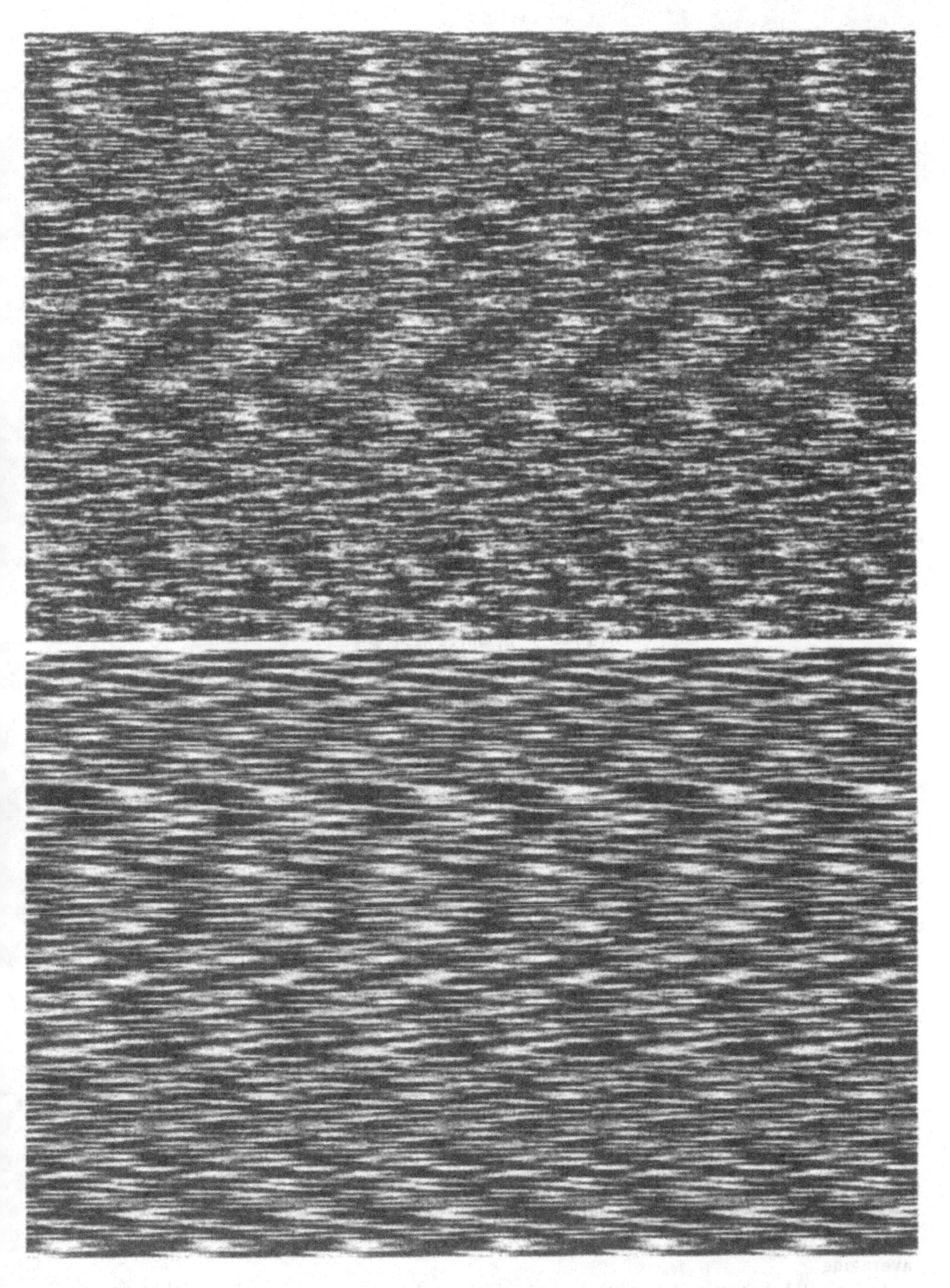

Figure 5. Independent lines of $1/f^2$ (white noise) and $1/f^3$ noise

Figures 4 and 5 present a series of autostereograms with independent line process having power spectra $1/f^n$ for $n = 0, 1, 2, 3$. Note the degradation in resolution of depth for $n = 3$. It is difficult to choose between $n = 1$ and $n = 2$ for overall quality. In the next Section we will further elaborate on the role of scale-space in stereo vision and in perceiving autostereograms in particular. There we will conclude that a 2D $1/f$ process is desirable; independent line processes of $1/f^2$ noise actually approximate this better than independent line processes of $1/f$ noise.

3: From Scale Space to Autocorrelation

So far we made the assumption that the visual system has a way to compute the ensemble averages of the horizontal line processes, the averages being needed to effectively evaluate the matching functions $\Lambda(x, \tilde{x})$ that lead to 3D interpretations. It was suggested that use could be made of the vertical direction to mimic ensemble averaging: By having different samples on different lines vertical averaging of line by line products could mimic ensemble averaging. The reader might object to this idea on the basis that not enough independent lines will enter the depth calculation to simulate ensemble averaging. It turns out that a simple frequency decomposition of the image, an operation which is widely believed to be performed by the visual system, can facilitate this process. This can be seen even in 1D. Consider, as before, an image line

$$I(x) = \sum_i a_i \sin(i w_0 x + \phi_i)\,.$$

Let us assume that we have independent lines of this form where the ϕ_i are independent and uniform and, for simplicity, the a_i are deterministic (i.e., the same from line to line). Let $I_i(x)$ denote $a_i \sin(i w_0 x + \phi_i)$. The calculation of the correlation takes the form

$$E(I(x)I(x+\tau)) = E(\sum_i I_i(x)I_i(x+\tau)) + E(\sum_{i \neq j} I_i(x)I_j(x+\tau)\,,$$

where the expectation is taken over the random phases of the sinusoids. The expectation eliminates the second term. However, if we wish to approximate the expectation by a sum over independent samples then, indeed, the number of samples required can be quite large. This is due to the relatively large number of essentially independent terms which need to be averaged out. However, if each eye-sensor were able to first extract I_i, by Fourier analysis, then the brain could compute $\sum_i I_i(x)I_i(x+\tau)$ directly thereby significantly reducing the number of independent samples required to approximate the ensemble average. Furthermore, since $I_i(x)I_i(x+\tau) = a_i^2 \cos(i w_0 \tau) + a_i^2 \cos(i w_0 (2x + \tau) + 2\phi_i)$, the second, spatially (i.e. x) dependent, term could also easily be removed by low pass filtering with a cutoff at or above $i w_0$. Note that a lower cut-off is not feasible since this would result in extreme loss of resolution in depth. (Recall that in reality all calculations are local.) Fourier analysis as indicated here is not a realistic assumption for the eyes but the point is this: band-pass decomposition of the images, correlation within bands, and low-pass filtering of the products prior to recombining can significantly reduce the burden of ensemble, i.e. vertical, averaging.

Let us elaborate a bit further on our ideal model. Suppose, as above, that the brain computes $I_i(x)I_i(x+\tau)$. It is reasonable to assume that averaging occurs before recombining over i. In this way we may conclude that the brain computes

$$\mathbf{P}^i(\tau) := E(I_i(x)I_i(x+\tau)) = \frac{1}{2}a_i^2\cos(iw_0\tau)$$

separately for each frequency band, i.e. (ideally) for each i. Now, let us postulate that, in order to adjust τ, $\mathbf{P}^i(\tau)$ is differentiated as a function of τ. Since the length scale associated with frequency $w_0 i$ should be $1/(w_0 i)$ we define

$$D^i(\tau) := \frac{1}{iw_0}\frac{\partial}{\partial\tau}\mathbf{P}^i(\tau) = -a_i^2\sin(w_0 i\tau)\,.$$

Note that this scaling is possible only if the averaging is done before recombining the various frequency bands, although the differentiation may occur after. Now, we will assume that the temporal derivative of τ, $\frac{\partial\tau}{\partial t}$ is computed by the brain by summing the length normalized contribution from different frequency bands. We will therefore define,

$$D(\tau) := \sum_i D^i(\tau) = -\sum_i a_i^2\sin(iw_0\tau)\,.$$

The question now is: what should we choose for a_i in order to obtain good behavior for the τ-adjustment rate. For simplicity, and to enable scale invariance, let us consider the possibility $a_i^2 \propto i^\beta$ for $i \geq 1$ and ask what to choose for β. Let $\omega_{\max}$ be the highest frequency actually used, then we have,

$$D(\tau) \simeq -(\text{const}\ )\int_{w_0}^{\omega_{\max}}\omega^\beta\sin(\omega\tau)\,d\omega = -(\text{const}\ )\tau^{-1-\beta}\int_{w_0\tau}^{\omega_{\max}\tau}u^\beta\sin(u)\,du$$

In the case of interest, $w_0\tau \ll 1$ and $w_{\max}\tau \gg 1$, and $-2 < \beta < 0$ we have $D(\tau) \simeq (\text{const})\,\tau^{-1-\beta}$. Given a global constraint on the energy in our images, arising, for example, from limitations in the rendering, β controls a trade-off between high resolution and the domain of convergence. This trade-off is balanced when we choose $\beta = -1$. Smaller values of β, i.e., $\beta < -1$ favor larger scales. Convergence slows as τ becomes small and resolution is reduced (because of noise.) For larger β, i.e., $\beta > -1$ we favor smaller scales, achieving high resolution but reducing the domain of convergence.

Once again, by invoking a scale invariance assumption we have concluded that a $1/f$ power spectrum is ideal for our basic patterns. But what properties should the 2D basic strip have ? We could, for example, let each horizontal line of the basic strip be an independent sample of a $1/f$ type noise process. The results of this idea are seen in Figure 4; if the eye processed each line independently and then, in the final "correlation" stage, averaged vertically over several lines, then, according to our analysis, this type of basic strip would be ideal. However, the eyes do not work in this, line by line, way. This is clearly indicated by the fact that a basic strip with 2D $1/f$ noise is significantly superior to line independent $1/f$ noise, as we shall see in the experiments.

A physiologically more plausible model would allow for the 2D filtering performed by the eyes prior to correlation. It is widely held that the visual system decomposes the images projected onto the eyes into various frequency bands. A standard model is isotropic band-pass filtering, i.e., the gain of the 2D frequency w depends only on $|w|$. For stereo processing there might well be some anisotropy but not the extreme version considered earlier. For simplicity we will study the effect of 2D filtering by considering the isotropic model.

Let us now try to understand our previous models from the 2D point of view. To begin with we shall return to the model of 1D correlation of the images with vertical averaging of the product. We will later adjust the model to add other elements. For notational convenience we will let I_τ be defined by $I_\tau(x,y) := I(x+\tau,y)$. Let $\mathcal{F}$ denote the Fourier transform, i.e.,

$$\mathcal{F}(I)(w) = \mathcal{F}(I)(w_1,w_2) = \int\int e^{i(w_1x+w_2y)} I(x,y)\,dx\,dy\,.$$

Let $G(w)\,(=\delta(w_2))$ denote the transfer function of (ideal) vertical averaging. Our simple model reduces to computing the following,

$$\mathcal{F}^{-1}(G\cdot\mathcal{F}(II_\tau)) = \int\int (e^{i\langle u,\vec{x}\rangle}\mathcal{F}(I)(u))^* G(w-u) e^{iw_1\tau}(e^{i\langle w,\vec{x}\rangle}\mathcal{F}(I)(w))\,dw\,du \tag{11}$$

where $\vec{x} = (x,y)$. We can introduce band pass filtering into this model as follows. Let $H^\sigma(w)$ denote the transfer function of some band-pass filter and assume $\sum_\sigma H^\sigma(w) = 1$. Then for each σ the eyes/brain may compute

$$\int\int (e^{i\langle u,\vec{x}\rangle} H^\sigma(u)\mathcal{F}(I)(u))^* G^\sigma(w-u) e^{iw_1\tau}(e^{i\langle w,\vec{x}\rangle} H^\sigma(w)\mathcal{F}(I)(w))\,dw\,du\,. \tag{12}$$

Here we have admitted the possibility that the filter G may depend on σ. In the case that $H^\sigma \simeq \delta(|w_1| - 2\pi/\sigma)$, i.e. ideal horizontal band-pass filtering, and G^σ introduces some low-pass filtering in the horizontal direction, then we reproduce the scenario described at the beginning of this section. Our 2D model would have H^σ represent 2D bandpass filtering in a band near $2\pi/\sigma$, i.e. H^σ passes frequencies $w = (w_1, w_2)$ with $|w| \simeq 2\pi/\sigma$. For simplicity we will assume H^σ is ideal, i.e.,

$$H^\sigma(w) = \begin{cases} 1 & w \in \Omega_\sigma \\ 0 & w \notin \Omega_\sigma \end{cases}$$

where Ω_σ is some annulus $\{|w| \simeq 2\pi/\sigma\}$. The filter G^σ should represent vertical averaging and also, perhaps, low pass horizontal filtering with cutoff near $|w| \simeq 2\pi/\sigma$.

We will study the effect of the 2D filtering via an illustrative example. We consider an image of the form

$$I(x,y) = \sum_i a_i \sin(iw_0x + \phi_{ik})$$

where the a_i are independent of y, k is determined by $y \in [k\epsilon/w_0, (k+1)\epsilon/w_0)$, and ϕ_{ik} are uniformly random in $[0, 2\pi]$ and, for now, independent for each i and k. This models an image in which pixels are vertically separated by ϵ/w_0 and each horizontal line is independent.

If we examine the 2D spectrum of such an image we find that, roughly speaking, energy a_i^2 is distributed uniformly in a strip around the line segment $\{w_1 = iw_0, w_2 \in (-w_0/\epsilon, w_0/\epsilon)\}$ and its reflection $\{w_1 = -iw_0, w_2 \in (-w_0/\epsilon, w_0/\epsilon)\}$. Thus, the energy associated with horizontal frequencies gets smeared, approximately uniformly, across a wide range of vertical frequencies.

For the image above, when G_σ is ideal vertical averaging we obtain

$$\mathcal{F}^{-1}(G_\sigma \cdot \mathcal{F}(I^\sigma I^\sigma_\tau)) = \frac{1}{2}\int_{\Omega_\sigma} |\mathcal{F}(I)(w)|^2 \cos(w_1\tau)dw \tag{13}$$

where I^σ denotes $\mathcal{F}^{-1}(H^\sigma \cdot \mathcal{F}(I))$. As before, we hypothesize that the eyes compute

$$D_\tau := \sigma \frac{\partial}{\partial \tau} \mathcal{F}^{-1}(G_\sigma \cdot \mathcal{F}(I^\sigma I^\sigma_\tau)) = -\frac{1}{2} \int_{\Omega_\sigma} |\mathcal{F}(I)(w)|^2 (\sigma w_1) \sin(w_1 \tau) dw$$

This reveals two weaknesses associated with independent lines. If the eyes do 2D band pass filtering as suggested above, then low (absolute) frequencies will not have enough energy in them and higher frequencies too much. This explains, in part, why 2D $1/f$ noise is superior as a basic strip to independent lines of $1/f$ noise. The second undesirable effect of having independent lines is that putting energy in frequencies (w_1, w_2) where $|w_2| \gg |w_1|$ is wasteful in the sense that stereo depth information is not carried by vertical components. Since the derivative above associated to $w = (w_1, w_2) \in \Omega_\sigma$ is, according to our model, scaled by $\sigma \simeq 1/|w|$ it is better, in the sense that the derivative will be larger, to have $w_1 \gg w_2$. The limiting case, $w_2 = 0$, is however undesirable since this requires phases ϕ_{ik} which do not vary over k, eliminating the tremendous value of vertical averaging. Thus we observe that there is a trade off between providing for the local averaging and maximizing horizontal signal.

A compromise is the following. We should let $\phi_{i,k}$ be positively vertically correlated, i.e. correlated in k, in a way which depends on i. We would like the randomness in the ϕ's to result in a 2D spectrum which is largely supported, roughly speaking, in the cone $|w_2| \leq |w_1|$. This can be achieved, for example, by letting ϕ_{ik} be a sample path from a random walk as follows,

$$\phi_{i,k+1} = \phi_{ik} + \gamma \xi_{i,k+1} / \sqrt{i}$$

where ξ_{ik} are i.i.d. uniform on $[-1, 1]$ (say) and where γ is an appropriate constant independent of i and k. In this way the 2D spectrum is also essentially $1/f$. The only remaining concern is whether vertical averaging can still effect cancellation of cross terms in the products $I^\sigma I^\sigma_\tau$. To see this we need to reconsider the derivation of equation (13). For simplicity let us assume that that horizontal frequencies are not spread across 2D bands. Then the phases of the cross terms take the form $\phi_{ik} \pm \phi_{jk}$. Thus, the cross term phases will behave like a random walk but the correlation length will be on the order of $1/i + 1/j$. Since, because of scale space filtering, we can assume $i \simeq j$ we see that the cross term phases fluctuate significantly more rapidly than the local length scale $\simeq 1/i$, and, therefore, vertical averaging would be able to effectively eliminate them.

Examples of images produced this way are given in Figures 6 and 7. Here we can see the role of vertical correlation. The figure at the bottom of Figure 6, a suitable vertically correlated $1/f$ noise pattern, is among the 'best' autostereograms we have been able to produce. Comparing Figure 6 with Figure 7 the superiority of $1/f$ to $1/f^2$ noise when the second dimension is properly taken into account is evident.

4: Discussion and Concluding Remarks

Autostereograms are natural generalizations of periodic patterns that were known to produce the so-called "wall paper phenomenon" documented in old books on visual perception. Ittelson, [Ittelson,60], describes this phenomenon as follows: "An observer stands a few feet distant from, and squarely facing, a wall covered with a regular, repeating pattern of small figures. By increasing the convergence of his eyes while observing the pattern on the wall, the observer will note that there are one or more amounts of convergence for

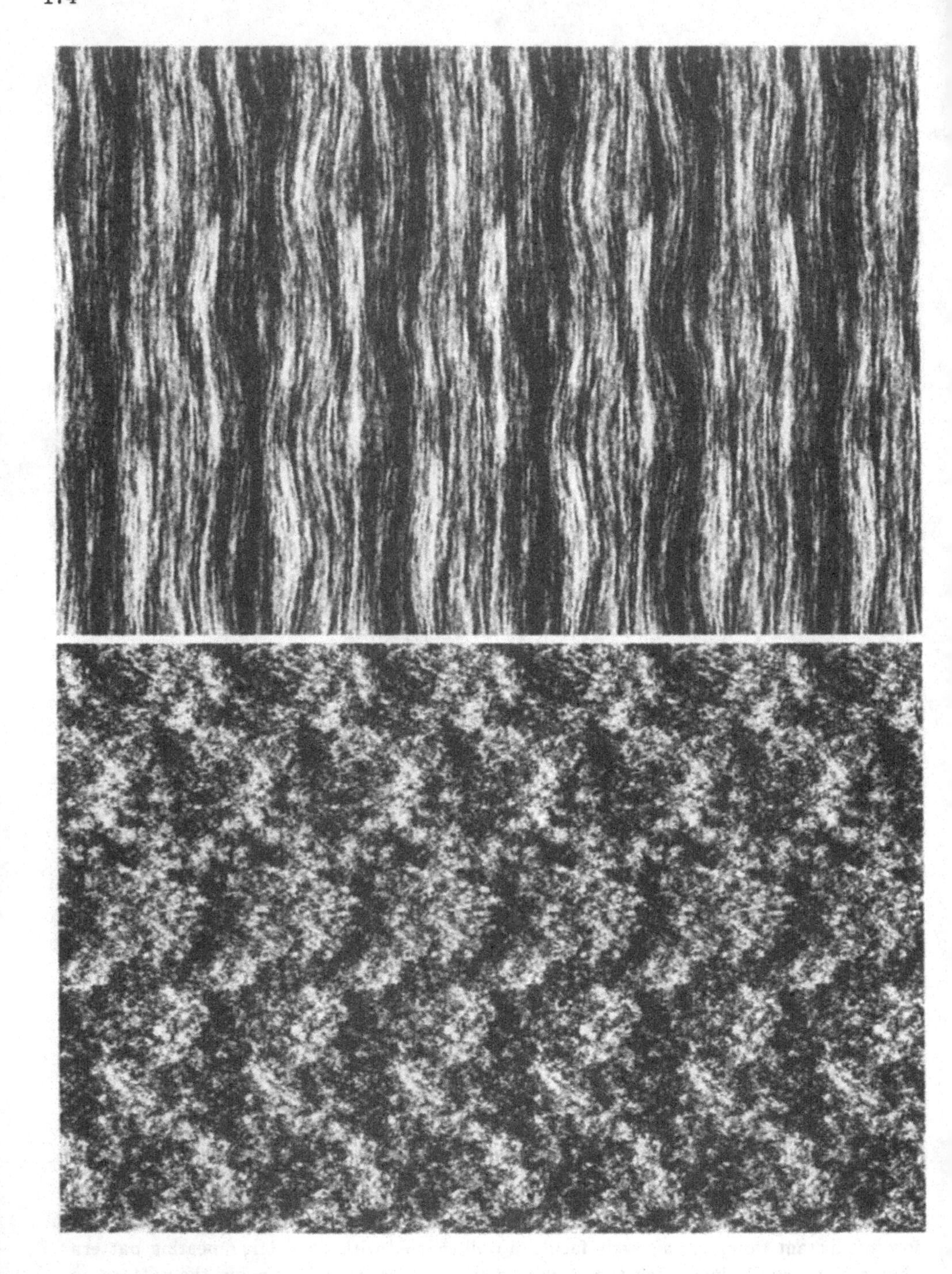

Figure 6. Patterns of $1/f$ noise with different vertical correlation

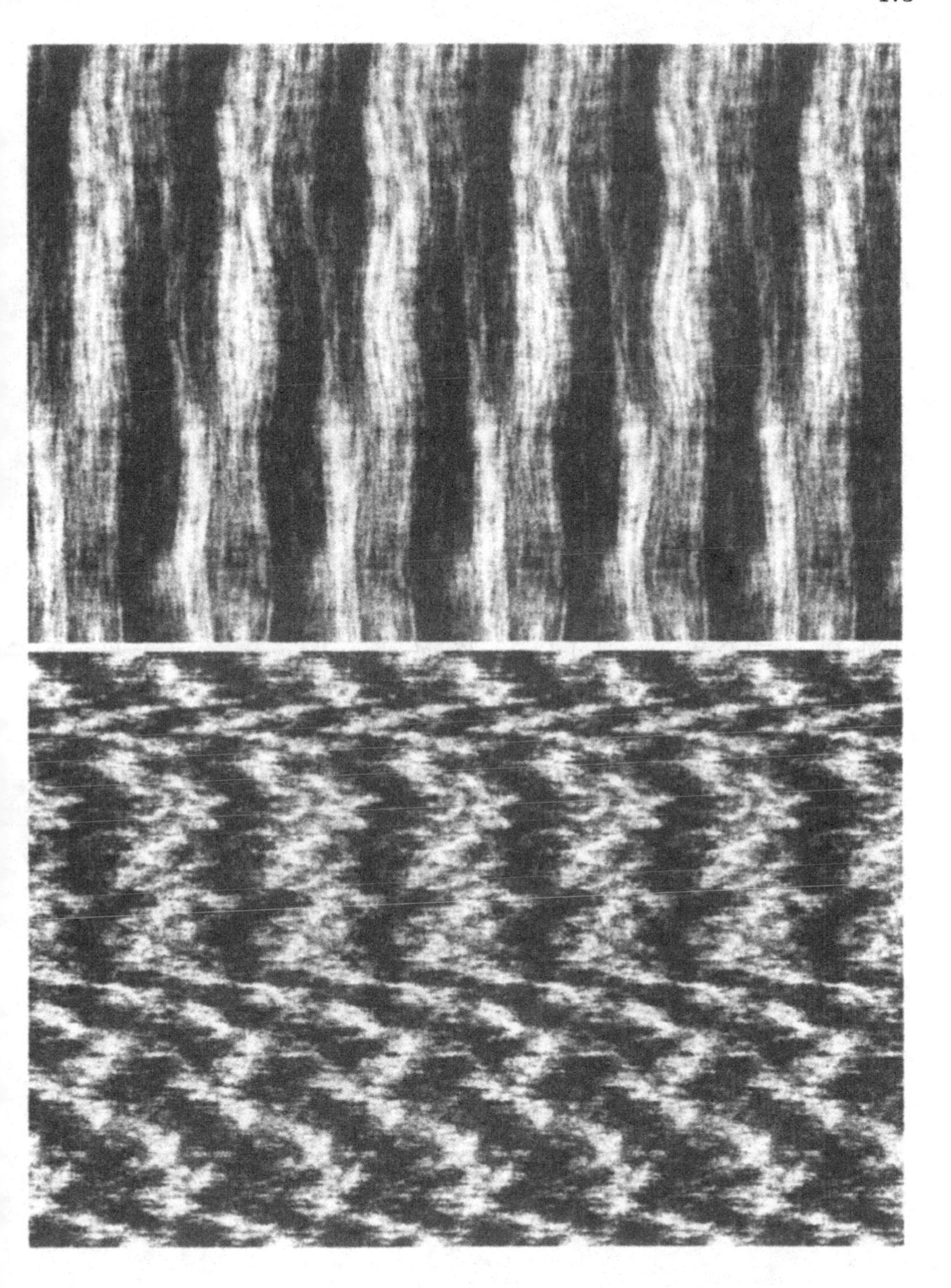

Figure 7. Patterns of $1/f^2$ noise with different vertical correlation

which fusion will be obtained, but with different, rather than the same, parts of the pattern fusing together. At the same time, the entire wall will appear to have moved nearer to the observer and become smaller. The same effect has been observed for a typewriter keyboard, postage stamps, and various other repeated figures."

In this short paper we explained the way autostereograms are produced for arbitrary depth profiles and posed the question of designing autostereograms for best visual interpretations. Then we analyzed possible simplistic mechanisms for depth recovery from the autostereographic images, and concluded with arguments that point toward $1/f$-noise processes as excellent generators for basic autostereogram patterns. Further work on this topic is currently under way, analyzing the relationships between the processes chosen to generate basic patterns and the ease of locking into the 3D interpretations, under a variety of further physiologically motivated stereo interpretation models.

Much deeper analysis is required to solve the problem of optimal autostereogram designs for complex models of stereo perception, but we believe that $1/f$ noise will turn out to be universally good for these purposes. Easily perceived autostereograms might one day be viable alternatives for the effective display of three dimensional surfaces and data.

References

[Julesz,64] B. Julesz, *Binocular Depth Perception without Familiarity Cues,* Science, Vol. 145, pp. 356-362, 1964.

[TylerChang,77] C. W. Tyler, J. J. Chang, *Visual echoes: the perception of repetition in random patterns,* Vision Research, Vol. 17, pp. 109-116, 1977.

[Tyler,83] C. W. Tyler, *Sensory Processing of Binocular Disparity,* in Vergence Eye Movements: Basic and Clinical Aspects, Butterworth, Boston, pp. 199-295, 1983.

[TylerClarke,90] C. W. Tyler, M. B. Clarke, *The Autostereogram,* Proceeding SPIE. Meeting on Stereoscopic Displays and Applications, Vol. SPIE 1256, pp. 182-197, 1990.

[ThiIngWit,94] H. W. Thimbleby, S. Inglis and I. H. Witten, *Displaying 3D Images: Algorithms for Single Image Random Dot Stereograms,* Computer, Vol. 27/10, pp. 768-774, 1994.

[TerTer,94] M. S. Terrel and R. E. Terrel, *Behind the Scenes of a Random Dot Stereogram,* American Math, Monthly, pp. 715-724, October 1994.

[Sperling,81] G. Sperling, *Mathematical Models of Binocular Vision,* SIAM-AMS Proc. Vol. 13, pp. 281-300, 1981.

[ArndtMallotBülthoff,95] P. A. Arndt,H. A. Mallot and H. H. Bülthoff, *Human stereovision without localized image features,* Biol. Cybern. 72, pp. 279-293, 1995.

[MagicEye,93] *N. E. Thing Enterprises, Magic Eye: A New Way of Looking at the World,* Michael Joseph Ltd, Penguin Group, 1993.

[Ittelson,60] W. H. Ittelson, *Visual Space Perception,* Springer Publishing Co, New York, pp. 123-127, 1960.

Shape Recovery from Stationary Surface Contours by Controlled Observer Motion

Liangyin Yu   Charles R. Dyer

Computer Sciences Department
University of Wisconsin
Madison, WI 53706

Abstract

The projected deformation of stationary contours and markings on object surfaces is analyzed in this paper. It is shown that given a marked point on a stationary contour, an active observer can move deterministically to the osculating plane for that point by observing and controlling the deformation of the projected contour. Reaching the osculating plane enables the observer to recover the object surface shape along the contour as well as the Frenet frame of the contour. Complete local surface recovery requires either two intersecting surface contours and the knowledge of one principle direction, or more than two intersecting contours. To reach the osculating plane, two strategies involving both pure translation and a combination of translation and rotation are analyzed. Once the Frenet frame for the marked point on the contour is recovered, the same information for all points on the contour can be recovered by staying on osculating planes while moving along the contour. It is also shown that occluding contours and stationary contours deform in a qualitatively different way and the problem of discriminating between these two types of contours can be resolved before the recovery of local surface shape.

1: Introduction

Natural objects are full of textures of all kinds, providing qualitatively different cues about surface shape. Different kinds of texture require different methods for analysis. One kind of surface texture, stationary surface contours, constrains surface shape in a way very different from "blob-like"texture. These stationary contours are one-dimensional curvilinear markings on the object surface, which, unlike occluding contours, do not "slide" across the surface as the vantage point changes [4] and, hence, only constrain the surface along a single dimension like a strip for a smooth surface [10]. Consequently, stationary contours have been studied mostly in the context of qualitative surface characterization [5, 9, 12, 14]. In contrast, since the observation of Barrow and Tenebaum [1] that occluding contours constrain surface orientation uniquely even from a single viewpoint, this kind of contour has been the focus of considerable research to quantitatively characterize the surface from one-dimensional curvilinear features.

The support of the National Science Foundation under Grant No. IRI-9220782 is gratefully acknowledged.

0-8186-7644-2 $5.00 © 1996 IEEE

Since both stationary and occluding contours are curvilinear features on the object surface, they constrain surface shape in a similar way, i.e., they tell us something about the tangent direction and degree the surface curves away from this direction. However, for a stationary observer, these two kinds of contours appear to be locally identical and, therefore, cannot be distinguished. On the other hand, the fact that an occluding contour slides across the object surface while a stationary contour is fixed on the surface present themselves quite differently to an active observer. This observation also makes the task of classifying contours an important problem for strategies that infer surface shape from contour [4, 11, 14].

For occluding contours, Giblin and Weiss [7] demonstrated that for certain specific motions, an active observer can recover surface shape from *known* occluding contours under orthographic projection. That result was extended by Cipolla and Blake [4] to arbitrary observer motion under perspective projection. They also devised a procedure for identifying the kind of contour after the surface shape was recovered. As Kutulakos and Dyer [11] pointed out, this detection procedure is hampered by both the measurement accuracy required and the necessity of recovering the surface shape first. Instead, they proposed an affine-invariant based *re-projection* approach to eliminate these problems. However, this method of recovering surface information from occluding contours is limited in areas where the surface shape contains occluding contours. In addition, to recover surface shape from occluding contour requires accurate measurement of observer motion, camera calibration, and is, in general, very sensitive to noise. The advantages of these methods come in two flavors: contour features can be extracted more easily and reliably, and they can strongly constrain the surface shape and characterize the surface directly. Both Koenderink [9] and Brady *et al.* [2] obtained results relating occluding contours and sign of Gaussian curvature. Brady *et al.* also showed how special curves such as occluding contours, asymptotes, and lines of curvature can provide strong constraints on surface shape. Kass, Witkin and Terzopoulos [8] proposed active contour models that can track image contours easily and this method was used by Cipolla and Blake [4] for tracking occluding contours.

For stationary contours, various qualitative properties were obtained and conjectured by Stevens [12], including the possibility that we might recover parts of the surface shape from contour deformation. Deformation of image contours in general was analyzed by Cipolla *et al.* [3, 4] by relating observer motion parameters to the deformation of image contours. They also showed how the sign of normal curvature can be determined from properties of projected image contours (e.g., inflection points). Zisserman *et al.* [14] proposed methods for qualitatively characterizing image contours. They showed, for example, how at least three different views of a contour are required to distinguish between a space curve and an occluding contour.

In this paper, we focus on the quantitative constraints imposed by stationary contours on surface shape. We show that an active observer can exploit these constraints and move deterministically to the *osculating plane* of a given point on the contour, and from there can recover the normal curvature of the surface as defined by the contour and the associated Frenet frame of the contour. Furthermore, when two non-collinear stationary contours intersect, and one of the principle directions is known, local surface shape can be completely recovered. This is both qualitatively and quantitatively different from the existing method of recovering surface shape from occluding contours in which a surface parameterization is obtained when the occluding contour slides across the surface [4, 7, 13].

In the first part of this paper we prove that the observer can explicitly choose its motion so that the projected curvature of a stationary surface contour monotonically decreases. The lower bound on this projected curvature is reached when the observer reaches the osculating plane defined by the surface contour. During the motion, the observer can either choose to fix the optical axis of the image plane or to rotate the optical axis so that the observed point on the contour is always on the optical axis. The latter is achieved through a combination of camera translation and rotation and is more natural.

However, when external references are available, choosing a fixed optical axis properly produces the shortest path to the destination. Both schemes are presented and results proved.

The second part of the paper describes how the Frenet frame can be recovered once the observer reaches the plane where the projected curvature for the given point on the contour reaches its minimum. In the process of reaching this position the observer can verify if the contour is indeed stationary. Our method for discriminating occluding from stationary contours differs from previous ones in that: (1) no motion parameters are used, and (2) it is applicable within areas where no occluding contours slide across the surface (i.e., elliptic concave parts of the surface). The recovered Frenet frame can be used by the observer to trace the stationary contour in order to recover the same information for all points on the contour (by always staying in the osculating planes). Furthermore, if there are points where two stationary contours intersect, the recovered Frenet frame can be used to parameterize the surface and this parameterization is unique if one of the principle directions (the direction where the normal curvature is either maximum or minimum at the given point) is known. This uniqueness is also true if more than two contours pass close to each other.

Finally, we show results of various recovered surfaces on a synthetic surface and the paths an observer actually takes to reach the osculating plane under purely translational motion, and under translation combined with rotational motion. The validity of the theory is further enforced by showing the result from recovering the surface of a vase sitting on a table.

2: Theoretical Framework

When a 3D surface is projected into a 2D image plane, the image formed is dependent on the lighting, surface properties, and the location of image plane. The problem of 3D shape recovery is to describe the surface in a parametric form from a set of 2D projected images.

In order to parameterize a surface locally, we need to designate two independent basis vectors and an origin. We also need three parameters to characterize how the distance (metrics) is measured on the surface and an additional three parameters to measure how the surface tangent turns away from the surface locally. Since these six parameters are not independent, but tied by a set of three *compatibility equations* [6], we need a minimum of three equations relating these parameters to completely characterize the local surface. The problem of recovering surface shape from contours is to derive these equations from a finite number of observations. In this section we describe the surface geometry and imaging model that will be used in the subsequent presentation.

In the following we will use ($\hat{x}$, $\hat{y}$, $\hat{z}$) to denote unit basis vectors in the 3D Euclidean coordinate frame (x, y, z).

2.1: Curves and Surfaces

Given a reference coordinate frame and a point P on a smooth surface S in space, the local shape of S at P is a set of parameterizations of the form $(x(u, v), y(u, v), z(u, v))$. Each parameterization differs from the others by a rigid transformation (compositions of translation and linear orthogonal transformations). The fundamental theorem of the local theory of surfaces asserts that the first and second fundamental forms uniquely define the surface up to a rigid transformation. These two forms, in turn, can be determined from the surface normal and first and second derivatives of the surface along two principle directions, where the surface curves most and least. This observation is the operational principle for our methods.

Given a point P on a 3D spatial curve, the Serret-Frenet equations relate an orthonormal frame $\{\hat{t}, \hat{n}, \hat{b}\}$, known as the *Frenet frame,* to two metric variables, κ and τ, known as *curvature* and

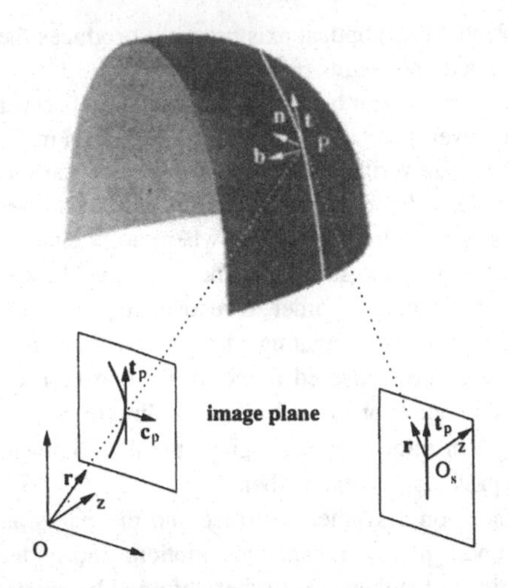

Figure 1. Locating the osculating plane for a stationary curve on a convex surface. Location O_s is where the observer reaches the osculating plane.

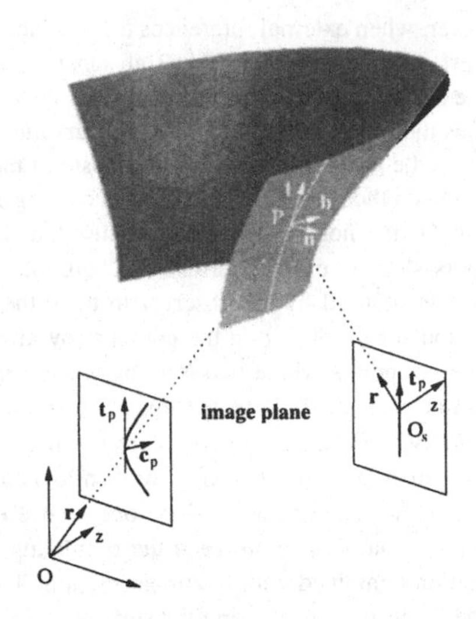

Figure 2. Locating the osculating plane for a stationary curve on a concave surface.

torsion:

$$\begin{aligned} \hat{t}' &= \kappa\, \hat{n} \\ \hat{n}' &= -\kappa\, \hat{t} - \tau\, \hat{b} \\ \hat{b}' &= \tau\, \hat{n}. \end{aligned}$$

For a stationary curve on an object surface, the normal of the curve is always uniquely defined if the curvature κ is not zero. If we determine an orientation for the curve and its tangent direction, the binormal of the curve is then determined by $\hat{b} = \hat{t} \times \hat{n}$. In the following we will use the convention of orienting the observer and the tangent $\hat{t}$ so that $\hat{x} \cdot \hat{t} > 0$ (see Figure 1). Hence the direction of $\hat{x}$ is an "upward" reference direction.

2.2: Contour Curvature under Projection

Consider the projection of a stationary curve segment C onto the image plane (Figure 1 and Figure 2). A point P on the curve is represented by the vector $\boldsymbol{r}$ in the observer frame. Let the Frenet frame at P be $\{\hat{t},\ \hat{n},\ \hat{b}\}$, and assume the observer is located at O and looks in the direction (viewing direction) $\boldsymbol{z}$. The image plane is at (0, 0, 1). The projected curvature κ_p of C_p at point P can be expressed in terms of the Frenet frame $\{\hat{t},\ \hat{n},\ \hat{b}\}$ as (see Appendix A)

$$\kappa_p = \frac{\kappa |\, \boldsymbol{r} \cdot \hat{b}|}{[(|\, \boldsymbol{r} \times \hat{t})|^2 - (\boldsymbol{r},\ \hat{t},\ \hat{z})^2)/(\boldsymbol{r} \cdot \hat{z})^2]^{3/2}}, \tag{1}$$

The advantage of Eq. (1) is the use of vector expressions in both the fixed Frenet frame and the observer frame, which is under control of the active observer. These forms are invariant to the observer's frame except for $\boldsymbol{r}$ and $\hat{z}$.

Since we haven't distinguished a stationary contour from an occluding contour at this point, the curve could be of either type. In the case of occluding contours, the expression κ_p actually reduces to the equation relating the *geodesic* curvature and the surface normal curvature [4, 9].

From Eq. (1) it can be seen that κ_p has minimum value 0 when $\boldsymbol{r} \cdot \hat{\boldsymbol{b}} = 0$. This is the case when the observer is in the plane defined by the binormal $\hat{\boldsymbol{b}}$, called the osculating plane, where the stationary contour C projects onto the image plane as a locally straight line. This is a well known result and we state it formally as follows:

Proposition 1. *Given a spatial curve on a smooth 3D surface and a point on the curve, the minimum projected curvature at this point is zero and will be observed when the observer is in the osculating plane containing this point.*

In the next section we show how an observer can reach this plane (that is, determine the binormal $\hat{\boldsymbol{b}}$ at a given point on curve C) from any location in space.

3: Moving to the Osculating Plane

The formulation above is in the frame of the observer, which we called the *observer frame.* By using the vector representation, we have some advantages when changing the observer frame during motion. However, it can be inconvenient when coordinate transformations are necessary. This difficulty can be circumvented by working in the *object frame* with the origin at P. These two frames can either be related through a pure translation or through a translation plus rotations. In the case of pure translation, the two frames can be transformed back and forth through a translation at any given location of the observer. We will call this case the *translation scheme.* More generally, the observer can choose to orient the observer frame with respect to the object frame in any convenient way and the two frames are related through an arbitrary rigid transformation at a given observer location. We will call this case the *rigid transformation scheme.* Both cases will be analyzed. In the following we will use $\boldsymbol{r}^*$ to denote the observer's location in the object frame.

3.1: Translation Scheme

When an active observer can control its motion so that only pure translation is performed (by using an external reference, for example), the observer and object frames are related by $\hat{\boldsymbol{x}}^* = \hat{\boldsymbol{x}}$, $\hat{\boldsymbol{y}}^* = \hat{\boldsymbol{y}}$, $\hat{\boldsymbol{z}}^* = \hat{\boldsymbol{z}}$. Consequently, $\boldsymbol{r}^* = -\boldsymbol{r}$. The observer may choose its frame arbitrarily as long as it satisfies $\boldsymbol{r} \cdot \hat{\boldsymbol{z}} > 0$ for the selected point on the surface, i.e., the viewing direction $\boldsymbol{r}$ generally "agrees" with the optical axis $\hat{\boldsymbol{z}}$. The following definition formalizes the statement that vectors $\boldsymbol{r}$ and $\hat{\boldsymbol{z}}$ are pointing generally *in the same direction.*

Definition 1. *Given a spatial curve C and the Frenet frame $\{\hat{\boldsymbol{t}}, \hat{\boldsymbol{n}}, \hat{\boldsymbol{b}}\}$ at a point on the curve, an observer frame is an* agreeable frame *for the point if for all possible observer movements, $\boldsymbol{r}$ and $\hat{\boldsymbol{z}}$ are in the same octant defined by the Frenet frame.*

Consider κ_p in the object frame as a scalar field of $\boldsymbol{r}^*$ (i.e., at any given point $\boldsymbol{r}^*$ in space, there is an associated field value $\kappa_p(\boldsymbol{r}^*)$) and consider the observer as a detector of the scalar field; that is,

$$\kappa_p(\boldsymbol{r}^*) = \frac{\kappa |\boldsymbol{r}^* \cdot \hat{\boldsymbol{b}}|}{[(|\boldsymbol{r}^* \times \hat{\boldsymbol{t}})|^2 - (\boldsymbol{r}^*, \hat{\boldsymbol{t}}, \hat{\boldsymbol{z}})^2)/(\boldsymbol{r}^* \cdot \hat{\boldsymbol{z}})^2]^{3/2}}. \tag{2}$$

Note that κ_p takes the same form in both the observer and the object frames. Conceptually, κ_p is a quantity to be observed in the observer frame, but in the object frame it is a scalar field defined in three-dimensional space.

Let $\boldsymbol{c} \triangleq \boldsymbol{r} \times \hat{\boldsymbol{t}} = (c_1, c_2, c_3)$. Let's go back to the observer frame and consider how various vectors project to the image plane and analyze their properties. The plane defined by $\boldsymbol{c}$ intersects the contour on the surface at point P and intersects the image plane along the direction of $\hat{\boldsymbol{t}}_p$, which is the projection of $\hat{\boldsymbol{t}}$ on the image plane (see Figure 1), and is given by

$$\hat{\boldsymbol{t}}_p = \frac{(c_2, -c_1, 0)}{(c_1^2 + c_2^2)^{1/2}}. \tag{3}$$

The orthogonal direction of $\hat{\boldsymbol{t}}_p$ is $\boldsymbol{c}_p$, which is also the projection of the normal of the plane (i.e., $\boldsymbol{c}$) on the image plane, and is given by

$$\boldsymbol{c}_p = (c_1, c_2, 0).$$

Let $\boldsymbol{r}_\perp$ be defined by $\boldsymbol{r}_\perp \triangleq (-c_2, c_1, (c_2 x - c_1 y)/z)$. Then $\boldsymbol{r}_\perp$ is orthogonal to $\boldsymbol{r}$, since $\boldsymbol{r} \cdot \boldsymbol{r}_\perp = 0$ and $\boldsymbol{c}_p$ is orthogonal to $\boldsymbol{r}_\perp$. Hence $\boldsymbol{r}_\perp$ is the normal of the plane spanned by $\boldsymbol{r}$ and $\boldsymbol{c}_p$.

In Appendix B, we show that the change of projected curvature κ_p in the direction of $\boldsymbol{c}_p$ takes the form

$$\nabla \kappa_p \cdot \boldsymbol{c}_p = \kappa_p \frac{\boldsymbol{c}_p \cdot \hat{\boldsymbol{b}}}{\boldsymbol{r} \cdot \hat{\boldsymbol{b}}}. \tag{4}$$

Note that κ_p is not a differentiable function at $\boldsymbol{r} \cdot \hat{\boldsymbol{b}} = 0$, but κ_p^2 is. This is the reason that at $\boldsymbol{r} \cdot \hat{\boldsymbol{b}} = 0$, $\nabla \kappa_p \neq 0$.

Since we want to locate the osculating plane, we should move in a direction that reduces κ_p until eventually reaching it. Since κ is bounded below, this is guaranteed if we can always find the desirable direction at any given point in space. The osculating plane is defined by $\boldsymbol{r} \cdot \hat{\boldsymbol{b}} = 0$ and this plane divides the space outside the object into two regions: $\boldsymbol{r} \cdot \hat{\boldsymbol{b}} < 0$ (region I) and $\boldsymbol{r} \cdot \hat{\boldsymbol{b}} > 0$ (region II). We will show that, in the translation scheme, the observer can deterministically move either in direction $\boldsymbol{c}_p$ or $-\boldsymbol{c}_p$ according to the region the camera is in, in order to reduce κ_p. In particular, we prove the following proposition in Appendix C (see Figures 1 and 2):

Proposition 2. *For a contour on a convex surface, if the observer chooses an* agreeable frame, *the direction of motion that reduces κ_p in the region defined by $\boldsymbol{r} \cdot \hat{\boldsymbol{b}} < 0$ is $\boldsymbol{c}_p$, and the direction of motion in the region where $\boldsymbol{r} \cdot \hat{\boldsymbol{b}} > 0$ is $-\boldsymbol{c}_p$. For a concave surface, the direction is reversed for each of the regions.*

3.2: Rigid Transformation Scheme

During active motion, the observer often needs to move by rotating as well as translating. One reason for this type of motion is to adjust the viewing direction so that the surface point being observed is in the direction normal to the image plane. That is

$$\hat{\boldsymbol{z}} = \frac{\boldsymbol{r}}{|\boldsymbol{r}|}.$$

We consider this case in this section.

Under the above condition, Eq. (1) takes the form

$$\kappa_p = \frac{\kappa |\boldsymbol{r} \cdot \hat{\boldsymbol{b}}|}{(|\boldsymbol{r} \times \hat{\boldsymbol{t}}|^2 / |\boldsymbol{r}|^2)^{3/2}} = \frac{\kappa |\boldsymbol{r} \cdot \hat{\boldsymbol{b}}|}{A^{3/2}},$$

where $A = |\, \boldsymbol{r} \times \hat{\boldsymbol{t}}|^2 / |\, \boldsymbol{r}|^2$. In the object frame, this becomes

$$\kappa_p(\boldsymbol{r}^*) = \frac{\kappa |\, \boldsymbol{r}^* \cdot \hat{\boldsymbol{b}}|}{(|\, \boldsymbol{r}^* \times \hat{\boldsymbol{t}}|^2 / |\, \boldsymbol{r}^*|^2)^{3/2}} = \frac{\kappa |\, \boldsymbol{r}^* \cdot \hat{\boldsymbol{b}}|}{[|\, \boldsymbol{c}^*|^2 / |\, \boldsymbol{r}^*|^2]^{3/2}}. \tag{5}$$

At each new camera position, assume the camera also makes a rotation so that the direction of $\hat{\boldsymbol{z}}$ is coincident with $\boldsymbol{r}$. The value of κ_p, then, depends only on $\boldsymbol{r}$, $\hat{\boldsymbol{t}}$ and $\hat{\boldsymbol{b}}$. Since the gradient vector needs to be computed in order to determine the dependency between the motion direction and the change of κ_p, and the object frame is the only one where we can perform the gradient operation, we have to use Eq. (5) directly in its general component form, that is,

$$\kappa_p = \frac{\kappa |\, \boldsymbol{r}^* \cdot \hat{\boldsymbol{b}}|}{[(c_1^{*2} + c_2^{*2} + c_3^{*2}) / (x^{*2} + y^{*2} + z^{*2})]^{3/2}}. \tag{6}$$

It then can be shown that

$$\nabla \kappa_p = \pm \frac{\kappa}{A^{5/2}} \left[A\, \hat{\boldsymbol{b}} - 3 \frac{(\boldsymbol{r}^* \cdot \hat{\boldsymbol{b}})}{|\, \boldsymbol{r}^*|^2} (\boldsymbol{c}^* \times \hat{\boldsymbol{t}}) - \frac{|\, \boldsymbol{c}^*|^2}{|\, \boldsymbol{r}^*|^2} \boldsymbol{r}^* \right]. \tag{7}$$

Since $\boldsymbol{c}^* \cdot (\boldsymbol{c}^* \times \hat{\boldsymbol{t}}) = 0$ and $\boldsymbol{c}^* \cdot \boldsymbol{r}^* = 0$ we have

$$\nabla \kappa_p \cdot \boldsymbol{c}^* = \pm \frac{\kappa}{A^{3/2}} (\boldsymbol{c}^* \cdot \boldsymbol{b}).$$

Using Eq. (5), we have

$$\nabla \kappa_p \cdot \boldsymbol{c}^* = -\nabla \kappa_p \cdot \boldsymbol{c} = \kappa_p \frac{\boldsymbol{c}^* \cdot \hat{\boldsymbol{b}}}{\boldsymbol{r}^* \cdot \hat{\boldsymbol{b}}}. \tag{8}$$

We can then prove a proposition similar to the one for the translation scheme:

Proposition 3. *For a contour on a convex surface, if the observer chooses its frame so that* $\hat{\boldsymbol{z}} = \boldsymbol{r} / |\, \boldsymbol{r}|$, *the direction of motion that reduces* κ_p *in the region defined by* $\boldsymbol{r} \cdot \hat{\boldsymbol{b}} < 0$ *is* $\boldsymbol{c}$, *and the direction of motion in the region where* $\boldsymbol{r} \cdot \hat{\boldsymbol{b}} > 0$ *is* $-\boldsymbol{c}$. *For a concave surface, the direction is reversed in each region.*

Note that in the observer frame, $\boldsymbol{c}$ is actually orthogonal to $\hat{\boldsymbol{z}}$ since $\hat{\boldsymbol{z}}$ and $\boldsymbol{r}$ are coincident. Hence $\boldsymbol{c} = \boldsymbol{c}_p$ and the above two propositions become identical. The only difference is that for the translation scheme $\boldsymbol{c}_p$ is always on the same plane, while in the rigid transformation scheme $\boldsymbol{c}$ translates and rotates as the observer approaches the osculating plane.

3.3: Discussion

We have shown that the observer can always move deterministically in either region outside the object surface to reach the osculating plane for a given marked point on the object surface. The direction of movement ($\boldsymbol{c}_p$) is always orthogonal to the projected tangent ($\boldsymbol{t}_p$). In the case of pure translation, the observer always moves within the established image plane and the point where it reaches the osculating plane will be on the intersecting line of the image plane and the osculating plane. Hence the length of path from the initial location to the osculating plane is determined by the initial viewing direction. Furthermore, without an external reference, it is very difficult to verify the motion being purely translational. This is not the case for the rigid transformation scheme since the surface mark itself is the external reference. The observer also has better control over the path

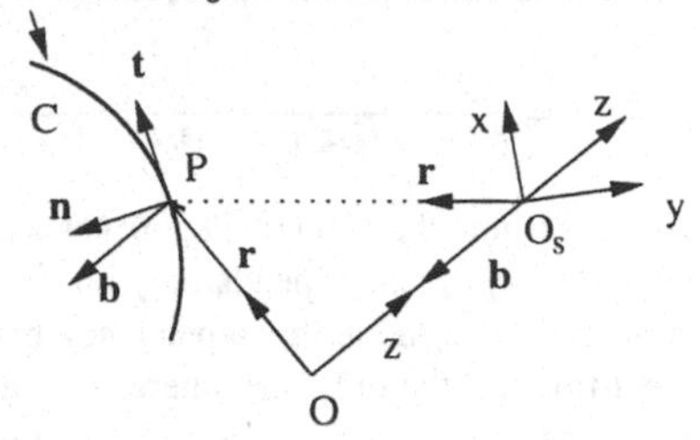

Figure 3. Recovery of Frenet vectors when moving away from the osculating plane along the binormal direction.

from the initial position to the osculating plane because the viewing direction can be guided by the reference rather than arbitrarily chosen. This difference is shown in Section 6.

On the other hand, when there are no well-defined stationary marks on the contour, pointwise correspondence across observations becomes a problem and the rigid transformation scheme cannot be used. However if the observer motion can be assured to be translational only (by external reference, for example), the plane formed by the initial observation direction ($\boldsymbol{r}$) and the direction of movement ($\boldsymbol{c}_p$) will intersect the object surface at the point P. In this case, the translation scheme will be the only one applicable.

Once in osculating plane the surface binormal $\hat{\boldsymbol{b}}$ can be recovered by

$$\hat{\boldsymbol{b}} = \pm \frac{\boldsymbol{r} \times \hat{\boldsymbol{t}}_p}{|\boldsymbol{r} \times \hat{\boldsymbol{t}}_p|}. \tag{9}$$

The sign of the above expression is determined by the local shape of the surface along the contour (negative for convex and positive for concave).

Propositions 2 and 3 also provide a way to determine the shape of the surface along the contour qualitatively, i.e., if it is a convex or concave surface strip [5, 10]. For example, if the projection of the contour is convex to the right (open to the left) and moving right decreases κ_p, then the surface strip is convex; otherwise it is concave.

Next we show how the rest of the Frenet frame can be recovered.

4: Frenet Frame Recovery

Once we have found the osculating plane, the recovery of the rest of the Frenet frame becomes possible. Consider Figure 3.

4.1: Curvature

If the observer chooses its frame so that $\hat{\boldsymbol{z}} = -\hat{\boldsymbol{b}}$, then as long as the observer translates only along $\hat{\boldsymbol{b}}$, we always have $\hat{\boldsymbol{t}} = (x', y', 0)$ and $\boldsymbol{c} = \boldsymbol{r} \times \hat{\boldsymbol{t}} = (-yz', zx', 0)$. Hence Eq. (15) becomes

$$\kappa_p = \frac{\kappa |\boldsymbol{r} \cdot \hat{\boldsymbol{z}}|}{(x'^2 + y'^2)^{3/2}} = \frac{\kappa |\boldsymbol{r} \cdot \hat{\boldsymbol{z}}|}{|\hat{\boldsymbol{t}}|^3} = \kappa |\boldsymbol{r} \cdot \hat{\boldsymbol{z}}|. \tag{10}$$

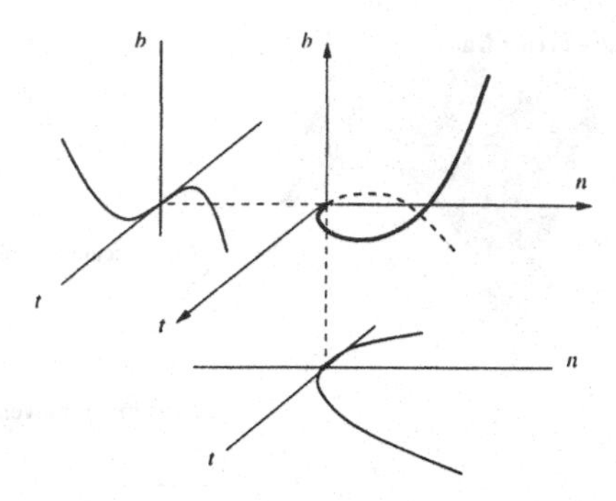

Figure 4. Curve projection onto planes defined by the Frenet frame.

Consequently, if the observer moves in the direction along $\hat{b}$ a distance d, the contour curvature κ at P is

$$\kappa = \frac{\kappa_p}{d}.$$

In this process, the exact measurement system used by the observer to measure the distance d is not important, as long as the projected curvature κ_p is measured against the same system. However, if the contour curvature κ is used to recover surface shape (see below), the measurement system has to be consistent with the metric used for the surface.

It should be noted that the requirement that the observer translate strictly along $\hat{b}$ implicitly assumes the existence of some external references. In another words, this action cannot be "intrinsic" and some kind of external reference must be used to accomplish this motion.

4.2: Tangent and Normal

From Eq. (3) we get

$$\hat{t}_p = \frac{(c_2, -c_1, 0)}{(c_1^2 + c_2^2)^{1/2}} = (x', y', 0). \tag{11}$$

Hence the components of $\hat{t}$ are identical to the components of $\hat{t}_p$ in the selected observer frame. This result gives us the tangent. Finally, from $\hat{n} = \hat{b} \times \hat{t}$ we get the last member of the Frenet frame, the normal.

Hence the Frenet frame can be recovered from the deformation of observables (the curvature of the projected contour) without knowing the depth z, which can be recovered from the triangulation of O_sOP in Figure 3 if external reference points can be found. Next we show that all the geometric metrics for the space curve C can be recovered.

4.3: Torsion

From the Serret-Frenet equations (Eq. 1), if the coordinate system is chosen to be the Frenet frame (i.e., $\hat{x}= \hat{t},\ \hat{y}= \hat{n},\ \hat{z}= \hat{b}$), we can derive the *local canonical form* of C :

$$\begin{aligned} x(s) &= s - \frac{\kappa^2 s^3}{6} + R_x \\ y(s) &= \frac{\kappa s^2}{2} + \frac{\kappa' s^3}{6} + R_y \end{aligned}$$

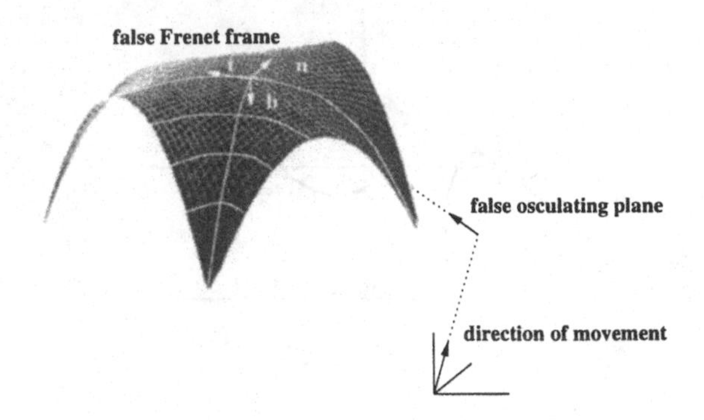

Figure 5. An occluding contour that appears to be a stationary contour.

$$z(s) = -\frac{\kappa \tau s^3}{6} + R_z.$$

The projections of C onto the $\hat{t} - \hat{n}$ and $\hat{t} - \hat{b}$ planes are shown in Figure 4. As can be seen in the equation, the local form of C on the $\hat{t} - \hat{b}$ plane is cubic and by estimating the third-order slope across the origin we can compute τ. This completely determines curve C at the point.

It should be noted that we cannot actually use the chosen $\hat{z}$ as our observer frame since the requirement $\hat{z} = \hat{b}$ will generally put the observer at a side view of the surface where the stationary contour coincides with the occluding contour. On the other hand, if we choose $\hat{z} = \hat{n}$ then the slope across the $\hat{x} - \hat{y}$ (i.e., $\hat{b} - \hat{t}$) plane is zero (i.e., $\kappa_p = 0$), making the estimation of τ unreliable. However, if the goal is to recover the surface geometry of the object, then recovering the curvature and the Frenet frame (actually $\hat{t}$) is sufficient.

5: Applications

5.1: Distinguishing Stationary Contours From Occluding Contours

There are qualitative differences between the deformations of a stationary and an occluding contour when the observer can move in controlled ways [11]. These differences enable us to discriminate between these two types of contours without first recovering the surface shape.

The deformation of a stationary contour occurs for two reasons. The first kind of deformation occurs during the process of locating the osculating plane, when the contour, as projected onto the image plane, deforms according to Propositions 2 and 3. The second kind occurs when the observer moves along the binormal $\hat{b}$ after the osculating plane is identified. In the second form, the contour does not actually deform locally (see Eq. (11)). There could be a surface and a motion that makes the deformation of the occluding contour behave like a stationary contour in the first form (see Figure 5). However, *all* occluding contours must deform when the observer moves along the (false) binormal because this direction is actually toward the object surface (see Eq. (9) and Figure 5).

In the previous sections we handled the case where there are no other markings on the surface. That is, there is only one contour with a marked point in the region of interest. In most practical cases, additional contours, texture or surface markings will make the task much easier. Nonetheless, we show that in this worst case, the deformation of projected curvature, κ_p, and the Frenet frame carry with them *intrinsic* information that allows us to distinguish stationary contours from other non-stationary ones.

5.2: Surface Shape Recovery From Multiple Contours

For a given local parameterization, the six parameters in the first and second fundamental forms that satisfy the three *compatibility equations* [6] completely determine the surface shape up to a rigid transformation. This is the fundamental theorem of the local theory of surfaces. Hence, there are three degrees of freedom that need to be fixed in order to determine the local surface shape. Since the value of the second fundamental form along a given direction equals the normal curvature of the surface along that direction, two intersecting stationary contours provide a local parameterization and two constraints for the three degrees of freedom. In this section, we consider how the third constraint can be obtained.

5.2.1: Surface Shape From Principle Curvature

From the differential theory of surface geometry, the two principle directions where normal curvature for the surface reaches extrema are orthogonal to each other. Any normal curvature κ_n at the point relates to these two extremal curvatures by *Euler's formula:*

$$\kappa_n = \kappa_1 \cos^2\theta + \kappa_2 \sin^2\theta,$$

where κ_1 and κ_2 are the two extreme curvatures and θ is the angle between the tangent for κ_n and one of the principle directions. If one of the stationary contours is along the principle direction, we can solve for both principle directions and the associated extreme curvatures. When neither of the two contours is along a principle direction but one of the principle directions can be determined by other means, the two Euler equations relating $\kappa_1, \kappa_2, \theta$ and the two known normal curvatures along the two contours enable us to solve the surface geometry completely.

Since principle directions are directions where the normal section of the surface is maximally or minimally curved, we need an active procedure capable of inspecting all directions around the surface point. *Shape from occluding contour* methods work well when the surface is elliptically convex in the neighborhood of the point, but fail otherwise.

5.2.2: Surface Shape From Another Contour

If we cannot determine any of the principle directions by examining the surface visually, they can still be determined algebraically. The result can be exact if there is another stationary contour passing through the intersection point where the two known contours intersect. In most cases, the shape can be estimated when additional contours pass through in the vicinity of the point. In the first case, we have three Euler equations to solve all required parameters. In the second case, since the third contour will intersect at least one of the first two contours, we can *parallel transport* [6, 10] the tangent and curvature along this third contour from the additional intersection point to the first intersection point and then solve the set of three Euler equations. The accuracy of this parallel transport depends on the curvature of the surface along which we make the parallel transport. For a locally cylindrical surface the result will be exact.

5.2.3: Mesh Representation

Stationary contours as visual cues are most effective in "meshed" representations of surfaces. For example, a meshed representation of a synthetic surface is shown in Figure 7. Intersecting lines with zero projected curvature along a particular direction give cues of a locally flat surface in that direction. Consequently, by assuming the surface is locally parabolic or cylindrical in one direction,

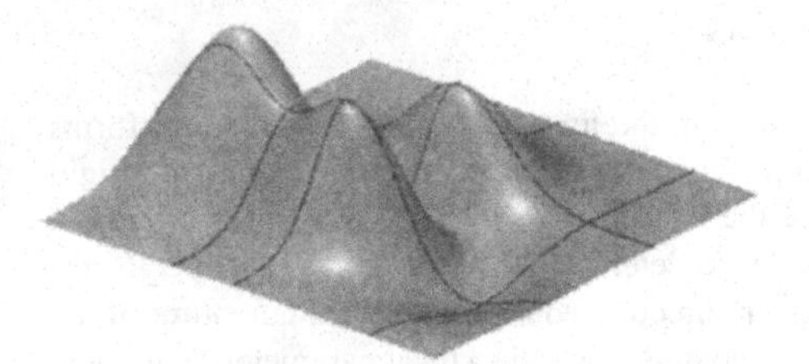

Figure 6. A synthetic surface with stationary contours.

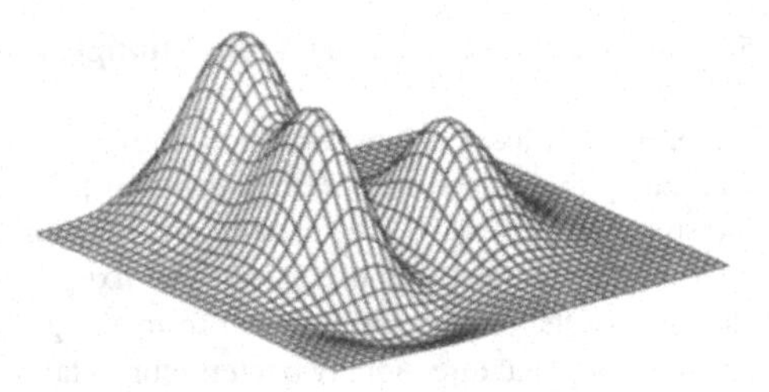

Figure 7. Mesh representation of the surface.

the mesh lines in the orthogonal direction provide the needed deformation as cues of surface shape in that direction. This is achieved through the implicit assumption that the orthogonal direction has constant curvature and the mesh lines in that direction provide a deformed sequence exactly as if the observer moves along that direction and observes the deformation of a stationary contour.

6: Experimental Results

We use both a synthetic surface and a ray-traced vase to show results of how the theory can be implemented. The surface in Figure 6 is populated with peaks, valleys and saddles and the dark contours on the surface are stationary contours on the surface and will be used by the observer to recover surface shape at the intersection points.

The recovery process involves camera motions to both the intersecting contours and, for each contour, recovering the tangent $\hat{t}$ and the curvature κ. From these two tangents, $\hat{t}_1$ and $\hat{t}_2$, the surface normal can be found as

$$\hat{N} = \frac{\hat{t}_1 \times \hat{t}_2}{|\hat{t}_1 \times \hat{t}_2|}.$$

The normal curvature κ_n of the surface along $\hat{t}_1$ and $\hat{t}_2$ can be recovered by applying the formula

$$\kappa_n = \kappa\, \hat{n} \cdot \hat{N},$$

where $\hat{n}$ is the normal vector in the Frenet frame for the contours.

Since we did not attempt to find the directions of the principle curvatures, it is assumed that these two contours are actually in the principle directions. Under this assumption, all six parameters of the first and second fundamental forms can be computed [6] and we can parameterize the local surface by using the two tangents as two basis vectors and the intersecting point as the origin. To illustrate, we overlay on the surface members in the family of quadratic functions that have the same parameters at the point in Figure 8. This is accomplished by using a quadratic Monge patch parameterization $(u, v, h(u, v))$ and solving for all the coefficients using the parameters from the two fundamental forms. The resulting functions are translated to the surface point and rotated according to the recovered surface normal $\hat{N}$. Locally, the quadratic functions exactly match the surface. Globally, the degree of fit will be determined by the variation of the normal curvatures along the principle curves passing by the surface point.

In Figure 9 we show the paths taken by the camera in the translation and the rigid transformation schemes from the same initial location. The scene is set up so that $\hat{z}$ is in the upward direction and $\{\hat{x}, \hat{y}, \hat{z}\}$ forms a right-handed system. At the surface point being tracked, the stationary contour

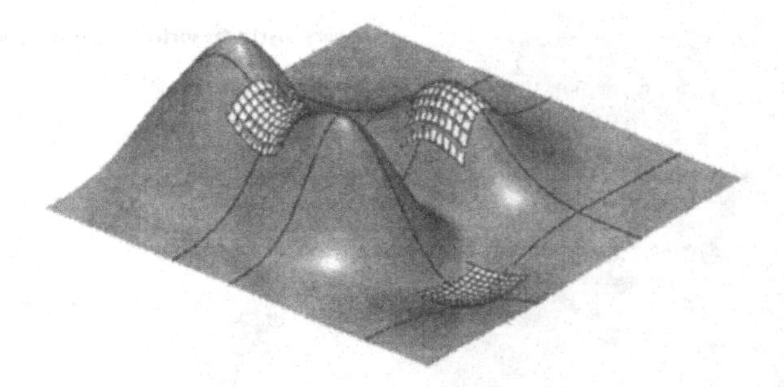

Figure 8. Synthetic surface with recovered elliptic and hyperbolic surface patches.

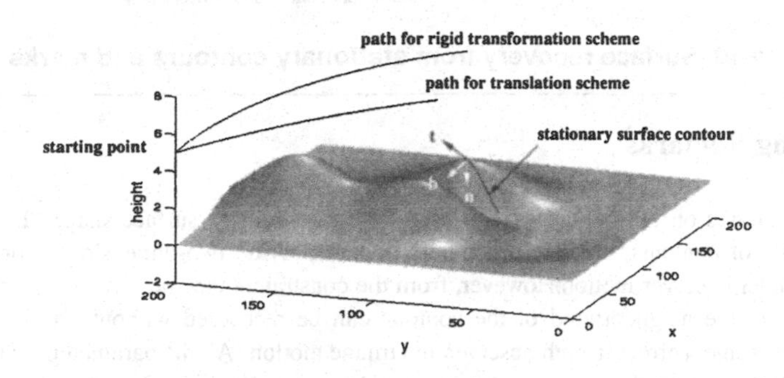

Figure 9. Paths produced by the translation scheme and the rigid transformation scheme.

moves in the direction of $\hat{t}$. The image plane for the translation scheme is arbitrarily set to be in the $-\hat{z}$ direction.

It can be seen that the path taken by the translation scheme is parallel to the x-y plane and intersects the osculating plane horizontally, i.e., the observer moves completely within the image plane. On the other hand, the rigid transformation scheme takes a more curved path because of the gradual turning of the viewing direction. In this particular case, the translation scheme actually reaches the osculating plane faster because of the position of the image plane. If the observer can ensure that its movement is purely translational by referring to external references, the path will be the shortest of the two schemes in all cases. However, this shortest path will bring the observer directly to the surface, i.e., the path intersects the osculating plane at the surface. This may not desirable in practical cases.

A more realistic example is shown in Figure 10 in which various stationary contours on a vase are tracked and the local surface shape at intersecting stationary marks is recovered. Complete shape recovery is possible only at these stationary marks. For points along a contour between two marks, interpolation is used to estimate the unknown curvature in the direction orthogonal to the contour.

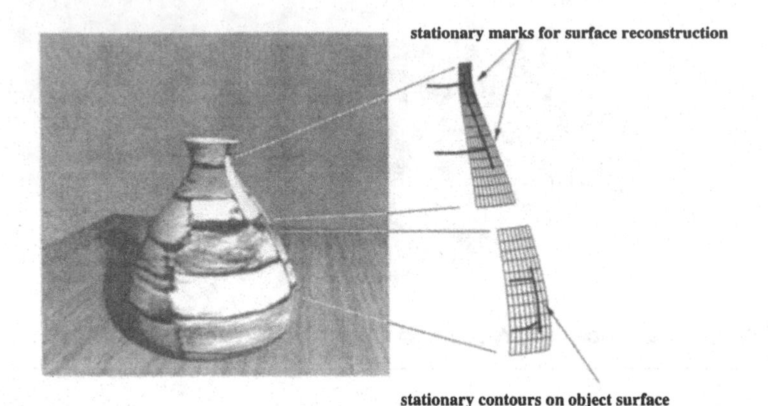

Figure 10. Surface recovery from stationary contours and marks.

7: Concluding Remarks

Curvilinear features on object surfaces are useful for constraining surface shape. In contrast to some other kinds of contours, stationary contours do not provide two-dimensional constraints on surface shape under observer motion.However, from the constraints they do provide, we have shown that the surface in the neighborhood of the contour can be recovered without knowing or being sensitive to measurement errors in both observer and image motion. All the parameters characterizing a stationary contour as a spatial curve can be recovered by controlled movement of the observer. In the process, the problem of discriminating between occluding contours and stationary contours is also solved. Another major contribution of this paper is the analysis leading to a method an observer can use to reach the osculating plane. Depending on whether external references other than the surface point are available or not, two different strategies for observer motion were presented. They perform differently but both enable the observer to move in directions that monotonically decrease the projected contour curvature on the image plane.

Contrary to the belief that stationary contours are only useful for acquiring qualitative surface information, we demonstrated that curvilinear features on a surface can be considered as "samples" of surface shape and as long as there are enough features in a region of interest, the shape can be recovered quite accurately. Suggestions are given concerning how to acquire information about the principle directions so that two intersecting surface contours can be used to constrain the local surface completely. Error analysis for the case when we can parallel transport a nearby point to the intersection point in order to solve the surface geometry is an important subject for future study.

References

[1] H. Barrow and J. Tenenbaum. Recovering intrinsic scene characteristics from images. In A. Hanson and E. Riseman, editors, *Computer Vision Systems*, pages 3–26. Academic Press, New York, 1978.

[2] M. Brady, J. Ponce, A. Yuille, and H. Asada. Describing surfaces. *Computer Vision, Graphics, and Image Processing*, 32:1–28, 1985.

[3] R. Cipolla. *Active Visual Inference of Surface Shape*. Springer-Verlag, Heidelberg, Germany, 1996.

[4] R. Cipolla and A. Blake. Surface shape from the deformation of apparent contours. *Int. J. Computer Vision*, 9(2):83–112, 1992.

[5] R. Cipolla and A. Zisserman. Qualitative surface shape from deformation of image curves. *Int. J. Computer Vision*, 8(1):53–69, 1992.

[6] M. P. do Carmo. *Differential Geometry of Curves and Surfaces*. Prentice-Hall, Englewood Cliffs, NJ, 1976.

[7] P. Giblin and R. Weiss. Reconstruction of surfaces from profiles. In *Proc. 1st Int. Conf. on Computer Vision*, pages 136–144, 1987.

[8] M. Kass, A. Witkin, and D. Terzopoulos. Snakes: active contour models. In *Proc. 1st Int. Conf. on Computer Vision*, pages 259–268, 1987.

[9] J. Koenderink. What does the occluding contour tell us about solid shape? *Perception*, 13:321–330, 1984.

[10] J. Koenderink. *Solid Shape*. MIT Press, Cambridge, MA, 1990.

[11] K. N. Kutulakos and C. R. Dyer. Occluding contour detection using affine invariants and purposive viewpoint control. In *Computer Vision and Pattern Recognition*, pages 323–330, 1994.

[12] K A. Stevens. The visual interpretation of surface contours. *Artifical Intelligence*, 17:47–73, 1981.

[13] R. Vaillant and O. Faugeras. Using extremal boundaries for 3-d object modeling. *IEEE Trans. Patt. Anal. Machine Intell.*, 14(2):157–173, 1992.

[14] A. Zisserman, A. Black, C. Rothwell, L. Van Gool, and M. Van Diest. Eliciting qualitative structure from image curve deformations. In *Proc. 4th Int. Conf. on Computer Vision*, pages 340–345, 1993.

Appendix A: Projected Curvature in the Observer Frame

Let the curve C on object surface be parameterized by its curve length s as $(x(s), y(s), z(s))$. The projection of C onto the image plane is a 2D curve C_p parameterized by $(\xi(t), \eta(t))$. From the imaging model for projection we have the standard projective equations:

$$\xi(s) = \frac{x(s)}{z(s)}, \qquad \eta(s) = \frac{y(s)}{z(s)}. \tag{12}$$

Since the natural parameter s of C becomes a general parameter t of C_p, the projected curvature κ_p of C_p is:

$$\kappa_p = \frac{|\xi'\eta'' - \xi''\eta'|}{[(\xi')^2 + (\eta')^2]^{3/2}}. \tag{13}$$

Because C is parameterized by curve length, from the definition of the Frenet frame we have:

$$\begin{aligned} \hat{t} &= (x', y', z') \\ \hat{n} &= \frac{1}{\kappa}(x'', y'', z'') \\ \hat{b} &= \hat{t} \times \hat{n} = \frac{1}{\kappa}(y'z'' - z'y'', z'x'' - x'z'', x'y'' - y'x'') = (b_1, b_2, b_3), \end{aligned} \tag{14}$$

where κ is the curvature of C at the point P (represented by $\boldsymbol{r}$ in the observer frame). Let $\boldsymbol{c} \triangleq \boldsymbol{r} \times \hat{t} = (c_1, c_2, c_3)$. Substituting the differentials in Eq. (12) into Eq. (13) and using Eq. (14), we have

$$\kappa_p = \frac{|c_1 x'' + c_2 y'' + c_3 z''|}{[(c_1^2 + c_2^2)/z^2]^{3/2}}. \tag{15}$$

The quantity $(c_1^2 + c_2^2)$ is the length of $\boldsymbol{c}_p$, where $\boldsymbol{c}_p$ is the projection of $\boldsymbol{c}$ onto the image plane and z is the component of $\boldsymbol{r}$ in the $\hat{z}$ direction. Using the component form of the cross product $\boldsymbol{c} = \boldsymbol{r} \times \hat{t}$ and Eq. (14) we can derive

$$c_1 x'' + c_2 y'' + c_3 z'' = \kappa\, \boldsymbol{r} \cdot \hat{b}. \tag{16}$$

The denominator of Eq. (15) can be rewritten as

$$\frac{(c_1^2+c_2^2)}{z^2}=\frac{|\,c|^2-(c\cdot\,\hat{z})}{(r\cdot\,\hat{z})^2}=\frac{|\,r\times\,\hat{t}|^2-(r,\,\hat{t},\,\hat{z})^2}{(r\cdot\,\hat{z})^2}, \tag{17}$$

where $(r,\ \hat{t},\ \hat{z})$ is shorthand for $r\times\,\hat{t}\cdot\,\hat{z}$. Hence Eq. (15) takes the vector form:

$$\kappa_p=\frac{\kappa|\,r\cdot\,\hat{b}|}{[(|\,r\times\,\hat{t})|^2-(r,\,\hat{t},\,\hat{z})^2)/(r\cdot\,\hat{z})^2]^{3/2}}. \tag{18}$$

Appendix B: Projected Curvature Gradient in the Object Frame

The scalar field $\kappa_p(r^*)$ given by Eq. (2) has steepest rate of change along the direction of its gradient, $\nabla\kappa_p$, and the change in an arbitrary direction r is given by $\nabla\kappa_p\cdot\,r$. Let

$$A=\frac{|\,r^*\times\,\hat{t}|^2-(r^*,\,\hat{t},\,\hat{z})^2}{(r^*\cdot\,\hat{z})^2}.$$

Then the gradient of κ_p in the object frame $\nabla\kappa_p(r^*)$ takes the form

$$\nabla\kappa_p(r^*)=\pm\frac{\kappa}{A^{5/2}}\left[A\,\hat{b}+3(r^*\cdot\,\hat{b})\frac{z'}{z^{*2}}(c_2^*,-c_1^*,-\frac{c_2^*x^*-c_1^*y^*}{z^*})\right]. \tag{19}$$

Note that the sign of the expression depends on the sign of $(r^*\cdot\,\hat{b})$ and that z' is the third component of $\hat{t}$, which is identical in both frames. Define

$$r_\perp^*\triangleq(c_2^*,-c_1^*,-\frac{c_2^*x^*-c_1^*y^*}{z^*}). \tag{20}$$

which, by its form, denotes a vector orthogonal to r^* since $r^*\cdot\,r_\perp^*=0$. Eq. (19) can then be expressed as

$$\nabla\kappa_p(r^*)=\pm\frac{\kappa}{A^{5/2}}\left[A\,\hat{b}+3(r^*\cdot\,\hat{b})\frac{z'}{z^{*2}}r_\perp^*\right]. \tag{21}$$

Since $r^*=-r$ and $r_\perp^*=-r_\perp$ we have $\nabla\kappa_p(r)=\nabla\kappa_p(r^*)$. This expresses the fact that the relative translational motion of object and observer is indistinguishable. But this does not carry over to rotational motion (see next section). Now let's consider the way κ_p changes along c_p^*, i.e., consider the expression $\nabla\kappa_p\cdot\,c_p^*$. Since $c_p^*\cdot\,r_\perp^*=0$ we have

$$\nabla\kappa_p\cdot\,c_p^*=\pm\frac{\kappa}{A^{3/2}}\,c_p^*\cdot\,\hat{b}.$$

Furthermore, using Eq. (1) and $c_p^*=-c_p$, $r^*=-r$, we have

$$\nabla\kappa_p\cdot\,c_p^*=-\nabla\kappa_p\cdot\,c_p=\kappa_p\frac{c_p^*\cdot\,\hat{b}}{r^*\cdot\,\hat{b}}=\kappa_p\frac{c_p\cdot\,\hat{b}}{r\cdot\,\hat{b}}. \tag{22}$$

Appendix C: Proof of Proposition 2

Proof. The space outside a convex surface is defined by $\boldsymbol{r} \cdot \hat{\boldsymbol{n}} > 0$. Consider region I where $\boldsymbol{r} \cdot \hat{\boldsymbol{b}} < 0$. Since $\boldsymbol{c} = \boldsymbol{r} \times \hat{\boldsymbol{t}}$ and $\hat{\boldsymbol{t}} = \hat{\boldsymbol{n}} \times \hat{\boldsymbol{b}}$, we have

$$\boldsymbol{c} = (\boldsymbol{r} \cdot \hat{\boldsymbol{b}})\, \hat{\boldsymbol{n}} - (\boldsymbol{r} \cdot \hat{\boldsymbol{n}})\, \hat{\boldsymbol{b}}. \tag{23}$$

The vector $\boldsymbol{c}_p$ is the projection of $\boldsymbol{c}$ onto the image plane and they are related through $\boldsymbol{c} = \boldsymbol{c}_p + (\boldsymbol{c} \cdot \hat{\boldsymbol{z}})\, \hat{\boldsymbol{z}}$. Applying Eq. (23), we have

$$\boldsymbol{c} = \boldsymbol{c}_p + \left[(\boldsymbol{r} \cdot \hat{\boldsymbol{b}})(\hat{\boldsymbol{n}} \cdot \hat{\boldsymbol{z}}) - (\boldsymbol{r} \cdot \hat{\boldsymbol{n}})(\hat{\boldsymbol{b}} \cdot \hat{\boldsymbol{z}})\right] \hat{\boldsymbol{z}}. \tag{24}$$

Hence

$$\begin{aligned}
\boldsymbol{c}_p \cdot \hat{\boldsymbol{b}} &= \boldsymbol{c} \cdot \hat{\boldsymbol{b}} - \left[(\boldsymbol{r} \cdot \hat{\boldsymbol{b}})(\hat{\boldsymbol{z}} \cdot \hat{\boldsymbol{n}}) - (\boldsymbol{r} \cdot \hat{\boldsymbol{n}})(\hat{\boldsymbol{z}} \cdot \hat{\boldsymbol{b}})\right](\hat{\boldsymbol{z}} \cdot \hat{\boldsymbol{b}}) \\
&= -(\boldsymbol{r} \cdot \hat{\boldsymbol{n}}) - (\boldsymbol{r} \cdot \hat{\boldsymbol{b}})(\hat{\boldsymbol{z}} \cdot \hat{\boldsymbol{n}})(\hat{\boldsymbol{z}} \cdot \hat{\boldsymbol{b}}) + (\boldsymbol{r} \cdot \hat{\boldsymbol{n}})(\hat{\boldsymbol{z}} \cdot \hat{\boldsymbol{b}})^2 \\
&= (\boldsymbol{r} \cdot \hat{\boldsymbol{n}})\left[(\hat{\boldsymbol{z}} \cdot \hat{\boldsymbol{b}})^2 - 1\right] - (\boldsymbol{r} \cdot \hat{\boldsymbol{b}})(\hat{\boldsymbol{z}} \cdot \hat{\boldsymbol{n}})(\hat{\boldsymbol{z}} \cdot \hat{\boldsymbol{b}}) \\
&\triangleq \alpha - \beta.
\end{aligned}$$

From Cauchy's inequality, we have

$$\hat{\boldsymbol{z}} \cdot \hat{\boldsymbol{b}} \leq |\hat{\boldsymbol{z}}||\hat{\boldsymbol{b}}| = 1.$$

Since the observer is in region I and observes a convex surface from an agreeable frame, we have $\boldsymbol{r} \cdot \hat{\boldsymbol{n}} > 0$ (convex), $\boldsymbol{r} \cdot \hat{\boldsymbol{b}} < 0$ (region I), and $\hat{\boldsymbol{z}} \cdot \hat{\boldsymbol{n}} > 0$, $\hat{\boldsymbol{z}} \cdot \hat{\boldsymbol{b}} < 0$ (agreeable frame). It follows that $\alpha < 0$ (the first part of the equation) and $\beta > 0$ (the second part of the equation). Hence $\boldsymbol{c}_p \cdot \hat{\boldsymbol{b}} < 0$. Consequently, for the observer in the agreeable observer frame, the change in κ_p in the direction of $\boldsymbol{c}_p$ is

$$\nabla \kappa_p \cdot \boldsymbol{c}_p = -\kappa_p \frac{\boldsymbol{c}_p \cdot \hat{\boldsymbol{b}}}{\boldsymbol{r} \cdot \hat{\boldsymbol{b}}},$$

which is always negative. Similar arguments hold for region II and for concave surfaces. □

Section 4

Object Recognition

Object Recognition Research: Matched Filtering becomes Bayesian Reasoning

Robert Hummel
Courant Institute of Mathematical Sciences
New York University

Abstract

One of the first things we learn in image processing is about matched filtering. The author learned about matched filtering from Azriel Rosenfeld, and also about convolutions, Fourier transforms, and many other foundational techniques. One of the goals of computer vision research, as practiced by the author and legions of students of Rosenfeld, has been to discover more sophisticated and robust methods for object recognition and image analysis. We begin by discussing some of that computer vision research and pattern matching, and the need for robustness. However, we view a large portion of computer vision as matched filtering, suitably embellished, often disguised. A number of example methods are shown to be very much related to some form of matched filtering. Typically, matched filtering arises when problems are formulated as L^2 optimization problems. We then consider a formulation of pattern matching as a Bayesian maximum likelihood computation, as opposed to a least square norm minimization. In this formulation, we discover that certain weighted functions that would ordinarily be invoked heuristically in order to account for noise and statistical variation can be derived and given precise meaning. We argue that the formulas that are derived have heuristically the right behavior, but will provide better performance than other heuristic formulas when applied to real applications.

1: Introduction

Computer vision research is beginning to show promise for commercialization. Small demonstration projects in research labs, particularly in robotics, manufacturing, and parts inspection, demonstrate capabilities that seemed unlikely a decade ago (and yet were promised to be around the corner two decades ago). Aerospace contractors are beginning to develop production automatic target recognition systems, and renewed interest in countering armored threats in regional conflicts has brought an infusion of research funding to image processing for target recognition. Now, imagery is pervasive in medicine, and medical image processing applications show great promise for improved health care.

It is now realistic to expect the performance of a vision system to provide practical, commercializable applications, and that the performance of systems be quantifiably related to the parameters, scenarios, and inputs (e.g., the sensors) in the system. The demonstrations must be extensible, and the benefits must be quantifiable. Marketing, support, modifications, and product life-cycles will be new challenges in computer vision applications. The success of moving a fundamental research field to this level of accomplishment is largely a testament to the perseverance of the founders of the field, and new dedication will

 0-8186-7644-2 $5.00

be required by subsequent researchers to respond to the new needs and new environment, largely influenced by the success of the original developers.

There is good reason to believe that narrowly-defined applications with sizable markets are finally amenable to computer vision technology. It is not so much that the techniques developed over the years have reached fruition, although in many cases they have, but rather that processor speeds, DSP technology, sensors, and reduced prices make the economics of simple techniques more viable. Further, price reductions have been dramatic. Many researchers recall using $50,000 to $100,000 attached image processing systems, and now more typically make use of favor of $5,000 SCSI image digitizers, while others are beginning to look at the $500 camera systems that can be attached to workstations, with the intention that demonstration systems use commonly available inexpensive hardware.

Are we prepared for the transition of computer vision research to computer vision marketing? Alas, academic researchers have little appreciation of the subtleties of the marketplace, let alone the range of applications and requirements of the customers. In the new world, where DSP chips can easily achieve 700+ MIPS, the goal is not sophisticated algorithms but rather simple systems that work well and fail gracefully and are easily adapted to new environments.

The methodologies that researchers utilize in order to develop demonstrations will need to be modified to account for the new goals. In particular, it is critical that the performance of a system be demonstrated over a range of potential inputs, and that robustness of a system form a critical component of its evaluation. The results are more important than the sophistication of the methods.

The remainder of this paper is more technical, but related to these sentiments. Our focus is on object recognition. The comments that follow argue generally for the use of Bayesian methods, in the context of pattern matching, for model-based vision applications. We assert that Bayesian methods lead to well-justified formulas, and reasonable formulations. The principle alternative, which has been a mainstay of computer vision algorithm development, is based on optimization, and generally leads to some form of matched filtering. Our thesis is that the two are not so different. Our point is that after years of research and considerable success, we are still filtering (after all these years). But that's not bad, if we use representations where the filters are more robust.

We begin with a discussion of classical matched filtering.

2: Matched Filtering

By matched filtering, we mean simply an inner product. So if $f(x, y)$ is an image, and $m(x, y)$ is a model or pattern that is being sought, we compute

$$\int f(x,y) \cdot m(x,y)dxdy$$

in order to determine the degree of match. If multiple patterns are sought, say $m_i(x, y)$, for $i = 1, \cdots n$, then n inner products are computed, and the largest wins.

Beginning with optical character recognition, and continuing with optical computing and target recognition, matched filtering holds an important appeal and promise. The technique is easily implemented, and has mathematical justification. Specifically, if one wants to find the best match in terms of the L^2 norm, then the minimum of $\|f - m_i\|^2$ is equivalent to maximizing $\langle f, m_i\rangle - (1/2)\|m_i\|^2$. If all the models have the same norm, then we have derived matched filtering, and if the models have different norms, then we have good justification for a minor modification to matched filtering.

Alas, it is doctrine in computer vision, and to a lesser extent in pattern recognition, that matched filtering does not work. Unless the number of models is minimal, noise and variability kills the technique. The doctrine is re-learned by successive generations of researchers, and is verified empirically and by theoretical analysis.

But, as we will see, variations of matched filtering reappear in many guises. The methodology of computer vision research seems to dictate that matched filtering form the key component of the matching engine, providing it is sufficiently embellished. The performance of these systems, as we all well know, is adequate for the examples that are published with the papers, but falls apart when the method is extended to more models or more complicated systems. The problem of dealing with noise remains the key difficulty in the application of computer vision technology. The source of the problem is that matched filtering can't deal with noise effectively, which has been known for decades.

At the same time, Bayesian techniques show great promise in alleviating problems with noise and inadequate matching engines. The difficulty with Bayesian techniques lie in the representation of the information that is to be analyzed. The representation is critical, but is not dictated by any theory. Accordingly, Bayesian networks or Bayesian reasoning systems have to be developed for each application, often in ad-hoc ways. A better methodology is required.

In the sections to follow, we will not solve the problem of formulation the representation in order to utilize Bayesian reasoning to break the log-jam of matched filtering. However, we will show how matched filtering itself can be reformulated and modified so as to admit a Bayesian interpretation, which should assist in providing new methodologies that lead to more robust performance.

3: Other guises for matched filtering

Matched filtering can be formulated as the task of maximizing a collection of vector dot products. If $\vec{x}$ is a vector representing the ordered pixel values, and $\{\vec{x}_i\}_{i=1}^n$ is a collection of target vectors, then the problem is to find the index that maximizes $\vec{x} \cdot \vec{x}_i$ over all possible i. Since the $\vec{x}_i$ will typically contain multiple translates of the same prototype pattern, many of the vector dot products can be efficiently implemented as convolutions. Since the $\vec{x}_i$ are not orthogonal, there can be considerable cross-talk. The vector $\vec{x}$ of pixel values may represent the result of an edge detector or other filter of raw sensor data, and may also incorporate multispectral and multisensor information.

Next, let us consider statistical pattern recognition. Although there are many forms, let us first consider the K-nearest neighbor classifier, with K equal to one. In this classifier, a feature vector $\vec{x}$ locates the prototype that is nearest, i.e., that minimizes $||\vec{x} - \vec{x}_i||^2$ over i. This is equivalent to maximizing

$$\vec{x} \cdot \vec{x}_i - \frac{1}{2}||\vec{x}_i||^2,$$

which we may write as $(\vec{x}, 1) \cdot (\vec{x}_i, b_i)$, where $b_i = -(1/2)||\vec{x}_i||^2$. Thus we once again get a vector dot product maximization, but in this case, the vector represents the feature values of a region of interest with an appended component that is always one, and the matched filters encode the prototype feature vectors, with an appended bias term.

What about classifiers with K greater than one? Perhaps there is something essential in better pattern recognition methods and more sophisticated classifiers that relieve them from the tyranny of matched filtering. For example, a two-class 5-nearest neighbor classifier might use a voting scheme among the five nearest neighbors to a test pattern, resulting in one of the two classes assigned to the test. However, the result of the voting scheme is that the multiparameter feature space is decomposed into regions that result in assignments to one class or another, and the resulting decomposition can be closely approximated by a 1-nearest neighbor classifier, providing enough prototypes are inserted. In other words, by changing the prototypes, a 1-nearest neighbor classifier can be used to approximate just about any static pattern recognition scheme.

4: Model-based vision

Next, let us consider a model-based vision application. Suppose that we are using an hypothesis-and-test approach, such as the Lowe SCERPO method [1] or successive methods. As some stage in the processing, there is a 3-D model m that is hypothesized to be present, and some number of parameters that have been estimated, so as to predict certain features. The system must determine other parameter values, and/or verify currently established values, updating parameters according to new features extracted from the image. There are two phases to this process: (1) matching scene features to model features that can be predicted in the image, but may be located along a multiparameter transformation orbit; (2) improving viewpoint parameters based on newly discovered feature matches.

The first of these two phases, matching scene features to model features, can be formulated as follows. We view scene features as vectors $\vec{y}_i$, which include position information in the scene, and can also include attribute information. Likewise, the hypothesized model is composed of a similar collection of features, say $\vec{z}_j$. We have a transformation T that depends on unknown parameter values $p_1 \cdots p_n$, we wish to determine parameter values such that the collection of transformed model features,

$$\{T(p_1 \cdots p_n; \vec{z}_j)\}$$

best approximates a subset of observed scene features

$$\{\vec{y}_i\}.$$

A metric is required in order to measure the distance, and a typical measure is a mean square metric in feature space. This metric is common due to its mathematical tractability—the parameter search is facilitated when the minimization can be performed assuming an L^2 metric. Nonetheless, the problem has many subtleties, which is familiar to researchers in computer vision as the *correspondence problem.* In various forms, the problem can be solved by subgraph isomorphism, relational graph matching, relaxation methods, analytic optimization, and by many other methods. In general, some of the parameter values may be fixed, others constrained, and others are completely undetermined, and certain correspondences of features may already be established, and new correspondences are sought. The metric measuring the degree of match may well be changed as matches are hypothesized. However, in any given iteration, new matches are sought, either along with providing additional constraints on parameters, or as a verification to already determined parameters, and the metric to evaluate correspondences is fixed for the iteration. Thus, if all p_i are

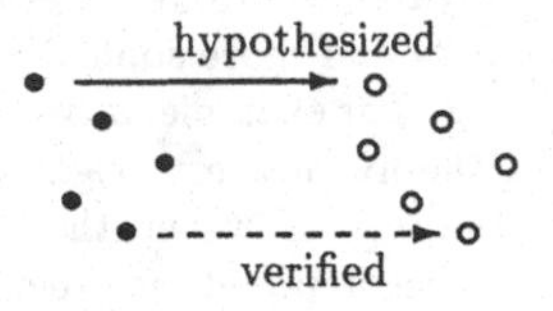

Establishing new matches

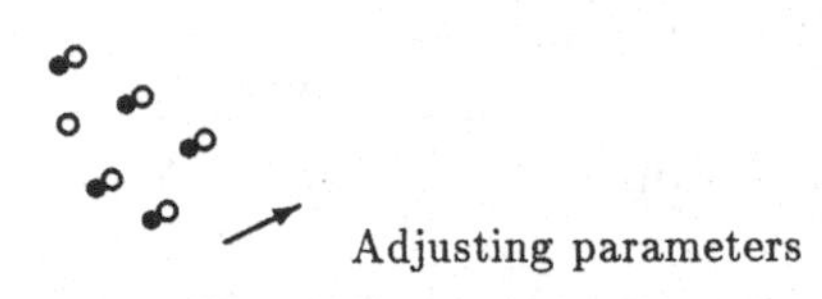

free, then we wish to find a subset S of the observed features, a map from the subset S matching $\vec{y}_i \in S$ to a subset of the model features $\vec{z}_{j_i}$, such that the minimum

$$\min_{p_1 \cdots p_n} \sum_{\vec{z}_i \in S} \|T(p_1 \cdots p_n; \vec{z}_{j_i}) - \vec{y}_{j_i}\|^2$$

is small relative to the number of matches in S. One formulation establishes a threshold fraction for each model, and declares a model to be recognized if there exists a matching of at least the specified fraction of model features to features in the scene such that the minimum parameter fit error (above) gives a sufficiently small value. Other ways of formulating minimization problems with constraints, including penalization methods, are discussed in [2].

There are many special cases, some of which we discuss later.

The second phase, viewpoint refinement, is formulated as follows. We have a collection of scene features $\vec{y}_i$, for $i = 1 \ldots k$, in correspondence with specific features in a hypothesized model, say $\vec{z}_{j_i}$, $i = 1 \ldots k$. Using a norm that can take into account the stated correspondences, we wish to find the optimal parameters $p_1 \cdots p_n$ in order to minimize the distance from

$$\{T(p_1 \cdots p_n; \vec{z}_{j_i})\}_{i=1}^{k}$$

to the set of features $\{\vec{y}_i\}_{i=1}^{k}$. In a sense, the second phase is the same as the first, except that the collection of features are now fixed and in correspondence, whereas the matching phase described above must posit correspondences as part of the solution.

Suppose we fix on L^2 norm. Then the problem is to find the $p_1 \cdots p_n$ minimizing

$$\sum_{i=1}^{k} \|T(p_1 \cdots p_n; \vec{z}_{j_i}) - \vec{y}_{j_i}\|^2.$$

The figure depicts the two problems: finding matches and parameter refinement.

Clearly, both phases are optimization problems, and can be solved by many different approaches to numerical optimization.

Suppose that T is linear in all its variables. This is a special case that is unrealistic, but nonetheless surprisingly common, and will be approximately true much of the time.

Clearly, a local linear approximation is often valid, and since iterative optimization operates locally, a linear assumption is not totally unuseful.

In that case, the solution to both phases, for a fixed model and fixed matching, is a linear minimization problem, of the form

$$\text{minimize } ||Ax - \vec{b}||^2 \text{ over } x$$

whose solution is given by the solution to the "normal equation," and can be solved by a simple iterative procedures; equivalently the minimum value can be obtained by evaluating $||[A(A'A)^{-1}A' - I]\vec{b}||^2$. Since $\vec{b}$ alone depends on the extracted scene features $\{\vec{y}_i\}$, and A depends on the current parameter values and the model features $\{\vec{z}_j\}$, the minimization is implementable as a vector dot product (i.e., a matched filter operation), such that the minimum value can be formulated as a vector product $\vec{x} \cdot \vec{x}_i$, where $\vec{x}$ depends on the values in the matrix A, and hence on the model features $\vec{z}_j$, and $\vec{x}_i$ depends on the vector $\vec{b}$, and hence on the $\vec{y}_i$'s, although the functional relation may not necessarily be simple.

Of course, for the first phase, the model parameters, the subset of model features, and the match to scene features are not fixed. Thus many innovative search strategies can be developed to perform the minimization in an efficient manner. However, here's a brute force method. We begin by enumerating all possible matches, over all possible subsets of model features, over all possible models, which we index by α. Further, for each α, some of the model parameters are fixed, and the others are free. For each such α, the minimization is a linear problem, as describe above, and leads to a vector product $\vec{x} \cdot \vec{x}_\alpha$ to obtain the value to be minimized (or maximized). If we find extrema over the many α, the result is posited recognitions.

For the second phase, no maximization is required over multiple α. Instead, a simple optimization is required. This can be solved by matched filtering, or some related technique, depending on the metric norms. In the case of the first problem, since the parameter values are unknown, the set of available parameters might have to be discretized, which can lead to errors. Indeed, the correspondence problem is interesting precisely because one wants to find a search strategy that is better than exhaustive search.

We see that at least one form of the hypothesis-and-test approach to model-based vision can be implemented as a sequence of matched filtering decisions. Of course, this formulation made a linear assumption, or at least a local linear assumption, of the transformation of features T, and the approach that has been outlined is brute-force, and likely to be improved in actual implementation. Nonetheless, our formulation shows the pervasiveness of matched filtering.

5: Unstable problems and unstable methods

At this point, we have said that:

- Matched filtering won't work; and
- All methods lead to matched filtering, perhaps in disguise.

This theme will continue, even as we discuss geometric hashing.

There is a seeming dilemma. If we really believe the points, then we should give up hope. Either, matched filtering is not so bad after all, as long as we get the representation right, or matched filtering is the wrong approach, and we must make sure that the methods that are used to solve the problems are different than matched filtering. After all, we have only shown that standard vision problems have formulations amenable to solution by matched filtering. We have not said that matched filtering is the *only* approach to solving vision problems.

The point is that problems, particularly optimization problems, can be unstable or stable, and if they are stable, it is still possible to fail to solve them due to the fact that the methods are unstable.

The computer vision community hopes that the following is true:

1. The computer vision problems as formulated above are stable problems;
2. The decades of failure of matched filtering approaches means that matched filtering is an unstable method for solving most of these problems; and
3. Stable approaches will soon be found, or have been found, and are yet to be fully promulgated.

But it could also be that the problems as formulated are unstable, and that a better formulation is required. At this point, it is customary to point to biological vision systems as an existence proof, and thus support for the proposition that the recognition problem must be stable. Thus, the implication is that the fault lies in the methods, although it could be that the fault lies in the formulation. In particular, the optimization formulation is not necessarily the only way to describe the matching problem.

6: Geometric Hashing

Geometric hashing [3] is a method for organizing pattern matching, and is related to the object matching formulation given in the previous section. However, by discretizing the space of parameter values for the transformations in a clever way, the enumeration of possible matches becomes more manageable. Geometric hashing as a search method is particularly attractive because (1) it is parallelizable, (2) efficient especially when dealing with large databases, and (3) permits easy adaptation.

Using independent features with attributes, an image scene is represented using a collection of vectors, $\{\vec{y}_i\}$, as before. Likewise, each model is represented as a collection of features, say for model m_k, the features are $\{\vec{z}_{k,i}\}_{i=1}^{n_k}$. The recognition should be independent to some class of transformations that can be applied to the models, such as translation, rotation, and perhaps skew transformations.

Accordingly, we define a *basis set* to be a minimal collection of features such that placing a basis set in one-to-one correspondence with another basis set (such as a basis set in the scene) determines a transformation in the class of transformations under which recognition must be invariant. For example, if the features are points in $\Re^2$, and if recognition is to be similarity (translation, rotation, and scale) invariant, then a pair of points establishes a basis.

For any particular basis, remaining features may be normalized with respect to the basis. So, if we choose a basis B in a model m_k, then all features $\vec{z}_{k,i}$ in m_k may be transformed in such a way as to move the basis to some pre-determined configuration. Likewise, given a basis B' in the scene, then all features $\vec{y}_i$ may be transformed so as to move the basis B' to the same pre-determined configuration. If model m_k occurs in the scene with basis B in the model in correspondence with the features B' in the scene, then after normalization,

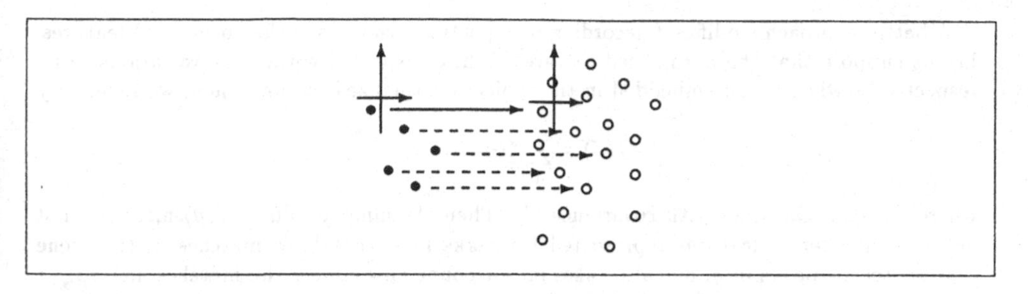

the normalized model features should occur within the normalized scene features in the same configuration.

Accordingly, rather than searching for transformation parameters by a numerical scheme, in conformance with the idea of enumerating over all possible matches indexed by α (as formulated in the previous section), we begin by expanding the model base, considering each model m_k and every basis B within m_k to be a prototype *pattern*. Thus the new collection of models, now called patterns, are pairs (m_k, B), subject to the condition that the basis B is a basis set of features in the model m_k.

To perform the search, we choose a basis B' in the scene, and normalize the features $\vec{y}_i$ with respect to the basis B'. The goal is to determine if any of the model/basis pairs (m_k, B) match the normalized scene. The question is the same as asking whether any of the models are embedded in the scene in such a way that the scene basis B' matches to any basis in the models. If all bases have been included in the enumeration, then it suffices that B' simply lies within an embedded model. The match problem is depicted in the figure.

For any particular model/basis (m_k, B), the question is very simple. We have a collection of normalized model features, which we call $\{\vec{x}_i\}$. We have a collection of normalized scene features $\{\vec{u}_j\}$. We want to find an embedding of $i \to j_i$ for a subset of the possible i such that each $\vec{x}_i$ lies "near" $\vec{u}_{j_i}$.

This problem can be formulated as an optimization problem, but let us consider a heuristic matched filtering approach directly. After all, we are attempting to find a match of the pattern $\{\vec{x}_i\}$ within $\{\vec{u}_j\}$. We now revert to a continuous formulation.

Let

$$f(\vec{x}) = \sum_i \delta(\vec{x} - \vec{x}_i),$$

and let

$$g(\vec{x}) = \sum_j \delta(\vec{x} - \vec{u}_j).$$

(Note that f is dependent on the model/basis pair (m_k, B), and so we might denote it as $f_{(m_k,B)}$. Nominally, the problem is to determine whether delta peaks in f match a subset of peaks in g. Heuristically, this can be determined by considering $\langle f, g\rangle$, although clearly some "smearing" of the peaks is required in order to account for normal noise and possible minor displacements. One way of doing this is to "blur" f by a gaussian,

$$\tilde{f} = f * G_C,$$

where G_C is a gaussian distribution with mean zero and covariance C. We then compute $\langle \tilde{f}, g\rangle$.

A better approach modifies f according to expected variations of the normalized features. Let us suppose that the normalized features $\vec{x}_i$ have expected covariance variations of C_i respectively, when found embedded in the typical (normalized) scene. Then, we redefine $\tilde{f}$ as

$$\tilde{f}(\vec{x}) = \sum_i G_{C_i}(\vec{x} - \vec{x}_i),$$

where G_{C_i} is a Gaussian with covariance C_i. Then the inner product $\langle \tilde{f}, g \rangle$ measures not only the number of features represented by peaks in $\tilde{f}$ that have matches in the scene represented by peaks in g, but also takes into account how closely the matches line up.

Clearly, we are back to matched filtering. Worse, the justification is heuristic. And it is not clear if the scheme will not work because it is matched filtering, or it won't work because the problem is still formulated poorly, or perhaps it will work if the features are robust, stable, and discriminative.

7: Bayesian Pattern Matching

This section shows that the fundamental pattern matching problem, formulated in the previous section, can be given a Bayesian formulation, leading to both a justification and a modification of the heuristically derived formula for matched-filtering -type recognition. Our optimistic conclusion is that matched filtering is probably okay after all, but that a non-optimization formula is needed for stability. Moreover, the right formula is critical, and the features and representation of the features is even more critical. This formulation is based on the thesis work of Isidore Rigoutsos [4]. See also [6] for more details.

The event that we hypothesize is that model m_k is present in the scene with basis B in m_k matching scene basis B'. We may denote this binary event as $E(m_k, B, B')$. A maximum likelihood formulation asks us to compute and maximize

$$\Pr(E(m_k, B, B') | \{\vec{u}_j\}),$$

where the $\vec{u}_j$ are the unnormalized scene features. Note that this formulation is slightly nonstandard, in that if the model m_k appears in the scene, then there may well be multiple pairs (B, B') for which the event is true, and which "discovers" the model. We only need find one such pair, if the model is present.

If the features are conditionally independent, then the above expression can be decomposed. The conditional independence assumption is a strong one, and imposes a large constraint on the kinds of features with which we can operate. The condition means that under the assumption that a particular model is present in a particular location, knowledge of a feature at one location (or with one set of parameters) has no influence on the probability density distribution of another feature. The condition is not at all true, for example, when dealing with edge elements (edgels), which tend to line up in lines, and are thus very much dependent.

Using various Bayesian manipulations, and using the independence assumption, one can derive

$$\log \left(\frac{\Pr(E(m_k, B, B')) | \{\vec{u}_j\}}{\Pr(E(m_k, B, B'))} \right) =$$

$$\log K + \sum_j \log \left(\frac{\Pr(\vec{u}_j | E(m_k, B, B'))}{\Pr(\vec{u}_j)} \right)$$

where the probability terms in the sum should be interpreted as density function evaluations. See [5] for details of these manipulations. If we assume that all prior probabilities of the events $E(m_k, B, B')$ are uniform, then we can simply strive to maximize the right hand side. The $\log(K)$ term will be independent of (m_k, B, B'), and so can be ignored.

Accordingly, let us examine carefully the summand terms on the right hand side of the equation above in light of the patterns $\{\vec{x}_i\}$ and $\{\vec{u}_j\}$.

The denominator $\Pr(\vec{u}_j)$ is very simple. This is simply the prior probability of a feature occurring with values $\vec{u}_j$, where the features are chosen randomly in a random scene. We will assume that the number of features extracted is fixed and known (although modeling the density using a non-homogeneous Poisson process might be more realistic). There is a background density distribution, say $\rho(\vec{x})$ on the feature space, and this probability is simply an evaluation of this density function. This function can be derived from a modeling of the image space and of the feature extraction process, or could be developed empirically.

The numerator is a more complicated story. This density function is different, because it is conditioned on the knowledge that event $E(m_k, B, B')$ is true. This means that we know that the normalized scene contains the features $\{\vec{x}_i\}$, for $i = 1 \dots n_k$, where there are n_k features in the model m_k. Let us suppose that we have s observed features. Let us further suppose that of the n_k features in the model m_k, it is reasonable to expect on average a certain amount of obscuration, and thus only $\beta \cdot n_k$ features will be observed from the model, and the remaining will be background clutter features. Finally, let us assume that each of the expected features $\vec{x}_i$ has an expected variation with covariance matrix C_i. Then the expected density distribution is given by

$$\tilde{\rho}(\vec{x}) = \frac{s - \beta n_k}{s} \cdot \rho(\vec{x}) + \frac{\beta}{s} \cdot \sum_{i=1}^{n_k} G_{C_i}(\vec{x} - \vec{x}_i).$$

Note that the function has total density one (assuming ρ has total density one), and that the delta masses at the locations of the pattern of the normalized model m_k have become Gaussian "bumps," as heuristically considered in the previous section. Accordingly, each summand has the form

$$\log\left(1 - \frac{\beta \cdot n_k}{s} + \frac{\beta}{s\rho(\vec{x})} \sum_i G_{C_i}(\vec{x} - \vec{x}_i)\right).$$

Plugging into the Bayesian formulation above, we see that the maximization of the likelihood is tantamount to computing, for each collection of $\{\vec{x}_i\}$ based on the model m_k and basis B, given the normalized scene features $\{\vec{u}_j\}$ based on the scene basis B',

$$\sum_{j=1}^{s} \log\left(\frac{s - \beta n_k}{s} + \frac{\beta}{s\rho(\vec{u}_j)} \sum_{i=1}^{n_k} G_{C_i}(\vec{u}_j - \vec{x}_i)\right).$$

This can be rewritten as

$$-s \log\left(\frac{s}{s - \beta n_k}\right) +$$

$$\sum_{j=1}^{s} \log\left(1 + \frac{\beta/\rho(\vec{x})}{s - \beta n_k} \sum_{i=1}^{n_k} G_{C_i}(\vec{u}_j - \vec{x}_i)\right).$$

And now for some magic. For any given $\vec{u}_j$ value, at most one of the values of $\vec{x}_i$ will lie close. Let's call the closest one i_j. Then in the second term, the Gaussian terms will all be essentially zero, except possibly for the term $G_{C_{i_j}}(\vec{x}_{i_j} - \vec{u}_j)$. Thus we may replace the sum over i with the single term i_j! This analysis, incidentally, is using a greedy approach to finding an association between predicted and extracted features; a different analysis is possible if we are interested in maximizing over all possible associations.

So we obtain the following formula. We fix B' in the scene, and obtain the collection $\{\vec{u}_j\}_{j=1}^{s}$. We have many prestored model/basis patterns, each one consisting of a collection of the form $\{\vec{x}_i\}_{i=1}^{n_k}$. The support level for a particular model/basis depends on finding the nearest model normalized feature $\vec{x}_{i_j}$ to each normalized scene feature $\vec{u}_j$. The support for the particular model/basis is then

$$-s\log\left(\frac{s}{s-\beta n_k}\right)+$$

$$\sum_{j=1}^{s}\log\left(1+\frac{\beta/\rho(\vec{u}_j)}{s-\beta n_k}G_{C_{i_j}}(\vec{u}_j-\vec{x}_{i_j})\right).$$

The bias term $-s\log(s/(s-\beta n_k))$, which we will denote by c_k, depends only on the model m_k, and not on the basis B chosen within model m_k. Accordingly, the support may be rewritten as

$$c_k + \langle \tilde{f}_{(m_k,B)}, g\rangle,$$

where g is a sum of delta functions at the locations $\{\vec{u}_j\}_{j=1}^{s}$, as before, and $\tilde{f}_{(m_k,B)}$ is again redefined, and discussed briefly below.

But, at this point, we may note that the pattern matching problem has once again been formulated at matched filtering! The collection of biased inner products should be maximized over all (m_k, B), and then separate collections may be evaluated and maximized for different B' (which effects the function g, i.e., $g = g_{B'}$). Accordingly, if Bayesian pattern matching provides a more stable performance than optimization methodologies, it is not because matched filtering is inherently inappropriate. Rather, there is still hope that the Bayesian approach is better because the change in formulation provides different formulas and different representations.

Finally, we define the Bayesian pattern-based matched filter $\tilde{f}_{(m_k,B)}$. For a feature vector $\vec{x}$, and $i(\vec{x})$ defined as the index of the nearest normalized vector among $\{\vec{x}_i\}$, i.e., the collection of feature vectors in m_k normalized according to the basis B, we then have

$$\tilde{f}(\vec{x}) = \log\left(1+\frac{\beta/\rho(\vec{x})}{s-\beta n_k}G_{C_{i(\vec{x})}}(\vec{x}-\vec{x}_{i(\vec{x})})\right).$$

By using fancy data structures, such as hash tables and k-D trees, or self-balancing trees, the computation of the multiple inner products can be made quite efficient. This is what we refer to as "geometric hashing." But it is not our purpose here to argue for indexing and/or hashing methods here. Instead, our point is that a Bayesian formulation leads to yet another matched filtering formula, but where the filter operates in a domain equivalent to the range space of a feature value (which we have viewed as a vector, because there can be multiple attributes to a single feature), and with a formula that is non-obvious and only somewhat intuitive. Indeed, the formula for $\tilde{f}$ above can be closely approximated by a sum of log terms (since if $\vec{x}$ is distant from a particular $\vec{x}_i$, then the Gaussian will essentially evaluate to zero, and the log term is thus also nearly zero), but this is slightly different than a sum of Gaussians, which was our heuristic formula from the previous section.

8: Comments

At this point, it would be typical and desirable to state that the revised formula derived in the previous section for Bayesian pattern matching yields much better results, which we can document with examples. However, if we did this, we would be guilty of exactly the kind of article criticized in the first section, where the results are given in a summary in the penultimate section. Moreover, the jury is still out. We believe that Bayesian pattern matching will work much better than traditional object recognition systems, given the same features, but this is simply an intuition.

But from the standpoint of methodologies, we have two points: (1) There are alternative formulations; (2) we still haven't gotten rid of matched filtering, although it is considerably transformed and disguised.

9: Summary and Conclusions

For many years, people have worked on feature extraction, with the conviction that stable, robust features are important. We now state here that stable, robust features are essential. Viewing recognition as a pattern matching problem, the patterns have to be representative of the objects, and the patterns are composed of extracted features. Our discussion above has been mostly about the matching engine. Although we have shown alternative methodologies and developed new matching formulas, because we always come back to matched filtering, it is entirely possible that the precise matching engine does not matter. We believe that it does matter, and that the Bayesian approach should lead to more robust performance. However, in no way does the alternative approach obviate the need for stable features. Indeed, if anything, since our Bayesian approach is based on matching patterns of extracted features, we have emphasized the need for stable features.

Acknowledgements

Thanks to Isidore Rigoutsos, with whom many of these ideas were formulated in the context of his thesis on Bayesian interpretations of geometric hashing. Many, many thanks to Azriel Rosenfeld, who started it all and gave me an interesting and happy career.

References

[1] D. Lowe, *Perceptual Organization and Visual Recognition,* Kluwer Academic Publishers, 1985.

[2] R. Hummel, B. Kimia, and S. Zucker, "Debluring Gaussian Blur," *Computer Vision, Graphics, and Image Processing* **38**, pp. 66-80, 1987.

[3] Y. Lamdan and H. Wolfson, "Geometric Hashing: A General and Efficient Model-Based Recognition Scheme," In *Proceedings of the 2nd International Conference on Computer Vision*, pages 238–249, 1988.

[4] I. Rigoutsos, *Massively Parallel Bayesian Object Recognition*, Ph.D. thesis, New York University, August, 1992.

[5] E. Charniak and D. McDermott, *Introduction to Artificial Intelligence*, Addison-Wesley, 1985.

[6] I. Rigoutsos, R. Hummel, "A Bayesian approach to model matching with geometric hashing," *Computer Vision and Image Understanding,* Vol. 62(1), pages 11–26, 1995.

Finding and Describing Objects in Complex Images

Michael F. Kelly and Martin D. Levine
Department of Electrical Engineering &
Centre for Intelligent Machines
McGill University
Montréal, Canada.

1: Introduction

This work presents a novel method which allows *coarse* object representations to be extracted directly from *unsegmented* gray-scale images. The explicit use of multiscalar sampling employing *annular symmetry operators* provides the basis for identifying symmetrical structures which are then used to construct object representations. It is suggested that the latter may be appropriate for a variety of practical tasks relying on visual information. Furthermore, it is conceivable that simple grouping operations similar to those demonstrated by annular operators may underly early stages of visual processing in biological organisms.

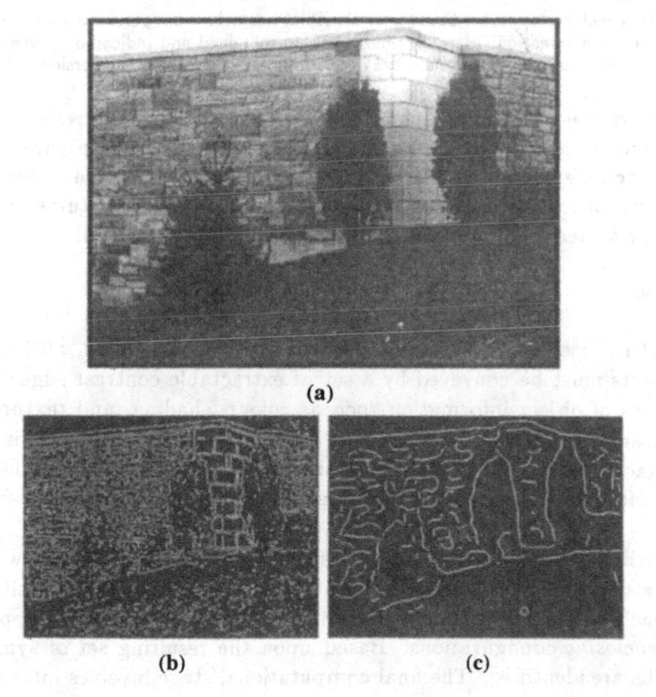

(a) (b) (c)

Figure1.
Tree image (a) and two edge maps obtained at small (b) and large (c) scales. Using this data, how can the predominant objects be located computationally?

Consider the image presented in Figure 1 which depicts three trees set against a wall. It is assumed that all relevant object information is captured by the set of contrast edges (or simply edges) that can be extracted from the image. While to a human observer the trees appear as distinct objects, it is not clear how this information may be extracted computationally from the associated edge maps. At small scales, the edge map presents a complex branching structure from which it is impossible to infer the tree outlines by a boundary-based traversal alone. For larger scales, many of the continuous paths in the edge map do not accurately represent the boundary of a single object. A common problem that further impedes boundary traversal approaches is the presence of gaps in the edge map that may occur at any scale.

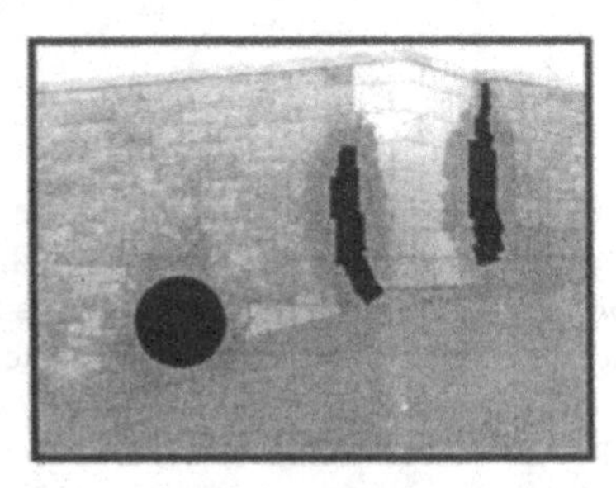

Figure2.
This work presents a multiscalar region-based analysis which identifies the 'perceptually significant' objects in images without human intervention. Here all three trees are obtained and indicated by black structures (two elongated *limbs*, and one round *blob*) that have been superimposed onto a bleached version of the original image.

A multiscalar region-based approach has been developed which addresses these problems. After the method is applied to the tree image, the results shown in Figure 2 are obtained. In addition to locating the objects, a coarse structural representation is also produced for each. In this example, two objects are modelled as long narrow structures (limb), while the third one is represented as an approximately circular structure (blob).

1.1: Overview

The method is based upon two fundamental assumptions. First, all information pertaining to objects must be conveyed by a set of extractable contrast edges. Hence, other potential sources of object information such as colour, shading, and texture are ignored. Second, it is assumed that all objects possess a degree of symmetry which is manifested in the set of extracted edges. The location of objects within image data is then based upon the detection and interpretation of the symmetrical relationships between sets of extracted edge data.

The approach is represented by the series of computational steps shown in Figure 3. Starting with a single gray-scale image as input, edge contours are first identified at multiple scales. For each contour map, annular symmetry operators are then applied to detect *symmetrical enclosing* configurations. Based upon the resulting set of symmetry points, blobs and limbs are identified. The final computational stage involves interpreting a set of extracted parts in order to derive representations for objects in the image.

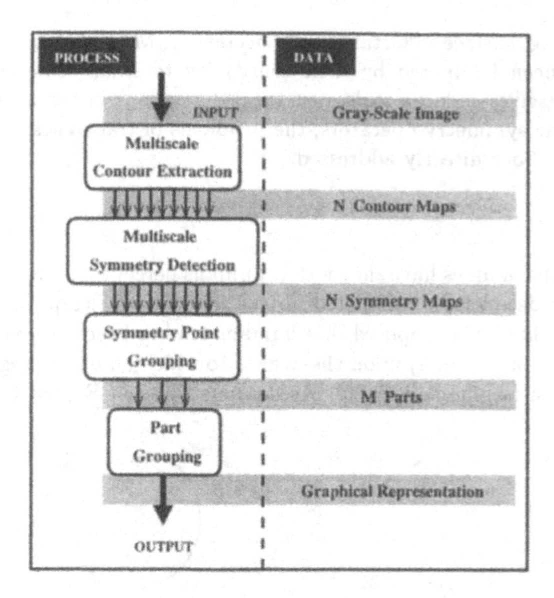

Figure3.
The method consists of four computational *processes* shown in the left-hand column. The data transferred between each process is indicated by the shaded strips labelled in the right-hand column.

2: Annular Symmetry Operators

A novel aspect of this work is the use of annular symmetry operators to detect the presence of symmetric enclosure over a range of scales. The use of annular 'masks' for extracting symmetry information was originally proposed in [29]. In this work, the authors chose to implement an annular sampling approximation based on a mixed wave-diffusion process, and the idea in its original form was not developed further. It will become clear that the annular sampling approach provides advantages over propagation-based methods since it leads to simplified descriptors, even in cases involving complex unsegmented image data.

The approach has been motivated by the importance that geometric symmetry has in mediating human perceptual phenomena [19][11]. In the field of computational vision, symmetry has also been widely recognized as a valuable tool - particularly for the explicit representation of two-dimensional image structures based on the "medial axis transform" of Blum [1]. There are numerous examples where related symmetry-based methods have also been employed (for a small but representative set see [10][26][28][24][16][33]). In addition to deriving representations for objects, there is evidence suggesting that extracted symmetry information can also be readily used as the basis for decomposing objects into simpler component parts [2][15].

Two limitations have traditionally prevented the application of symmetry-based techniques under practical imaging conditions. The first relates to the segmentation of image objects prior to symmetry analysis. Generally, this has required that objects be represented by closed curves corresponding to the object's occluding contour. However, in practice it is often impossible to directly acquire closed edge-based object contours due to a variety of factors such as lighting, surface reflectance, and occlusion. Many previous symmetry-based methods have also been hampered by their sensitivity to small scale contour variations which manifest themselves as large scale structural distortions in the axial representation. By employing annular symmetry operators, the problems of fragmentation and small scale contour variation are both directly addressed.

2.1: Enclosure

Numerous classical examples have shown that humans perceptually associate disjoint sets of planar curves in order to infer completed object structure [18][19]. For these perceptual *grouping* processes, it has been proposed that humans tend to base their perception of inside and outside (or figure and ground) upon the degree to which an object region is *enclosed* or *surrounded* by a set of fragments [27][25]. As demonstrated in Figure 4, the phenomenon

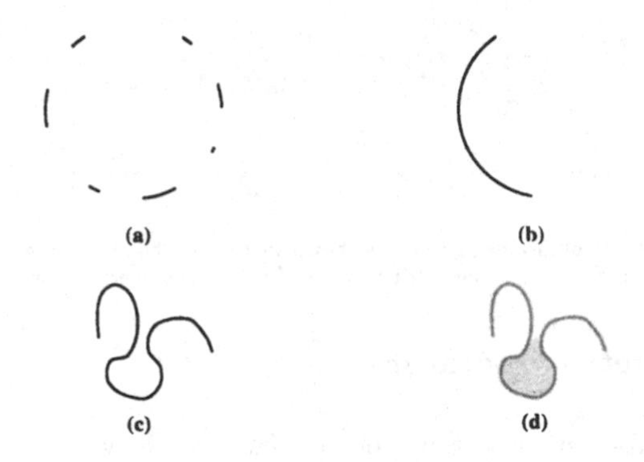

Figure4.
(a) A circular contour with several gaps elicits a strong figure/ground inference to a human observer. (b) Another circular contour having the same arc length elicits a much weaker figure/ground inference. It is the spatial *distribution* of contour fragments (not the combined arc length) which appears to be the most perceptually relevant quantity. This spatial distribution is referred to as *enclosure*. (c) A curve, demonstrating several enclosed regions, gives rise to two opposing figure/ground interpretations. The region in (c) corresponding to the shaded region in (d) is perceived as an independent structure - even though there is no principled means for topologically closing the contour locally based on curve endpoints or by inferring smooth tangential continuation.

of enclosure appears to depend more upon the *distribution* of contour fragments than it does on the total combined arc length of a bounding contour. Previous methods have quantified the *closure* of a curve by attempting to artificially close a set of fragments - for example by filling the gaps between fragment endings [9][6]. However, perceptually enclosed regions can occur even in settings when it is not clear how to topologically close a contour based on information such as curve endings and tangent directions (see Figure 4(c,d)). Annular operators provide a means of addressing this problem by providing a mechanism for identifying and analyzing enclosed regions that are isolated as a function of spatial scale.

2.2: Detecting Symmetric Enclosure

Each annular symmetry operator is located by its center point c, and has a spatial extent bounded by two concentric circles of different radii centered at the point c (see Figure 5). Here the term *scale* refers to the *size* of annular operator for which an enclosing configuration is detected. This contrasts with previous work where scale has often been defined in terms of smoothing processes [17][32].

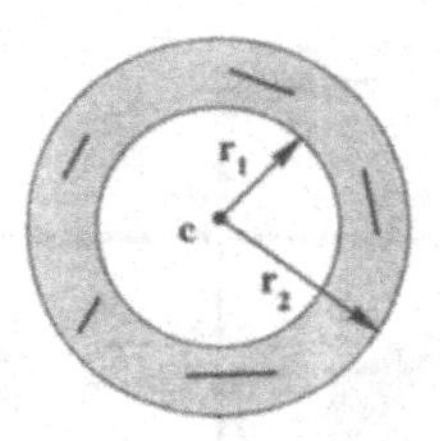

Figure5.
(a) A set of contour fragments falling within the annular region defined by concentric circles of radius r_1 and r_2 centered at the point c. The parameter κ defines the annular width: $\kappa = \frac{r_2}{r_1}$. In practice this ratio is kept fixed for all operators in order to maintain sampling consistency with scale.

To apply an annular operator at a given scale requires that contour fragments first be extracted at a scale which is proportional to the size of the annulus employed. For this, a grouping process similar to [21] has been used. Edges extracted at a single small spatial scale are grouped at a series of progressively larger scales so as to infer the presence of longer and (relatively) smoother contours at larger scales [12].

For contour fragments falling within a given annular region, the local contour orientation is used to define a binary compatibility function. A local polar coordinate system (r, θ) is defined within the annular band relative to the center point c and having unit vectors $(\hat{r}, \hat{\theta})$. The compatibility function assigns a TRUE value for those angular positions θ where the contour has a local orientation value that is parallel (within a tolerance) to the $\hat{\theta}$ direction. The function assigns a FALSE value otherwise. The resulting function is referred to as an *angular profile*. It is used to compute enclosure values by looking for *break angles* corresponding to the angular run-lengths associated with FALSE intervals in the function. Each of these n intervals is represented by its angular extent θ_i. Enclosure is inferred for a given annular operator if $\max[\theta_i] < \tau$, where τ is a fixed angular threshold (typically set to $\tau = \frac{4\pi}{3}$). If enclosure is inferred, the degree of enclosure is quantified as follows:

$$e = 1 - \frac{1}{4\pi^2} \sum_{i=1}^{n} (\theta_i)^2 \tag{1}$$

A similar nonlinear expression has been used to quantify the degree of *closure* for fragmented planar curves based upon psychophysical results obtained using human subjects performing visual search tasks [6]. In the current work, since only curve segments falling within an annular region are considered, this sampling strategy isolates enclosing structures as a function of spatial scale. The separation and subsequent analysis of enclosing structures as a function of scale provides one of the key advantages associated with the annular approach (see Sections 2.3 & 2.4).

For enclosing structures containing gaps, an *orientation* value is assigned based upon the gap location relative to the annular center (see Figure 6). This information is critical since it permits the differentiation of overlapping limb structures in complex settings (see Section 3.4).

Once an enclosing contour configuration is detected and quantified, the presence of enclosure at a given spatial location and operator scale is indicated by a *symmetry point* which

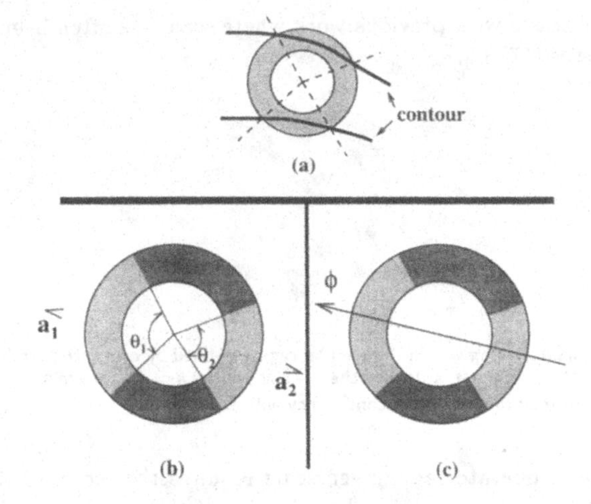

Figure6.
(a) A contour configuration detected by an annular operator. Dashed lines indicate positions where the contour passes outside the annular region. (b) Two break angles are associated with regions where no contour points are found (light shaded). These are represented by the angular intervals θ_1 and θ_2 measured relative to the annular centre. The direction of the bisector for each break interval defines the break angle direction (a_1 and a_2). (c) The average ϕ of the two break angle directions is used to represent the enclosure orientation.

corresponds to the operator's center point c. In practice, an economical (discrete) sampling strategy [14][13] is used to detect symmetric contours at any spatial location within the image. A range of operator scales is selected which permits the detection of any symmetric structure whose scale s falls within a continuous range $s_1 \leq s \leq s_2$ (see Figure 7).

2.3: Properties I. Structures at Different Scales

There is a direct relationship between the use of annular operators and the more traditional wave propagation approaches that have been used to identify symmetry points [29][1][15]. As the width of its annular band shrinks to zero, a given annular operator could *in principle* be used as a *co-circularity* detector - in the same way that bitangency has often been used to define the symmetry points resulting from isotropic wave propagation interactions. However, when compared to previous wave propagation methods, annular operators have distinct advantages for analysing objects with internal structures since they provide a practical means for approximating wave "flow through" [1]. This becomes particularly important in the analysis of objects when structures exist over a range of different scales (see Figure 8).

For previous symmetry methods that have used wave propagation, the process is most commonly terminated at points where two wave crests meet. This form of wavefront "quenching" or "shock formation" is often visualized in terms of the simple grass-fire analogy originally proposed by Blum [1]. In many cases, the extraction of symmetry information using flow through is necessary since it permits the interaction between contours in cases when there are intervening contour structures (see Figure 8(a)). Past methods which have employed flow through (either explicitly or implicitly) have computed the full *symmetry set* for simple objects [3][29][23]. However, like the medial axis transform [1], these flow

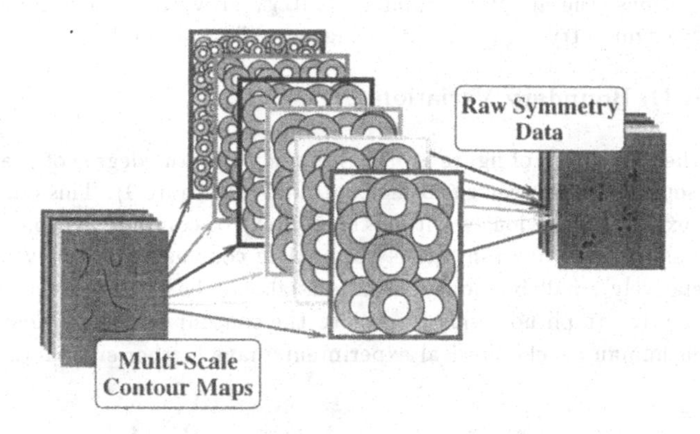

Figure7.
For each contour map a corresponding set of annular operators is employed to detect symmetric enclosures at a given scale. All operators applied at one scale have the same size. Starting with a set of N contour maps, the operation results in a set of N symmetry maps. Each map contains the set of symmetry points which correspond to the annular centers where enclosure was detected (see [13] for details beyond the scope of this paper).

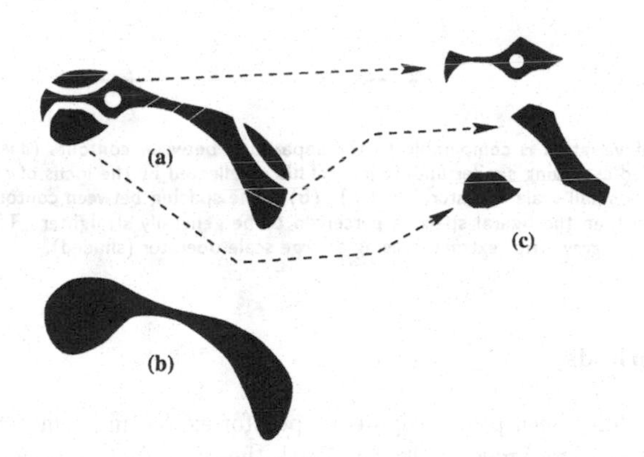

Figure8.
Object structure manifested at a variety of different scales. (a) There is a global perception of structure which is similar to that shown in (b). Note that the associated contour in (a) is fragmented and that numerous internal structures exist (c).

through methods are generally quite sensitive to boundary variations and hence produce complex symmetric structures which are difficult to interpret. Due to a reduced sensitivity to boundary variations, the annular sampling strategy provides a more robust means for approximating the symmetry set generated using wave flow through.

2.4: Properties II. Boundary Variation

For humans, the perception of figure is often influenced by the degree of spatial variation and the relative separation between enclosing contours (see Figure 9). This phenomenon has been noted and examined previously within the context of computer vision [15][7][23][28]. These observations have led to an intuitively appealing conclusion - namely that for larger scale regions, relatively small boundary variations should be treated as a kind of 'noise' based upon its relative amplitude with respect to the contour separation distance. Recent findings based on human psychophysical experiments have further substantiated this view [4].

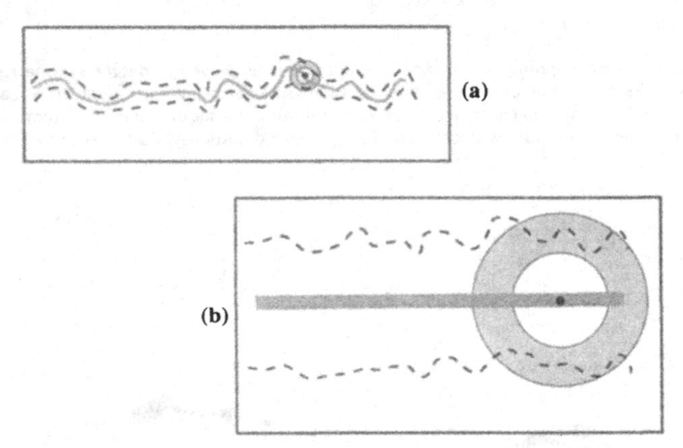

Figure9.
(a) If the amplitude of variation is comparable to the separation between contours (dashed), the intervening figural space is perceived as having similar undulations. This is reflected by the locus of symmetry points (gray curve) extracted using a small scale operator (shaded). (b) If the spacing between contours is large relative to the boundary variation, then the figural space is perceived to be generally straighter. This is reflected by the locus of symmetry points (grey strip) extracted using a large scale operator (shaded).

2.5: Related Methods

Several methods have been previously developed for extracting symmetries directly from the *grey-scale surface* of an image [31][8][5]. With these, extraction is performed as a function of scale - where the term *scale* is used to describe the amount of smoothing applied to the grey-scale surface prior to symmetry extraction. This approach has been demonstrated to work well under certain conditions (for example, uniformly dark objects on a uniformly light background). However, its application is often limited due to the significant structural distortions [17][32] that result in either the annihilation or the spatial translation of contours as a function of smoothing-scale. This effect is particularly dramatic in cases involving *contrast reversals*, which occur quite commonly along the boundary of imaged objects (see Figure 10). Since the annular method employs a purely spatial definition of scale (as opposed to one based on smoothing), it is possible to detect and represent enclosed regions in cases involving contrast reversals (see Figure 16).

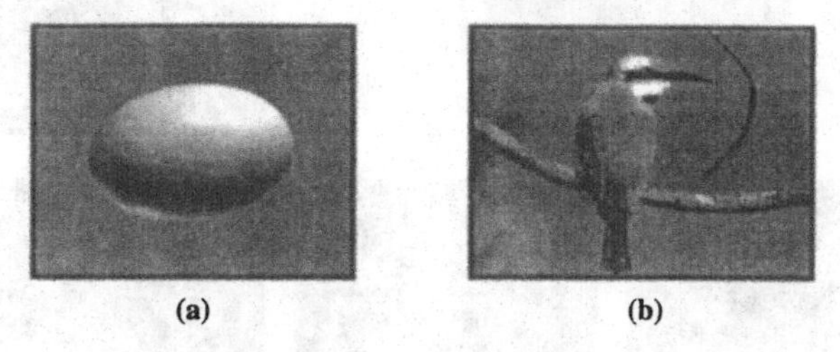

Figure10.
Objects with boundary contrast reversal. (a) An artificial image of an ellipsoid having uniform reflectance. (b) A natural image. In both cases, the gradient direction along the object boundary is variable - with its direction alternating between outward and inward pointing as the boundary is traversed.

There is significant similarity between the application of annular operators and the 'core' based analysis proposed in [4]. Both methods provide a convenient and robust means for approximating the symmetry set generated using wave propagation with flow through. In the current work, the annular sampling approach has been adopted for several practical reasons. First, it is amenable to a discrete analysis which leads to efficient implementations employing a minimum number of operators to adequately sample a given image [13][14]. Second, annular operators lend themselves more directly to the (nonlinear) computation of *enclosure* (See Section 2.1). One final advantage is that the approach provides a direct means for assigning an *orientation* to enclosed structures. This facilitates the *grouping* of symmetry information in complex settings involving multiple object structures (see Section 3.4). An example demonstrating the multiscale extraction of symmetry points for a simple binary image is presented in Figure 11.

3: Interpreting Symmetry Data

After extracting raw symmetric-enclosure data, the result is a set of points that fall within a three-dimensional scale space $\mathcal{S}$. A given point is identified by its *spatial position* (x, y) and the *scale* (s) at which symmetric enclosure is detected. The goal is to perform progressive grouping operations upon this data in order to infer the presence of gross object structures such as those shown in Figure 1.

3.1: Blobs and Limbs

It is assumed that two general types of symmetric structures can be used to represent the set of objects under consideration. *Limbs* are associated with enclosed object regions that are elongated. *Blobs* are associated with object regions which are less directed and more "globular". Similar object primitives have previously been shown to be useful for the representation of both three-dimensional [30][22] and two-dimensional [33] objects. The goal of interpreting symmetry data is to arrive at graphical representations for object structures in terms of these primitive elements. Blobs and limbs are detected by locating specific signatures within the set of extracted symmetry points.

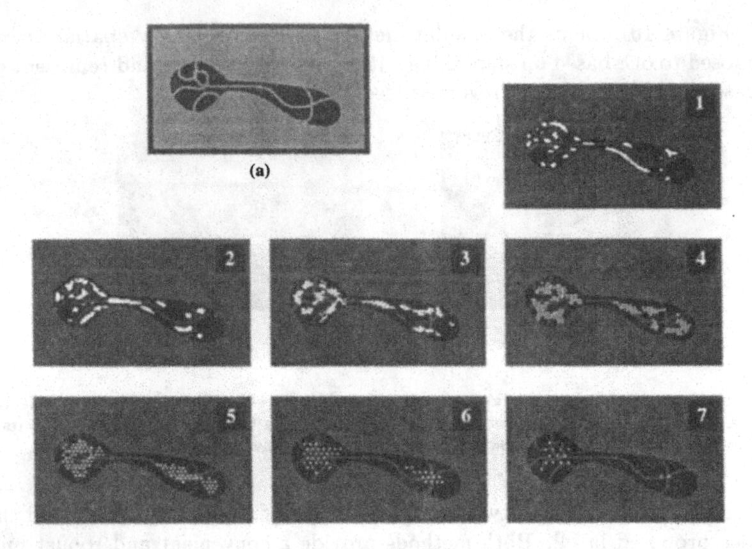

Figure11.
Image used to demonstrate the part extraction phase (a) followed by extracted symmetry points for a series of 7 consecutive (discrete) scales. Symmetry points are superimposed onto the original image and are shown as white dots. Ground intensity has been reduced to enhance the visibility of symmetry points. Note that the discrete spatial sampling becomes sparser at larger scales.

3.2: Symmetry Point Volumes

To simplify our discussion, annular sampling is described within a continuous[1] three-dimensional space $\mathcal{S}$. The presence of symmetric image structures is manifested through characteristic symmetry point distributions which define continuous *volumes* within the space $\mathcal{S}$. For approximately circular contour structures that have no predominant elongation in any one spatial direction, this information is reflected in the $\mathcal{S}$ space by a volume which demonstrates an approximately spherical form (see Figure 12(a)). For image structures that have significant spatial elongation, we find that the resulting $\mathcal{S}$ space volume has a cylindrical form (see Figure 12(b)). The differences which characterize these symmetry volumes provides the basis by which blobs and limbs are identified and represented.

3.3: Part Representations

For a given symmetry volume, simple descriptors are required to represent the associated blob or limb. A useful choice of descriptor is the equivalent medial axis [1] which can be abstracted directly from symmetry point volumes. A single point in the $\mathcal{S}$ space is chosen to represent a blob in terms of its spatial position and scale of enclosure (see Figure 12(a)). For limbs, a locus of symmetry points defining a space curve is used to represent its structure (see Figure 12(b)). Since these descriptors are directly related to the medial axis, they both provide the same degree of descriptive power.

[1] In practice, annular sampling is performed by implementing a *discrete* sampling within this continuous space [12][13].

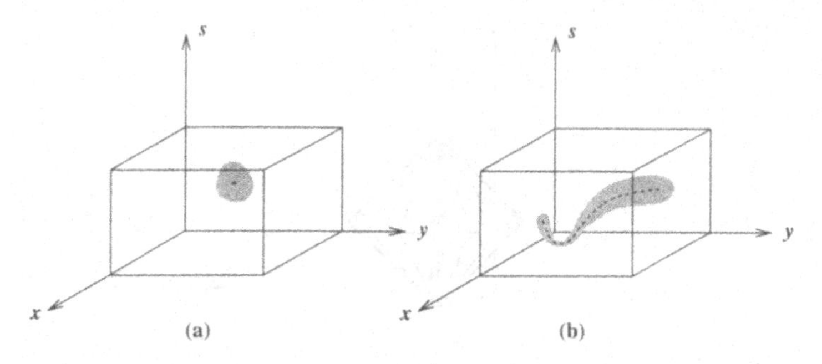

Figure12.
(a) For spatially non-elongated parts (blobs), the set of extracted symmetry points defines a volume which has an approximately spherical form within the $\mathcal{S}$ space. (b) For spatially elongated parts (limbs), the set of symmetry points defines a volume which has an approximately cylindrical form within the $\mathcal{S}$ space. Note that the girth of the limb volume varies as a function of the scale s at which enclosure is detected. Desirable representations take the form of a single point in the case of blobs (a), and a space curve (dashed) in the case of limbs.

3.4: Extracting Representations

Starting with raw symmetry data, the search for blobs and limbs is performed independently within the $\mathcal{S}$ space. The extraction of representations depends upon locating the symmetry volumes described in Figure 12. In each case, parts are extracted sequentially based upon a measure of ***physical support***. It is assumed that parts associated with large amounts of contour data are less likely to be formed accidentally. Thus, those having the longest contour arc length are extracted first. Once a part has been extracted, all symmetry points which define the associated symmetry volume are deleted from the set of candidate points, and the process continues by searching for the next part. By extracting parts in order of decreasing physical support, we minimize the chance of corrupting non-accidental structures through the deletion process. The specific processes employed for extracting blob and limb representations is outlined below.

BLOBS: These are found by locating instances of continuous line segments in the $\mathcal{S}$ space that run parallel to the s axis. For each of the symmetry points comprising the path locus, the enclosure function e is required to exhibit a maximum within a local spatial neighbourhood. When implementing this process using discrete sampling, the local maxima in e are first detected at each scale independently. Paths are then constructed by identifying sets of spatially coincident maximal points within consecutive sampling scales (see Figure 13). The ability to construct such a path provides evidence that a non-elongated part (a blob) exists at a given point. Having found such a path, its midpoint is then used to represent the entire blob.

LIMBS: To infer the presence of limbs, a search is performed which attempts to find the path having the greatest physical support while maintaining a high degree of overall smoothness along the path. During this process, paths are constructed such that a set of *continuity constraints* are satisfied. These constraints ensure that each path is continuous in terms of the local spatial coordinates (x, y), the local scale (s), and the local *enclosure orientation* values (ϕ). Once an optimal path is found, the limb is represented by the associated locus of symmetry points.

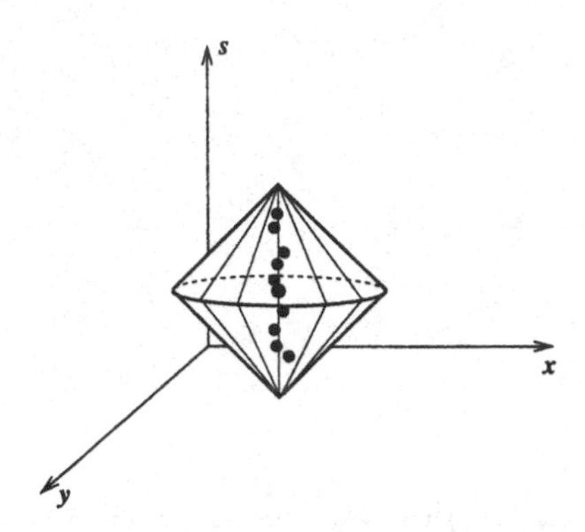

Figure13.
Within the $\mathcal{S}$ space, the set of symmetry points generated by a perfectly circular contour defines a symmetry volume which is bounded by two right circular cones joined at their bases [12]. For a blob, the extent of symmetry points within the $\mathcal{S}$ space will be approximately the same as for a circle. The detection of a blobs requires that a series of spatially localized enclosure extrema (black dots) be found such that they form a continuous path parallel to the s axis.

3.5: Perceptual Significance

Once a set of blobs and limbs has been extracted, all are then ranked in terms of their relative *perceptual significance.* Parts which are either very large or which have a high contrast with respect to the background are assigned a high perceptual significance value [12].

3.6: Examples

The set of extracted symmetry points for a simple binary object was shown in Figure 11. After identifying a set of parts, the blobs are shown in Figure 14 and the limbs in Figure 15. These demonstrate how multiple representations are extracted in cases involving internal structure. Results for a natural image were illustrated earlier in Figure 2 which depicts the three most perceptually significant parts extracted from the image (Figure 1).

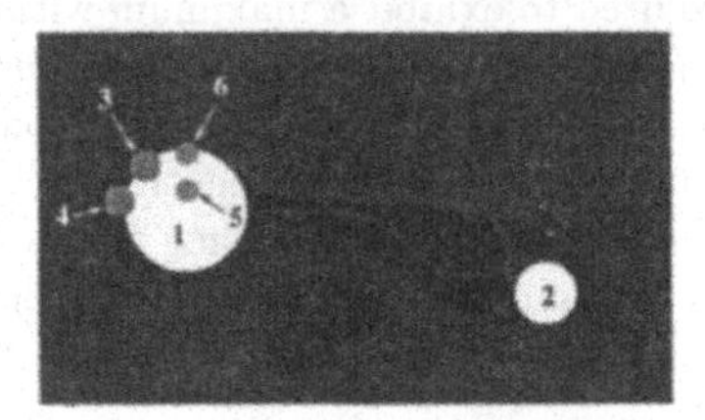

Figure14.
Darkened version of the original image (Figure 11(a)), with blobs overlayed as filled circles. Circles are shaded either white or gray to aid visualisation. The size of circles is proportional to the scale at which enclosure was detected. Numbers indicate an ordering of the extracted parts - starting with the most perceptually significant part (1).

In Section 2.5 it was noted previously how contrast reversal often occurs for object boundaries imaged under natural conditions. Figure 16(b) demonstrates the structural

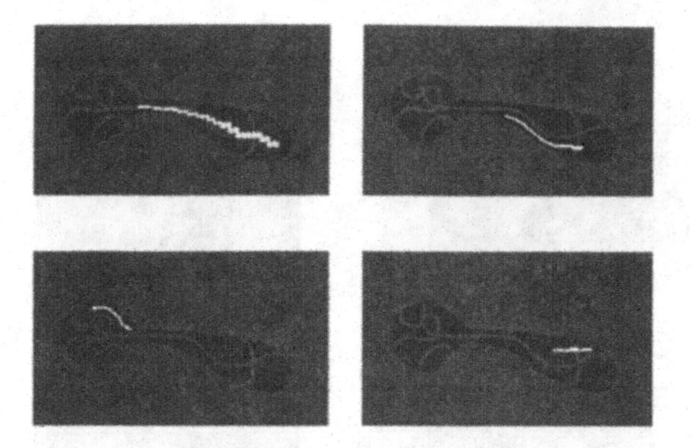

Figure15.
The first four extracted limbs shown in order of decreasing perceptual significance from top to bottom, left to right. Note how the highest ranking part captures the global limb symmetry, while the other parts depict various internal structures. Limbs are represented by a series of filled squares, where each is centered about a symmetry point. The size and orientation of the squares reflects the scale and orientation of the enclosing structure.

degradation that results if smoothing is employed. The application of previous approaches which rely on smoothing - either as part of a surface- [31][8][5][20] or edge-based [4] approach - will clearly be limited in these cases. The extracted limbs and blobs obtained with our method are shown in Figure 16(c,d). Much of the object structure is coarsely captured by this set of parts.

4: Constructing Graph-Based Representations

The grouping of symmetry points produces a set of blobs and limbs. The final computational stage in Figure 3 combines these parts in order to form descriptors for *objects* in the scene. The result of this process is a *graph* - where nodes correspond to object parts and edges represent the spatial relationships between the various parts.

4.1: Finding Objects

Starting with the combined set of blobs and limbs, grouping begins with the most *perceptually significant* part. This *principal part* is used to initiate the search for other parts associated with the same object. This produces a graph which represents the object. Once this process is completed for the first object, all associated parts are deleted from the candidate list. The process is then repeated for all other objects in the image - each time starting with the most perceptually significant part that remains after deletion.

4.2: Joined Parts

Once a principal object part has been identified, other parts associated with the same object are found by attempting to detect material *joins* between parts. Since the method does not explicitly employ boundary-based information, there is no knowledge of the high

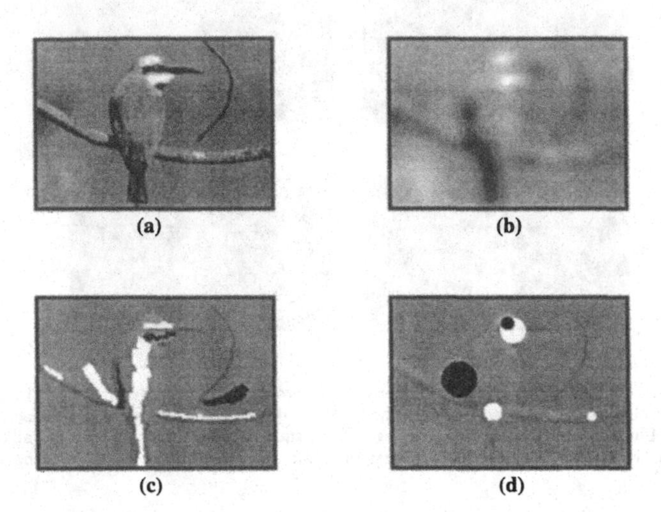

Figure16.
Contrast Reversal. (a) Original image from Figure 10 and (b) a blurred version. (c) Set of extracted limbs superimposed onto a bleached version of the image (black or white to differentiate individual structures). (d) Set of extracted blobs superimposed onto a bleached version of the image (black or white to differentiate individual structures).

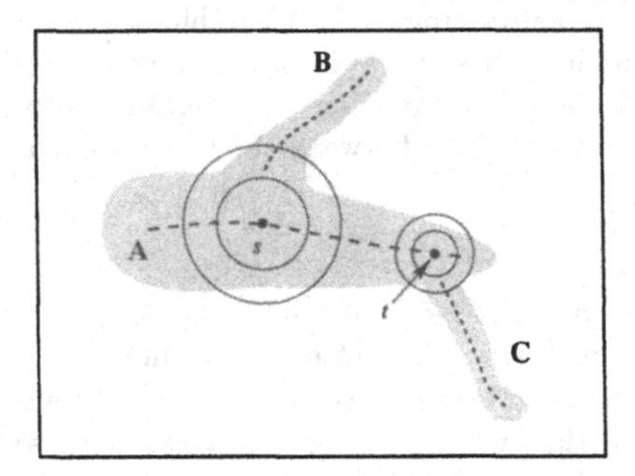

Figure17.
An object (filled) is decomposed into the three limb parts A, B, and C represented by paths (dashed) extracted from the raw symmetry data. The relationships between the constituent parts are found by first choosing the most perceptually significant part A as the root of an object graph. A search is then performed which traverses the set of points representing the part A. At centers s and t, the associated annular regions are shown by concentric circles. Note that the parts B and C are represented by paths which end within the annular regions centered at s and t, respectively. Since these annular regions define the scales at which enclosure was detected (for A) at the points s and t, material continuity is inferred between the principal part A and the secondary parts B and C.

curvature points (or corners) that are often used to indicate the presence of part junctions[2]. However, based on the locus of points which comprises a part representation, it is possible to predict when part joins will most likely occur (see Figure 17).

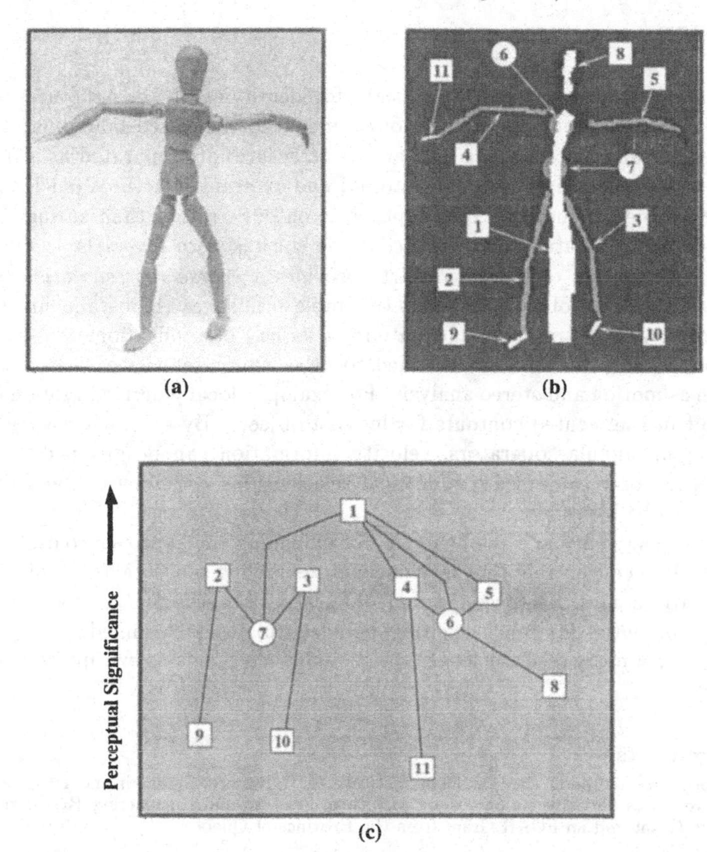

Figure18.
(a) Original image. (b) Set of joined parts labelled in order of decreasing relative perceptual significance. (c) Resulting graph that represents the joins between parts. Limb 1 extending between the doll's legs arises due to symmetries between the outer leg contours. Since the method does not explicitly employ contour curvature information, it is impossible to determine which components of 1 are associated with only the doll's torso[2].

A graph is used to capture the spatial relationships between the set of joined object parts. For a given primary part, other *secondary parts* that are joined to it are represented in terms of their spatial relationships with the primary part. The process of finding connected parts continues recursively for all other parts that are connected to any of the object parts (see

[2]While this is an obvious drawback of the method in its current form, we believe that a coarse approximation of the regional scale is often necessary before the appropriate choice of scale for a boundary-based approach can be made. The possibility of a hybrid approach - incorporating both region and boundary features - has been examined elsewhere [12]. The identification of enclosed regions permits the establishment of *stable reference points* and also permits the selection of an *appropriate scale* before attempting curvature analysis.

Figures 18). Once all parts associated with the first object are located, another principal part is found, and the process is repeated for the new (second) object.

5: Conclusion

Annular symmetry operators provide a means for identifying object structures from fragmented contour data without requiring prior segmentation or closed bounding curves. By using a multiscalar annular sampling framework, structures are separated as a function of scale. This permits the differentiation of internal and external structures (such as texture) from the overall object structure. By adopting a contour-, rather than surface-based approach, objects can be located and described when contrast sign reversals occur along the object boundary. The set of extracted parts provides a *coarse representation* of imaged objects. A description of objects in terms of simple primitives (blobs and limbs) has an appealing simplicity, and it may be adequate for a variety of applications.

The method can also be directly extended to other computer vision tasks that rely on grouping, such as motion and stereo analysis. For example, locally derived motion estimates generally result in fragmented contours for moving objects. By employing a region-based analysis employing annular operators, velocity information can be grouped across large spatial extents in order to detect symmetrical relationships - thereby enabling individual moving objects to be identified.

Based upon its simplicity and the potential for extending the approach to different visual grouping tasks, it is conceivable that mechanisms resembling multiscalar annular sampling may underly early stages of processing in biological vision systems. In the future, there is a significant potential for much exciting and fruitful research aimed at exploring the implications of symmetry and enclosure as grouping mechanisms within the context of biological vision.

Acknowledgements

The authors would like to thank the Canadian Institute for Advanced Research and PRECARN for its support. This work was partially supported by a Natural Sciences and Engineering Research Council of Canada Strategic Grant and an FCAR Grant from the Province of Quebec.

References

[1] H. Blum. Biological shape and visual science (part i). *Journal of Theoretical Biology*, 38:205–287, 1973.

[2] H. Blum and R. N. Nagel. Shape description using weighted symmetric axis features. *Pattern Recognition*, 10:167–180, 1978.

[3] M. Brady and H. Asada. Smoothed local symmetries and their implementation. Technical Report A.I.Memo 757, Massachusetts Institute of Technology AI Lab, 1984.

[4] C. A. Burbeck and S. M. Pizer. Object representation by cores: Identifying and representing primitive spatial regions. *Vision Research*, 35:1917–1930, 1995.

[5] J.B. Burns, K. Nishihara, and S.J. Rosenschein. Appropriate-scale local centers: a foundation for parts-based recognition. In *Proceedings of ARPA*, 1994.

[6] J. Elder and S. Zucker. A measure of closure. *Vision Research*, 34(24):3361–3369, 1994.

[7] D. S. Fritsch, S. M. Pizer, B. S. Morse, D. H. Eberly, and A. Liu. The multiscale medial axis and its applications in image registration. *Pattern Recognition Letters*, 15:445–452, 1994.

[8] J. M. Gauch and S. M. Pizer. The intensity axis of symmetry and its application to image segmentation. *IEEE Transactions on Pattern Analysis and Machine Intelligence*, 15(8):753–770, 1993.

[9] D. Jacobs. *Recognizing 3-D Objects Using 2-D Images.* PhD thesis, Massachusetts Institute of Technology, Cambridge Mass., 1992.

[10] T. Kanade. Recovery of the three-dimensional shape of an object from a single view. *Artificial Intelligence*, 17:409–460, 1981.

[11] L. Kaufmann and W. Richards. Spontaneous fixation tendencies for visual forms. *Perception and Psychophysics*, 5(2):85–88, 1969.

[12] M. F. Kelly. *Annular Symmetry Operators: A Multi-Scalar Framework for Locating and Describing Imaged Objects.* PhD thesis, McGill University, Montréal, Canada, 1995.

[13] M. F. Kelly and M. D. Levine. A sampling strategy using multi-scale annular operators. Technical Report TR-CIM-93-20, Center for Intelligent Machines, McGill University, Montréal, Canada, 1993.

[14] M. F. Kelly and M. D. Levine. Annular symmetry operators: A method for locating and describing objects. In *Proceeding of the* 5th *International Conference on Computer Vision*, pages 1016–1021, Cambridge, 1995.

[15] B. Kimia. *Toward a Computational Theory of Shape.* PhD thesis, McGill University, Montréal, 1990.

[16] B. B. Kimia, A. R. Tannenbaum, and S. W. Zucker. Shapes, shocks, and deformations, I: The components of shape and the reaction-diffusion space. *International Journal of Computer Vision*, 15:189–224, 1995.

[17] J. J. Koenderink. The structure of images. *Biological Cybernetics*, 50:363–370, 1984.

[18] K. Koffka. *Principles of Gestalt Psychology.* Harcourt, Brace and World, New York, 1935.

[19] W. Köhler. *Gestalt Psychology.* New American Library, New York, 1947.

[20] T. Lindeberg. Detecting salient blob-like image structures and their scales with a scale-space primal sketch: A method for focus of attention. *International Journal of Computer Vision*, 11(3):283–318, 1993.

[21] D.G. Lowe. Organization of smooth image curves at multiple scales. *Int. J. Comp. Vision*, 3:119–130, 1989.

[22] D. Marr and K. H. Nishihara. Representation and recognition of the spatial organization of three-dimensional structure. *Proc. Royal Society of London (ser. B)*, 200:269–294, 1978.

[23] R. Mohan and R. Nevatia. Perceptual organization for scene segmentation and description. *IEEE Transactions on Pattern Analysis and Machine Intelligence*, 14(6):616–634, 1992.

[24] K. Rao and R. Nevatia. Describing and segmenting scenes from imperfect and incomplete data. *CVGIP: Image Understanding*, 57:1–23, 1993.

[25] I. Rock. *The Logic of Perception.* MIT Press, Cambridge Mass., 1983.

[26] A. Rosenfeld. Axial representations of shape. *Computer Vision, Graphics, and Image Processing*, 33:156–173, 1986.

[27] E. Rubin. *Visuell wahrgenommene Figuren.* Copenhagen, 1921.

[28] E. Saund. Symbolic construction of a 2-d scale space image. *IEEE Transactions on Pattern Analysis and Machine Intelligence*, 12(8):817–830, 1990.

[29] G. L. Scott, S. C. Turner, and A. Zisserman. Using a mixed wave/diffusion process to elicit the symmetry set. *Image and Vision Computing*, 7(1):63–70, 1989.

[30] L. Shapiro, J. Moriarty, P. Mulgaonkar, and R. Haralick. Sticks, plates and blobs: a three-dimensional object representation for scene analysis. *AAAI*, 80, 1980.

[31] S. Wang, A.Y. Wu, and A. Rosenfeld. Image approximation from gray scale "medial axes". *IEEE Transactions on Pattern Analysis and Machine Intelligence*, 3(6):687–696, 1981.

[32] A. P. Witkin. Scale space filtering. In *Proceedings of the* 8th *International Joint Conference on Artificial Intelligence*, pages 1019–1022, 1983.

[33] S. C. Zhu and A. L. Yuille. Forms: a flexible object recognition and modelling system. In *Proceeding of the* 5th *IEEE International Conference on Computer Vision*, pages 465–472, Cambridge, Mass, 1995.

Networks that Learn for Image Understanding

Tomaso Poggio and Kah-Kay Sung

Abstract

Learning is becoming a central problem in trying to understand intelligence and in trying to develop intelligent machines. This paper describes some recent work on developing machines that learn in the domains of vision and graphics. We will introduce an underlying theory which connects function approximation techniques, neural network architectures and statistical methods. While these techniques have limitations, one can overcome these limitations by using the idea of virtual examples. We shall describe some learning-based systems we have developed that recognize objects, in particular faces, find specific objects in cluttered scenes, and produce novel images under user control. Finally, we will discuss about the implications of this research on how the brain might work. [1]

1 Introduction

Learning and vision are two very broad fields of research. In this paper we will introduce an underlying theory which connects supervised learning, function approximation techniques, neural network architectures and statistical methods. We shall make a few general observations about supervised learning approaches in vision. Then we will briefly describe some learning-based image understanding (IU) applications developed at the MIT Center of Biological and Computational Learning. Finally, we will discuss about the implications of this research on how the brain might work.

Vision systems that *learn and adapt* represent one of the most important future directions in IU research. This reflects an overall trend – to make intelligent systems that do not need to be fully and painfully programmed. It may be the only way to develop vision systems that are robust and easy to use in many different tasks.

Building systems without explicit programming is not a new idea. Duda and Hart [10] is still one of the best textbooks. However, there are extensions of the classical pattern recognition techniques, extensions often called Neural Networks. Neural Networks have provided a new metaphor – learning from examples – that makes statistical techniques more attractive. As a consequence of this new interest in learning, we are witnessing a renaissance of statistics and function approximation techniques and their applications to new domains such as vision.

[1] This paper describes research done within the MIT Center for Biological and Computational Learning in the Department of Brain and Cognitive Sciences, and at the MIT Artificial Intelligence Laboratory. This research is sponsored by grants from the Office of Naval Research under contracts N00014-91-J-1270 and N00014-92-J-1879; by a grant from the National Science Foundation under contract ASC-9217041 (funds provided by this award include funds from DARPA provided under the HPCC program). Additional support is provided by the North Atlantic Treaty Organization, ATR Audio and Visual Perception Research Laboratories, Kodak, Daimler Benz and Siemens AG. Tomaso Poggio is supported by the Uncas and Ellen Whitaker Chair. Kah-Kay Sung is currently a lecturer at the Department of Information Systems and Computer Science, National University of Singapore.

0-8186-7644-2 $5.00 © 1996 IEEE

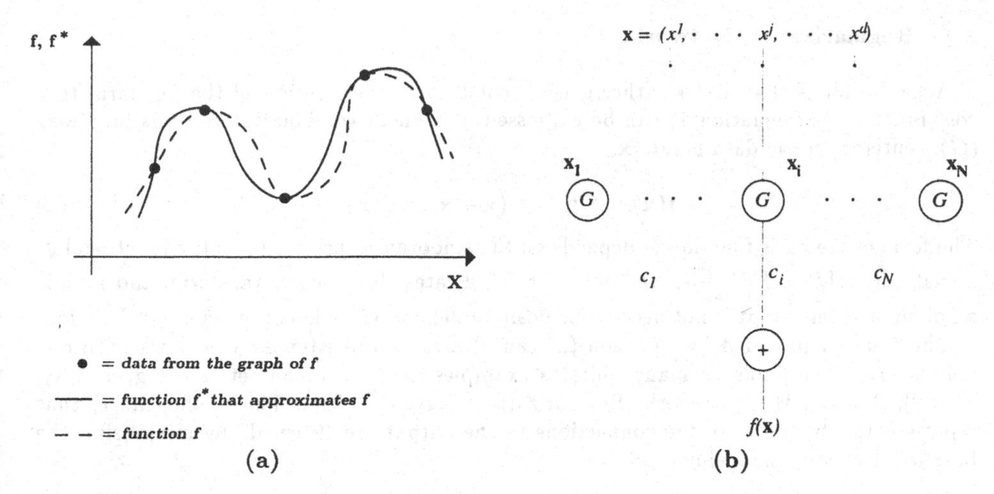

Figure 1. **(a):** Learning-from-examples as multivariate function approximation or interpolation from sparse data. Generalization means estimating $f^*(x) \approx f(x)$, $\forall x \in X$ from the examples $f^*(x_i) = f(x_i)$, $i = 1, \ldots, N$. **(b):** A Regularization Network. The input vector $\mathbf{x}$ is d-dimensional, there are N hidden units, one for each example $\mathbf{x}_i$, and the output is a scalar function $f(\mathbf{x})$.

2 Learning-from-Examples as Multivariate Function Approximation

In this section we will concentrate on one aspect of learning: *supervised learning.* Supervised learning, or *learning-from-examples*, refers to systems that are trained, instead of programmed – by a set of examples, that is input-output pairs $(\mathbf{x}_i, \mathbf{y}_i)$. At run-time these systems would hopefully provide a correct output for a new input not contained in the training set. One can formulate *supervised learning* as a regression problem of approximating a multivariate function from sparse data, i.e. the examples (Figure 1(a)). Generalization means estimating the value of the function for points in the input space where data is not available.

Once the ill-posed problem of learning-from-examples has been formulated as a function approximation problem, an obvious approach to solving it is *regularization.* Regularization imposes a constraint of smoothness on the space of approximating functions by minimizing the cost functional:

$$H[f] = \sum_{i=1}^{N}(y_i - f(\mathbf{x}_i))^2 + \lambda\phi[f], \tag{1}$$

where the stabilizer ϕ is typically a measure of smoothness of the solution f.

The functional regularization approach can also be regarded from a slightly more general probabilistic and Bayesian perspective. In particular, as Girosi, Jones and Poggio [15] [14] (see also Poggio and Girosi [23],[22] and Wahba [28]) describe, a Bayesian approach leads to the maximum *a posteriori* (MAP) estimate of $P(f|\mathcal{D}) \propto P(f)\ P(\mathcal{D}|f)$, where the set $\mathcal{D} = (\mathbf{x}_i, \mathbf{y}_i)_{i=1}^{N}$ consists of the input-output pairs of training examples. *Iff* the noise model is additive and Gaussian (i.e. $P(\mathcal{D}|f)$ is Gaussian) and the prior $P(f)$ is a Gaussian distribution of a linear functional of f, then the MAP estimate, that is the f maximizing $P(f|\mathcal{D})$, is equivalent to the f that minimizes equation (1).

2.1 Regularization Networks

A key result is that under rather general conditions, the solution of the regularization cost functional in equation(1) can be expressed as a linear combination of basis functions (G), centered on the data points $\mathbf{x}_i$:

$$f(\mathbf{x}) = \sum_{i=1}^{N} c_i G(\mathbf{x} - \mathbf{x}_i) + p(\mathbf{x}), \qquad (2)$$

The form of the basis function G depends on the smoothness prior, that is the functional ϕ. Specifically, $\phi[f] = \int \frac{|\tilde{f}(\mathbf{s})|^2}{\tilde{G}(\mathbf{s})} d\mathbf{s}$, the operator $\tilde{}$ indicates the Fourier transform, and $p(\mathbf{x})$ is a polynomial term that is not needed for positive definite G such as the Gaussian function.

The solution provided by equation (2) can always be rewritten as a network with one hidden layer containing as many units as examples in the training set (see Figure 1(b)) [23] [8]. We call these networks Regularization Networks (RN). The coefficients c_i that represent the "weights" of the connections to the output are "learned" by minimizing the functional H over the training set.

2.2 Radial Basis Functions

An interesting special case arises for radial ϕ. In this case the basis functions are radial and the network is

$$f(\mathbf{x}) = \sum_{i=1}^{N} c_i G(\|\mathbf{x} - \mathbf{x}_i\|) + p(\mathbf{x}).$$

Legal G's include the Gaussian, the multiquadric, thin-plate splines and others [22]. In the Gaussian case, these RBF networks consist of units, $G(\|\mathbf{x} - \mathbf{x}_i\|)$, each tuned to one of the examples, $\mathbf{x}_i$, with a bell-shaped activation curve. The examples, $\mathbf{x}_i$, which are the unit centers, behave like *templates*.

"Vanilla" RBFs can be generalized to the case in which there are fewer units than data, and the centers $\mathbf{x}_i$ are to be found during the learning phase of minimizing the cost over the training set. A weighted norm can also be introduced instead of the Euclidean norm and again the associated matrix can be found during the learning phase. These generalized RBF networks have sometimes been called HyperBF networks [22].

3 Three observations

Given this general framework for looking at the problem of learning-from-examples, three observations are relevant.

3.1 Regularization Provides a General Theory

The first point is summarized in Figure 2: *several representations for function approximation and regression as well as several Neural Network architectures can all be derived from regularization principles with somewhat different prior assumptions on the smoothness of the function space (the stabilizer ϕ).* They are therefore quite similar to each other.

In particular, the radial class of stabilizer is at the root of the techniques on the left branch of the diagram: RBF can be generalized into HyperBF and into some kernel methods and various types of multidimensional splines. A class of priors combining smoothness and additivity [15] is at the root of the middle branch of the diagram: additive splines of many different forms generalize into ridge regression techniques, such as the representations used in Projection Pursuit Regression [12], hinges [7] and several multilayer perceptron-like networks (with one hidden layer).

3.2 The Learning-from-Examples Framework almost implies a View-based Approach to Object Recognition

The second general point is that the simplest version of an example-based approach to object recognition – and, as we will see later, to graphics – is equivalent to a view-based approach to recognition. Consider a learning-from-examples network trained to classify different views of a 3D object against views of other distractor objects. Suppose the network has a Gaussian Radial Basis Function form. Then, as discussed by Logothetis, et al. [19] [20], each hidden unit is view-tuned to one of the example views, whereas the output can be view invariant but object specific (if a sufficient number of examples and units are available).

3.3 Prior Information is needed to Leverage the Example Set

The third and last observation has to do with sample complexity – how many "training examples" are needed for a supervised learning scheme to reasonably solve a specific problem.

Niyogi and Girosi [18] have recently proven the following result, which extends to Radial Basis Functions results due to Barron [2] for multilayer perceptrons:

Theorem 1 *Let $f \in H^{m,1}(R^d)$, where $H^{m,1}(R^d)$ is the space of functions whose partial derivatives up to order m are integrable, and let $m > d$, with m even.*

Let $f_{n,N}$ be a Gaussian basis function network, with n coefficients, n centers and n variances, that has been trained on a set of N data points. Then, with probability $1-\delta$, the generalization error is bounded by the following inequality:

$$E[(f - f_{n,N})^2] < O\left(\frac{1}{n}\right) + O\left(\left(\frac{nd\ln(nN) - \ln\delta}{N}\right)^{\frac{1}{2}}\right)$$

The plot of Figure 3 summarizes the main result of the theorem. Generalization error on new data, not used for training, depends on the number of examples N and on the complexity of the network, described here by the number of its free parameters n. Notice that for each N there is an optimal n. In many cases, especially when N is not very large, the choice of a good n is often more important in practice than the choice of one regression or classification architecture vs. another.

The main point here is however a simpler one: very often we are forced to work in the region of the plot with very small N, i.e., there is not enough training data. This, we believe, is one of the most basic limitations of all the nonparametric learning-from-examples schemes. What can be done? Poggio and Vetter [24] have pursued the idea of exploiting *prior information* about the specific problem at hand to generate *virtual examples* and thus expand the size of the training set, that is N. We shall describe specific instances of this approach for face recognition and face detection.

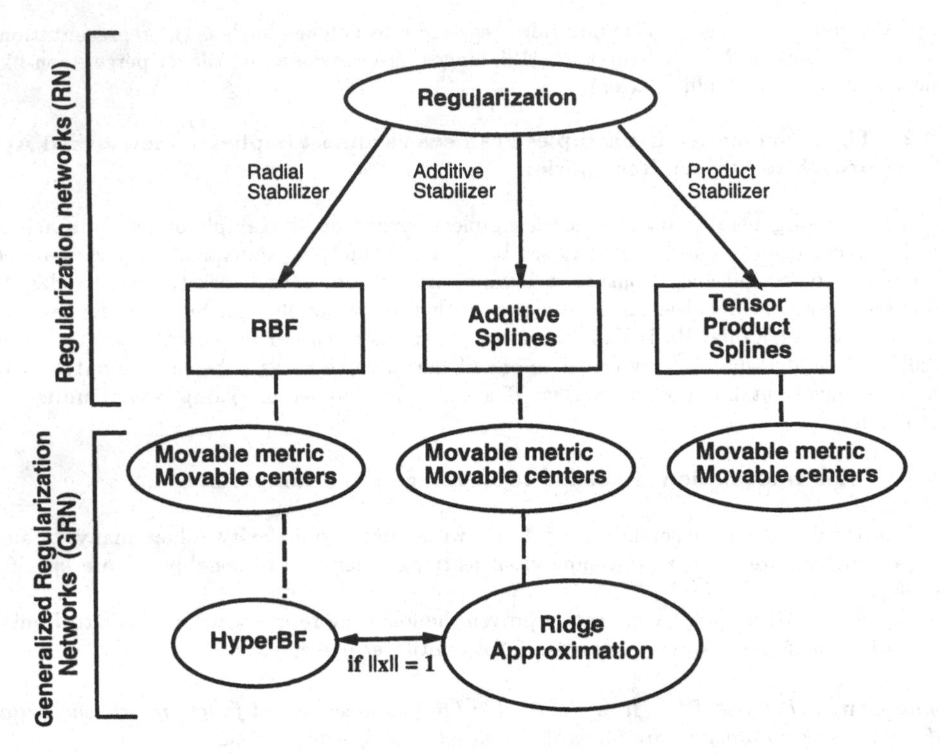

Figure 2. Several classes of approximation schemes and corresponding network architectures can be derived from regularization with the appropriate choice of smoothness priors, associated stabilizers and basis functions. From Girosi, Jones, and Poggio (1995).

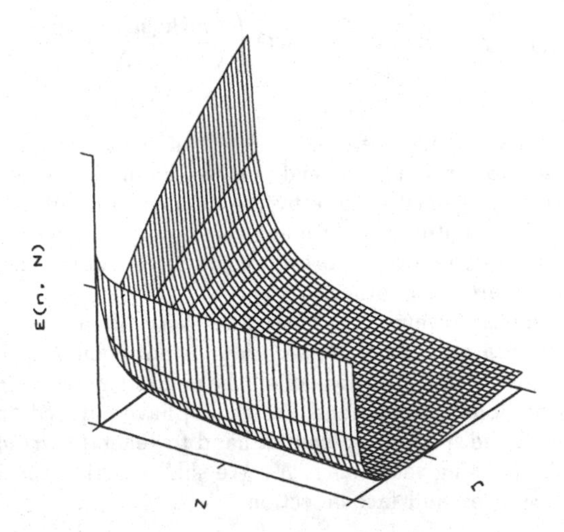

Figure 3. An illustration of the theorem. From Niyogi and Girosi (1994).

4 Learning-from-Examples for Image Understanding

We now describe three specific projects in our group that use the learning-from-examples techniques for approaching problems in computer vision and computer graphics, using human faces as the class of 3D objects.

4.1 Example-based Image Analysis and Image Synthesis

In the image analysis problem, a learning module is trained to associate input images to output parameters – such as the classification label for the object in the image and possibly pose and expression parameters associated with it. After training with a set of images and corresponding parameters, the network is expected to generalize correctly; that is, to classify new views of the same object, recognize another object of the same class, or estimate its pose/expression parameters.

In the "inverse" problem of image synthesis, a learning module is used to associate input parameters to output images. This module synthesizes new images and represents a rather unconventional approach to computer graphics. It takes several real images of a possibly non-rigid 3D object, and creates new images by generalizing from those views, under the control of appropriate pose/expression parameters assigned by the user during the training phase.

4.1.1 The Analysis Network

Given a set of example images labeled in pose and expression space (see Figure 4(a)), a RBF network (see equation (2)) can be trained to estimate the pose-expression parameters of new images of the same person. The network input, or our image representation, is a geometrical representation of (x, y) locations of facial features as pixel-wise correspondence with respect to a reference image [3], here chosen to be the upper left image in Figure 4(a). There are two parameters in the output, representing the amount of smile and rotation of the face. The network is constructed using four Gaussian RBF hidden units, one for each of the example images. After "training" with the four examples the network can "generalize" to new images of the same person, estimating the associated rotation and smile parameters. Examples of input images and their estimated parameters are shown in Figure 5(b).

4.1.2 The Synthesis Network

In the synthesis case the role of inputs and outputs are exchanged. Figure 5(a) shows the network [3] used to synthesize images of a specific person's face under control of two pose-expression parameters, the same ones used by the analysis network. For network output, since our image representation uses pixel-wise correspondence with respect to a reference image, the number of outputs, q, is twice the number of pixels in the image (each pixel has an x and y coordinate). The output geometry produced by the network is then "rendered" by applying 2D warping operations to the example image "textures".

Figure 4(b) shows some example images generated by the synthesis network. The four examples used here are the same as for the analysis case and are shown at the four corners of the figure. All the other images are "generalizations" produced by the network after the "learning" phase.

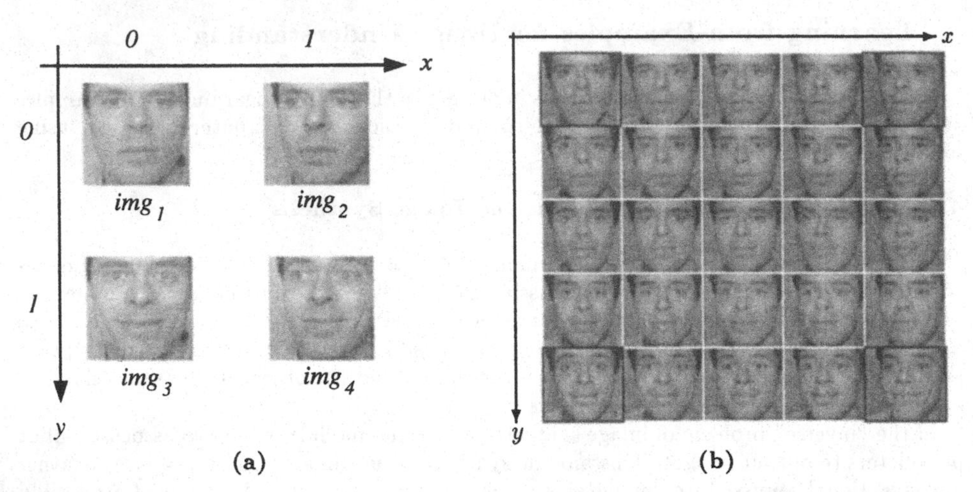

Figure 4. **(a)**: In our demonstrations of the example-based approach to image analysis and synthesis, the example images img_1 through img_4 are placed in a 2D rotation-expression parameter space (x, y), here at the corners of the unit square. For analysis, the network learns the mapping from images (inputs) to parameter space (output). For synthesis, we synthesize a network that learns the inverse mapping, that is the mapping from the parameter space to images. From Beymer et al. (1993). **(b)**: Multidimensional image synthesis using the example images bordered in black. All other images are generated by the synthesis network. From Beymer et al. (1993) .

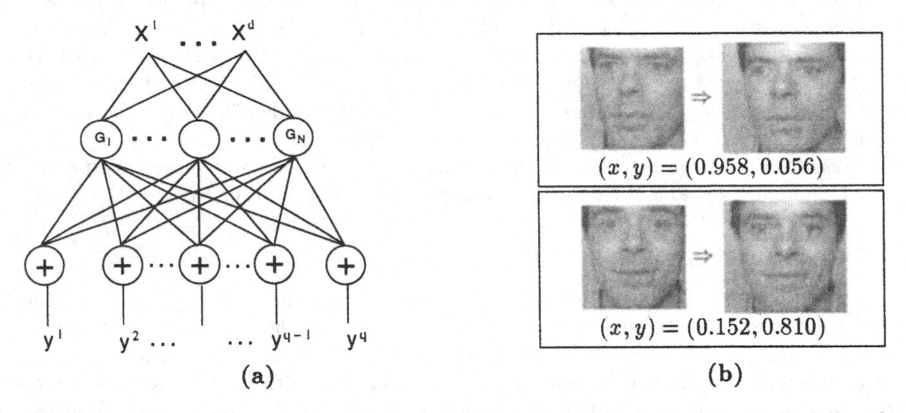

Figure 5. **(a)**: The network used to synthesize images from pose inputs. It differs from the Regularization Network in Figure 1(b) simply by having more outputs, q in this case. Since we use a pixel wise representation for our output images, q is proportional to the number of pixels in the image. **(b)**: In each of the boxed image pairs, the novel input image on the left is fed into an analysis RBF network to estimate rotation x and expression y. These parameters are then fed into the synthesis module of figure 4(b) that synthesizes the image shown on the right. This figure can be regarded as a very simple demonstration of very-low bandwidth teleconferencing: only two pose parameters need to be transmitted at run-time for each frame. From Beymer et al. (1993) .

4.1.3 Very-low Bandwidth Video E-Mail

The analysis and the synthesis network can each be used independently for a variety of applications. A particularly interesting application is very-low bandwidth video e-mail and video-conferencing. Figure 5(b) shows a demonstration of the basic idea. An analysis network trained on a few views of the specific user analyzes new images during the session in terms of a few pose and expression parameters. These few parameters are sent electronically for each frame and used at the receiver site by a similar synthesis network to reconstruct appropriate new views. This model-based approach can achieve in principle very high compression.

4.2 Face Recognition

We will now summarize our work over the last 4 years on face recognition, first covering two view-based face recognition systems that rely on several example views per person. We will then discuss work on synthesizing virtual example views to deal with situations where only one example view (also called model view) per person is available.

4.2.1 The Brunelli-Poggio frontal face recognition system

Following the face recognition work of Baron [1], Brunelli and Poggio [9] use a template-based strategy to recognize frontal views of faces. From an example model view, faces are represented using templates of the eyes, nose, mouth, and entire face. A normalized correlation metric is used to compare the model templates against the input image.

During recognition, input images are geometrically registered with the model views by aligning the eyes, which normalizes the input for the effects of translation, scale, and image-plane rotation. On a data base of 47 people, 4 views per person, the recognition performance is 100% when two views are used as model views and the remaining two views are used for testing. Using an algorithm very similar to the Brunelli-Poggio system, Gilbert and Yang [13] have also developed a fast PC-based face recognition system that uses custom VLSI chips for correlation.

4.2.2 The Beymer pose-invariant face recognition system

In a view-based extension of the Brunelli-Poggio system, Beymer [5] [6] developed a pose-invariant face recognizer that uses 15 views per person, views that cover different out-of-plane rotations (see Figure 6). As in the previous system, translation, scale, and image-plane rotation are factored out by first detecting eyes and nose features and then using these features to register the input with model views. To recognize a new input, the image is matched against all model views of all people and the best match is reported. The matching step performs normalized correlation using eyes, nose, and mouth templates. On a data base of 62 people and 10 test views per person, the system obtained a recognition rate of 98%.

4.2.3 Face Recognition with Virtual Views

As discussed previously, the key problem for the practical use of learning-from-examples schemes – and for non-parametric techniques in general – is often the limited size of the training set. Recently, we have seen how one can exploit prior knowledge of symmetry properties in 3D objects to synthesize additional training examples [24]. The idea is to

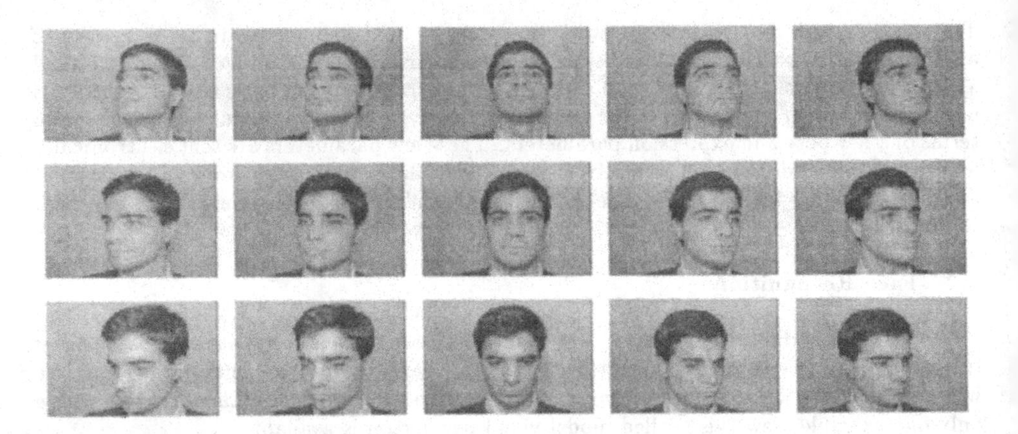

Figure 6. The pose-invariant, view-based face recognizer uses 15 views to model a person's face. From Beymer (1993) .

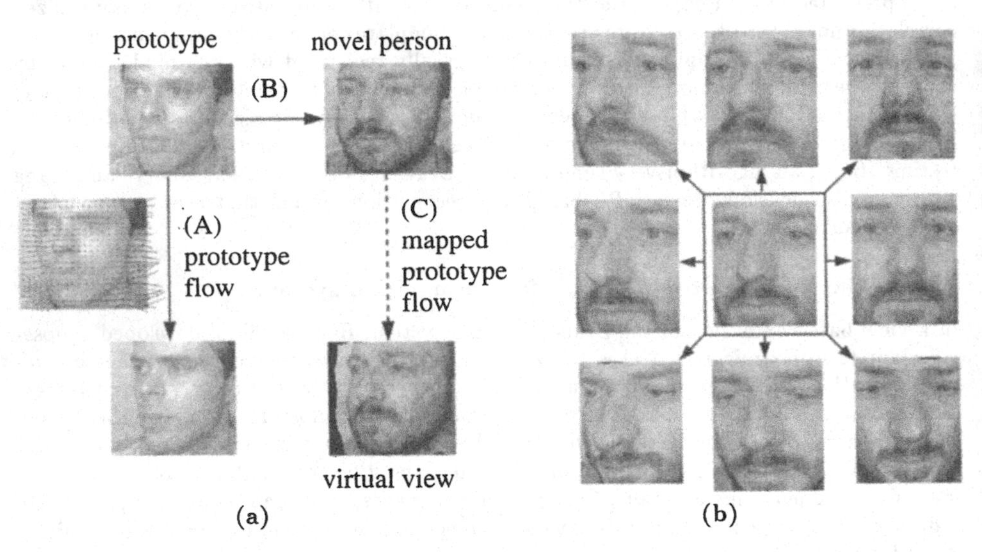

Figure 7. (a): In parallel deformation, (A) a 2D deformation representing a transformation is measured by finding correspondence among prototype images. In this example, the transformation is rotation and optical flow was used to find a dense set of correspondences. Next, in (B), the flow is mapped onto the novel face, and (C) the novel face is 2D warped to a "virtual" view. From Beymer and Poggio (1995b) . **(b):** A real view (center) surrounded by virtual views derived from it using parallel deformation. From Beymer and Poggio (1995b) .

learn class-specific transformations from example objects, and synthesize virtual views of new objects in the same class from a single view.

Figure 7(a) [4] shows one way of generating virtual face views by applying a transformation "learned" from prototypes of the same class (faces). Called *parallel deformation* [21], a 2D deformation measured on the prototype faces is mapped onto the novel face. The mapped deformation then drives a 2D warping of the novel face to the virtual view. Figure 7(b) shows the result of applying this technique to produce several rotated *virtual views* of the same face from a single real image.

The virtual views generated in this way have been used as model views in the pose-invariant recognizer described earlier to achieve recognition rates that are likely to compare well to human performance in the same situation [4].

4.3 Example-based Face Detection

Our last project deals with the general problem of object and pattern detection in cluttered pictures. It has been often said that this problem is even more difficult than the problem of isolated object recognition [16]. We have developed [27] [25] an example-based learning approach for locating vertical frontal views of human faces in cluttered scenes. The approach uses learning-based methods to complement human knowledge for capturing complex variations in face patterns.

4.3.1 Learning a Distribution-based Face Model

The first learning task is to build a distribution-based model for describing possible variations in face image appearances (see Figure 8). We use a representative sample of face patterns to approximate the distribution of frontal face views in a normalized 19×19 pixel image vector space. We also use a carefully chosen sample of "face-like" non-face patterns to help localize the boundaries of the face distribution. The final model consists of a few (6) elongated Gaussian "face" clusters that coarsely describe the frontal face pattern distribution in the 19×19 pixel image vector space, and a few (6) "non-face" clusters that explicitly carve out regions in the image vector space that do not correspond to face views. One can interpret these model clusters as a set of view-based "face" and "non-face" prototypes in a HyperBF object recognition network architecture.

4.3.2 Learning a Similarity Measure for Identifying Face Patterns

To search for faces at different image locations and scales, our system resizes each square image window to 19×19 pixels and matches the resized pattern against the distribution-based face model. The matching stage computes a "difference" feature vector of 12 directionally dependent distances between the resized pattern and the 12 model centers. Here, we also use example-based learning techniques to train a network classifier that identifies face patterns from non-face windows, based on their "difference" feature vector measurements with the model. For this stage, our experiments have shown that the choice of one network architecture vs. another is not very crucial.

4.3.3 Virtual Example Generation and Example Selection

Figures 9 and 10(a) show some face detection results by our system. For high quality CCD images, the system has a 96.3% face detection rate with an extremely low false alarm rate. The system's robust performance was largely due to the way we used virtual views to

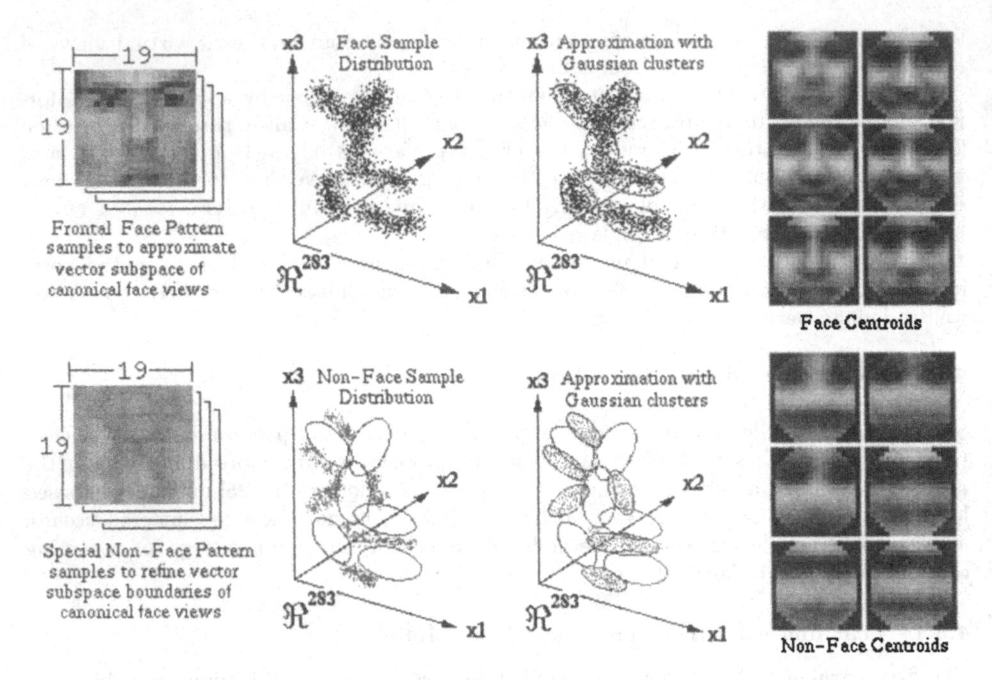

Figure 8. Our distribution-based face model. **Top Row:** We use a representative sample of frontal face patterns to approximate the distribution of frontal face views in a normalized 19×19 pixel image window vector space. We model the face sample distribution with 6 multi-dimensional Gaussian clusters whose centers are as shown on the right. **Bottom Row:** We use a carefully chosen sample of non-face patterns to help localize the boundaries of the face distribution. We also model the non-face sample distribution with 6 multi-dimensional Gaussian clusters. The final model consists of 6 Gaussian "face" clusters and 6 "non-face" clusters.

significantly increase the number of face training patterns. We generated 3000 virtual face examples by applying some simple image transformations to the 1000 real face views we had in our training database. For practical reasons, we also had to constrain the number of non-face patterns in our training data set without unnecessarily sacrificing the quality of negative examples. To do this, we used an incremental scheme for selecting high quality non-face training examples [25] [26]. The scheme starts with a small number of non-face patterns in the training database, and gradually adds new non-face patterns that the current system wrongly classifies as faces.

5 Object Recognition — Psychophysical and Physiological Connections

As shown in the previous sections, the example-based approach is successful in practical problems of object recognition, object detection, image analysis and image synthesis. We conclude by asking whether a similar approach may be used by the human brain. Over the last four years, psychophysical experiments have indeed supported the example-based

Figure 9. Some face detection results. From Sung and Poggio (1994) .

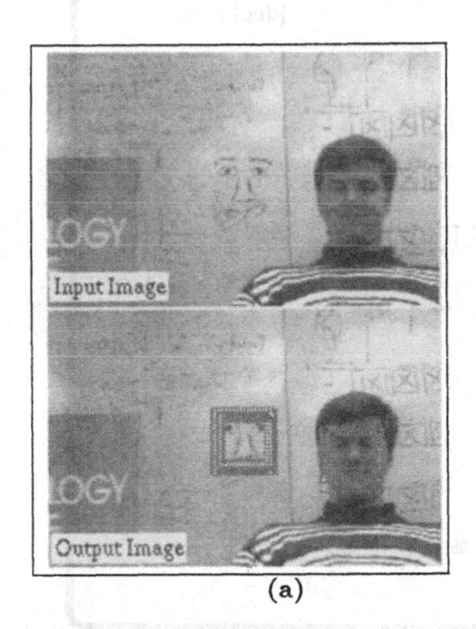

(a)

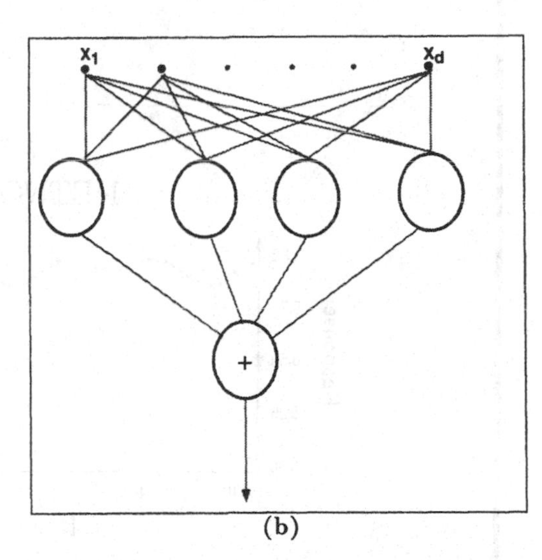

(b)

Figure 10. (**a**): More face detection results. The same system finds real faces as well as hand drawn faces. From Sung and Poggio (1994). (**b**): A RBF network with four units each tuned to one of the four training views shown in the next figure. The tuning curve of each of the unit is also shown in the next figure. The units are view-dependent and selective, relative to distractor objects of the same type.

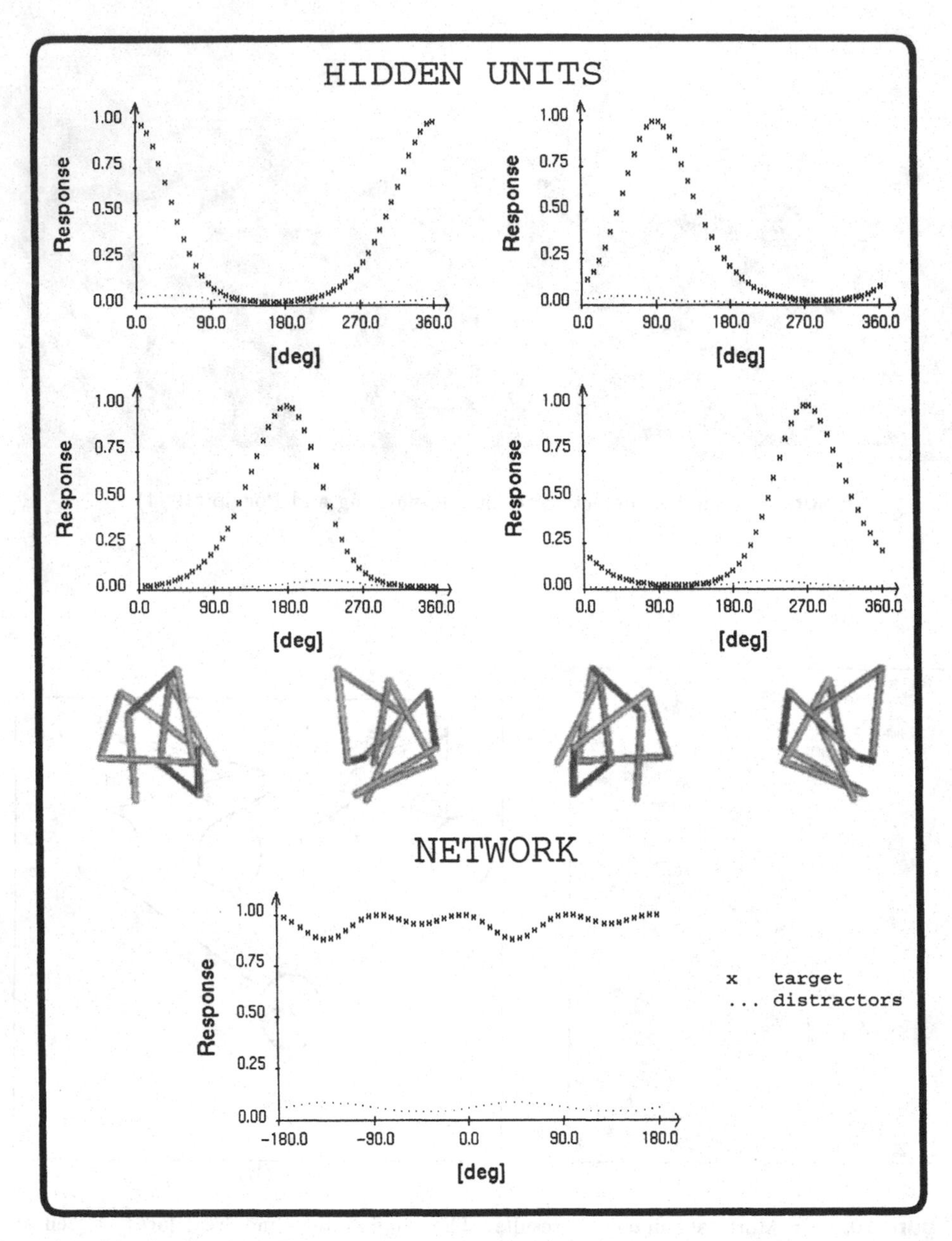

Figure 11. Tuning of each of the four hidden units of the network of the previous figure for images of the "correct" 3D objects. The tuning is broad and selective: the dotted lines indicate the average response to 300 distractor objects of the same type. The bottom graphs show the tuning of the output of the network after learning (that is computation of the weights c): it is view-invariant and object specific. Again the dotted curve indicates the average response of the network to the same 300 distractors. From Vetter and Poggio (unpublished).

and view-based schemes that we suggested as one of the mechanisms for object recognition. Very recently, physiological experiments have also provided a suggestive glimpse on how neurons in IT cortex may represent objects for recognition. The experimental results seem to agree surprisingly well with our model [17].

Figure 10(b) shows our basic module for object recognition. Classification or identification of a visual stimulus is accomplished by a network of units. Each unit is broadly tuned to a particular view of the object. We refer to this optimal view as the center of the unit. One can think of each unit center as a template to which the input is compared. The unit is maximally excited when the stimulus exactly matches its template but also responds proportionately less to similar stimuli. The weighted sum of activities of all the units represents the output of the network.

Consider how the network "learns" to recognize views of the object shown in Figure 11. In this example the inputs of the network are the (x, y) positions of the vertices of the wireframe object in the image. Four training views are used. After training, the network consists of four units, each one tuned to one of the four views as in Figure 11. The weights of the output connections are determined by minimizing misclassification errors on the four views and using as negative examples views of other similar objects ("distractors").

The graphs show the tuning of the four units for images of the "correct" object. The tuning is broad and centered on the the training views. Somewhat surprisingly, the tuning is also very selective: the dotted line shows the average response of each of the unit to 300 similar distractors [11]. Even the maximum response to the best distractor is in this case always less than the response to the optimal view.

Despite its gross oversimplification, our model relates well to some recent psychophysical and physiological findings, in particular to the existence of view-tuned and view-invariant neurons in the primate IT cortex, and to the shape of psychophysically measured recognition fields. We refer interested readers to the following pieces of work by Logothetis and co-workers for further details [17], [19] [20].

References

[1] Robert J. Baron. Mechanisms of human facial recognition. *International Journal of Man Machine Studies*, 15:137–178, 1981.

[2] A.R. Barron. Approximation and estimation bounds for artificial neural networks. *Machine Learning*, 14:115–133, 1994.

[3] D. Beymer, A. Shashua, and T. Poggio. Example based image analysis and synthesis. AIM-1431, AI Lab., MIT, 1993.

[4] David Beymer and Tomaso Poggio. Face recognition from one example view. In *Proceedings of the International Conference on Computer Vision*, pages 500–507, Cambridge, MA, June 1995.

[5] David J. Beymer. Face recognition under varying pose. AIM-1461, AI Lab., MIT, 1993.

[6] David J. Beymer. Face recognition under varying pose. In *Proceedings IEEE Conf. on Computer Vision and Pattern Recognition*, pages 756–761, Seattle, WA, 1994.

[7] L. Breiman. Hinging hyperplanes for regression, classification, and function approximation. *IEEE Transaction on Information Theory*, 39(3):999–1013, May 1993.

[8] D.S. Broomhead and D. Lowe. Multivariable functional interpolation and adaptive networks. *Complex Systems*, 2:321–355, 1988.

[9] R. Brunelli and T. Poggio. Face recognition: Features versus templates. *IEEE Transactions on Pattern Analysis and Machine Intelligence*, 15(10):1042–1052, 1993.

[10] R. O. Duda and P. E. Hart. *Pattern Classification and Scene Analysis.* Wiley, New York, 1973.

[11] S. Edelman and H. H. Bülthoff. Orientation dependence in the recognition of familiar and novel views of 3D objects. *Vision Research*, 32:2385–2400, 1992.

[12] J.H. Friedman and W. Stuetzle. Projection pursuit regression. *Journal of the American Statistical Association*, 76(376):817–823, 1981.

[13] J.M. Gilbert and W. Yang. A real-time face recognition system using custom VLSI hardware. In *IEEE Workshop on Computer Architectures for Machine Perception*, pages 58–66, New Orleans, LA, December 1993.

[14] F. Girosi, M. Jones, and T. Poggio. Priors, stabilizers and basis functions: From regularization to radial, tensor and additive splines. AIM-1430, AI Lab., MIT, 1993.

[15] F. Girosi, M. Jones, and T. Poggio. Regularization theory and neural networks architectures. *Neural Computation*, 7:219–269, 1995.

[16] A. Hurlbert and T. Poggio. Do computers need attention? *Nature*, 321:651–652, 1986.

[17] N.K. Logothetis and J. Pauls. Psychophysiological and physiological evidence for viewer-centered object representations in the primate. *Cerebral Cortex*, 5:270–288, 1995.

[18] P. Niyogi and F. Girosi. On the relationship between generalization error, hypothesis complexity, and sample complexity for radial basis functions. AIM-1467, AI Lab., MIT, 1994.

[19] N.K. Logothetis, J. Pauls and T. Poggio. View-dependent object recognition by monkeys. *Current Biology*, 4:401–414, 1994.

[20] N.K. Logothetis, J. Pauls and T. Poggio. Shape representation in the inferior temporal cortex of monkeys. *Current Biology*, 5:552–563, 1995.

[21] T. Poggio and R. Brunelli. A novel approach to graphics. AIM-1354, AI Lab., MIT, 1992.

[22] T. Poggio and F. Girosi. Networks for approximation and learning. *Proceedings of the IEEE*, 78(9), September 1990.

[23] T. Poggio and F. Girosi. Regularization algorithms for learning that are equivalent to multilayer networks. *Science*, 247:978–982, 1990.

[24] T. Poggio and T. Vetter. Recognition and structure from one 2D model view: observations on prototypes, object classes and symmetries. AIM-1347, AI Lab., MIT, 1992.

[25] K. Sung. *Learning and Example Selection for Object and Pattern Detection.* PhD thesis, Massachusetts Institute of Technology, Cambridge, MA, 1995.

[26] K. Sung and P. Niyogi. Active learning for function approximation. In *Advances in Neural Information Processings Systems 7*, pages 593–600, MIT Press, 1995.

[27] K. Sung and T. Poggio. Example-based learning for view-based human face detection. In *Proceedings from Image Understanding Workshop*, Monterey, CA, 1994.

[28] G. Wahba. *Splines Models for Observational Data.* Series in Applied Mathematics, Vol. 59, SIAM, Philadelphia, 1990.

A Comparative Study of Three Paradigms for Object Recognition - Bayesian Statistics, Neural Networks and Expert Systems. *

J. K. Aggarwal, Joydeep Ghosh, Dinesh Nair and Ismail Taha
Computer and Vision Research Center
The University of Texas at Austin, Austin, TX, USA
email: aggarwaljk@mail.utexas.edu

Abstract

Object recognition, which involves the classification of objects into one of many a priori known object types, and determining object characteristics such as pose, is a difficult problem. A wide range of approaches have been proposed and applied to this problem with limited success. This paper presents a brief comparative study of methods from three different paradigms for object recognition: Bayesian, Neural Network and Expert Systems.

1 Introduction

Recognizing 3-dimensional (3D) objects from 2-dimensional (2D) images is an important part of computer vision [1]. The success of most computer vision applications (robotics, automatic target recognition, surveillance, etc.) is closely tied with the reliability of the recognition of 3D objects or surfaces. The study of object recognition and the development of experimental object recognition systems has a great impact on the direction and content of research pursued by the computer vision community. Thus, it is not surprising that a plethora of paradigms, algorithms and systems have been proposed over the past two decades towards this problem. However, a versatile solution to this problem still evades the reach of even the best researchers, with only partial solutions and limited success in constrained environments being the state of the art. In fact, some researchers hold that it is not possible to design an object recognition system that is functional for a wide variety of scenes and environments and is still as efficient as a situation-specific system.

The difficulty in obtaining a general and comprehensive solution to this problem can be attributed to the complexity of object recognition in itself, as it involves processing at all levels of machine vision: lower-level vision, as with edge detection and image segmentation; mid-level vision, as with representation and description of pattern shape, and feature extraction; and higher-level vision, as with pattern category assignment, matching and reasoning (figure 1). The success of an object recognition system depends upon succeeding at all these levels. The task is made difficult by several factors, such as not knowing how many objects are there in an image, the possibility that the objects may be occluded, the possibility that unknown objects appear in the image, motion of the object, and variations

*This work was supported by the Army Research Office under Contracts DAAH 049510494 and DAAH 0494G0417.

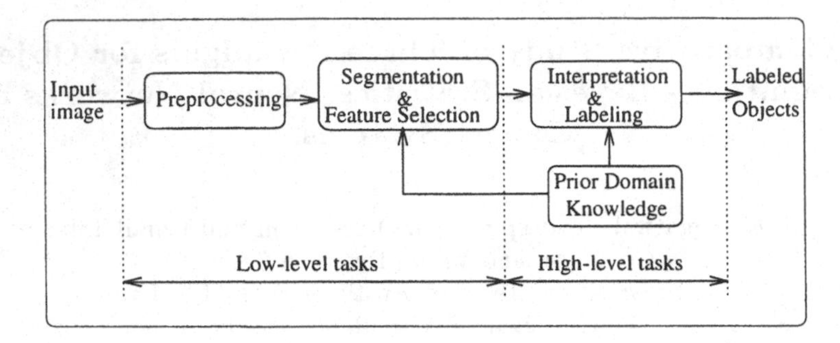

Figure 1. Levels of processing in an object recognition system.

in the sensing environment and the limits and accuracy of the sensor. In applications such as automatic target recognition (ATR), where targets have to be recognized in difficult outdoor scenes and adverse conditions, additional factors such as noise in the form of clutter, and deliberate mis-information such as camouflage, mislead the recognition system making the recognition process more difficult.

Recognition is the process of finding a correspondence between certain features in the image and similar features of the object model [2]. The most important issues involved in this process are: (a) identifying the type of features to use, and (b) determining the procedure to obtain the correspondence between image and model features. The reliability and efficiency of an object recognition system directly depends on how carefully these are addressed. Generally, recognition follows a bottom-up approach where the processing flows from the left to right in figure 1 and features extracted from an image are classified into one of many object types. However, attempts have also been made to approach the problem using a top-down perspective, where recognition is performed by determining if one of the many known objects appear in the image.

1.1 Model-Based Recognition

In this type of recognition, a 3D model(s) of the object(s) to be recognized is available. The 3D model contains concise and complete information about the object in terms of shape descriptions [3], object parts information, relationship between object parts, etc. The 3D structure of an object is frequently represented by CAD models [4], where volume-based representations of the object is built using primitives such as generalized cones, generalized cylinders and spheres. Typically, recognition involves extracting 3D information from the image and comparing it with the model features [4], or deriving a 2D description from the image and then comparing it with 2D projections of the model. In using the former method, the sensing device should be able to provide 3D information in some form (such as range data or depth information using a stereo setup) which can then be compared with the model. In the latter case, the task is a more difficult because the effects of self-occlusions and perspective must be considered, and the projection direction needs to be determined. In [5] the 3D structure of an object is constructed using an observed sequence of silhouettes. During matching, the 3D structure of the unknown object is constructed from different image views and more views are added to the construction process until features extracted from the object matches one of the object models. A comprehensive survey of model-based vision systems using dense-range images is presented in [6] and a recent survey is found in [10].

1.2 View-Based Object Recognition

View-based object recognition is often referred to as *viewed-centered* or *2D* object recognition, because direct information about the 3D structure (such as a 3D model) of the object is not available; the only a priori information is in the form of representations of the object viewed at different angles (aspects) and distances. Each representation (or characteristic view) describes how the object appears from a single viewpoint, or from a range of viewpoints yielding similar views. There is evidence showing that object recognition by humans is viewer-centered rather than object-centered [7].

The characteristic views may be obtained by building a database of images of the object or may be rendered from a 3D model of the object [8], [9]. Matching in this case is simpler than in model-based recognition because it involves only a 2D/2D comparison. However, the space requirements for representing all the characteristics views of an object tends to be considerable. Also, the number of model features to search among increases, because each characteristic view can be considered to be a model. Methods to reduce the search space have been addressed by grouping similar views [10] [11].

Broadly speaking, there are two ways to approach this problem. The first is based on matching salient information, such as corner points, lines, contours etc., that has been extracted from the image to the information obtained from the image database [1], [12]. Based on the best match, the object is recognized and its pose estimated. The second approach extracts translation, rotation and scale invariant features (such as moment invariants, Zernike moments or Fourier descriptors) from each image and compares them to the features that have been extracted from example images of all the objects. The comparison is usually done in the form of a classification operation.

1.3 The Three Paradigms

In order to build a system that can achieve success in an realistic environment, certain simplifications and assumptions about the environment and the problem being tackled are generally made. This process of simplification introduces uncertainties into a problem which may create inaccuracies/difficulties in the reasoning abilities of a system if these uncertainties are not represented and handled in a suitable manner. Some ways of dealing with uncertainty are by using: (1) methods that employ non-numerical techniques, primarily non-monotonic logic, (2) methods that are based on traditional probability theory, (3) methods that use neo-calculi techniques such as fuzzy logic, confidence factors and Dempster-Shafer calculus to represent uncertainties, and (4) approaches that are based on heuristic methods, where the uncertainties are not given explicit notations but are instead embedded in domain-specific procedures and data structures. In this section we review three paradigms that are commonly used for handling uncertainties in realistic systems: Bayesian statistic, neural networks and expert systems.

1.3.1 Bayesian Statistics

Bayesian methods provide a formal means to reason about partial beliefs under conditions of uncertainty [13] [14]. Within this formulation, propositions (A) are given numerical parameters ($P(A|K)$) whose values signify the degree of belief given some knowledge (K) about the current environment or problem, and the parameters can be combined and manipulated according to the rules of probability. If the knowledge K remains unchanged then the degree of belief about a proposition is often represented by $P(A)$. In the Bayesian formalism, belief measures obey the axioms of probability theory: The essence of Bayesian techniques lies in the inversion formula,

$$P(A|B) = \frac{P(B|A)P(A)}{P(B)}, \tag{1}$$

which states that the belief about any proposition A based on some evidence B can be computed by multiplying our previous belief about A by the *likelihood* that B will be true given that A is true. The denominator $P(B)$ is a normalizing constant. $P(A|B)$ is often called the posterior probability and $P(A)$ is often referred to as the prior probability. Bayes' theorem indicates that the overall strength of belief in a hypotheses A should be based on our previous knowledge and the observed evidence (B) and is obtained as a product of these two factors. In order to apply Bayes' theorem we need to have an estimate of the prior probabilities and also the underlying *likelihood* distributions. Depending on the application, different methods are used to determine these factors. Prior probabilities are usually estimated as the percentage of occurrence of the proposition over a period of time. The *likelihoods* are often estimated by making an assumption that simplifies the relationship between the hypothesis and the evidence. A commonly used assumption is that the evidence and the hypothesis are related by a normal (Gaussian) distribution. Other attractive features of the Bayesian approach are: (1) the ability to pool evidence from different sources while making a hypothesis and (2) amenability to recursive and incremental computation schemes, specially when evidence is accumulated in a sequential manner.

1.3.2 Neural Networks

Artificial neural networks are inspired by biological neural networks. Artificial neural networks (ANNs) are highly parallel networks of simple computational elements (nodes) [15]. Each node performs operations such as summing the weighted inputs coming into it and then amplifying / thresholding the sum. The properties of the nodes, their interconnection topology (number of layers and number of nodes per layer), the connection strengths between pairs of nodes (weights) and the method used to update these weights (learning rule) characterize a neural network. Neural networks are data-driven, adaptive models that are mainly used for function approximation (classification) problems and optimization problems. This is normally done by defining an objective function that represents the status of the network and then trying to minimize this function using semi- or non-parametric methods to iteratively update the weights. Learning in a neural network is usually performed using two distinct techniques: supervised and unsupervised. In supervised learning the network is presented with both the input and the desired output for each input, and learning takes place to determine the weight structure which best realizes this input/output relationship. In unsupervised learning the network is presented only with the input data and the network uses statistical regularities in the data to group it into categories. Generally, neural networks are often not used as stand-alone systems but as parts of larger systems as preprocessing or labeling / interpretation units.

1.3.3 Knowledge Based Approaches

Artificial Intelligence (AI) techniques have proven to fit well in high-level tasks that require reasoning capabilities and prior domain knowledge representation. A typical AI system has two main components, as shown in Figure 2: (1) *A knowledge base* component which includes general facts about the application domain as well as task specific knowledge, and (2) *a*

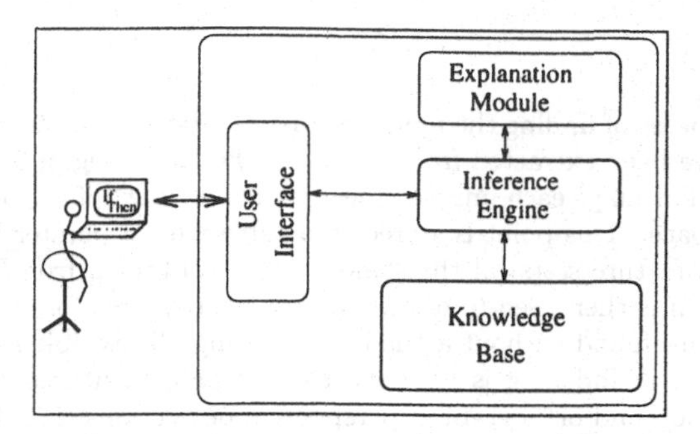

Figure 2. Main components of a knowledge-based system

control strategy such as an *inference engine* which controls the reasoning or search process. The knowledge base component of an AI system can be represented either as a set of procedures or in a declarative (i.e., non-procedural) fashion. Propositional logic, predicate calculus, decision trees, production rules, semantic nets, frames and slots, fuzzy logic and probabilistic logic are some of the commonly used knowledge representation techniques in the AI field. Though top-down (or goal-driven) and bottom-up (or data-driven) are the most commonly used control strategies, many successful AI systems use a hybrid top-down and bottom-up control strategy.

In this article we shall largely restrict ourselves to expert systems. Our focus will be on expert systems, and see how this paradigm has been used as a tool in designing and developing powerful object recognition systems.

2 Bayesian Statistics in Object Recognition

Bayesian statistics have been used at various stages of the object recognition process to provide a firm theoretical footing as well as to improve performance and incorporate error estimates into the recognition problem. The biggest advantage of a Bayesian (or probabilistic) framework is in its ability to incorporate uncertainty elegantly into the recognition process. Bayesian approaches also provide an error estimate with its decision, which gives another perspective for analyzing systems. Other advantages of using a Bayesian framework are [16]:

1. Modeling assumptions such as priors, noise distributions, etc., need to be explicitly defined. However, once the assumptions are made, the rules of probability give us an unique answer for any question that is posed.
2. Bayesian inference satisfies the likelihood principle where decisions are based on data that is seen and not on data that might have occurred but did not.
3. Bayesian model comparison techniques automatically prefer simpler models over complex ones (the **Occam's razor principle**).

On the other hand, since Bayesian decisions depend heavily on the underlying modeling assumptions, they are very sensitive to the veracity of these assumptions. Usually, to get a good description of the model a fairly large and representative amount of data is required.

Bayesian statistics have been used in object recognition for indexing, model matching and incorporating neighborhood relations under different contexts to some degree of success. Some representative applications of this framework are summarized in this section.

2.1 Indexing

Indexing is the process of finding the model from a database of models that best matches the features that have been extracted from an image. For indexing, a feature set(s) (index vector) is identified that maps each unique object model (or part of a model) into a distinct point in the index space. This point is stored in a table with a pointer back to the object model. At run time, feature set(s) of the same type are obtained from the image to form an index vector, which is then used to quickly access nearby pre-stored points. Thus a set of possible matches are found without actually comparing all possible image/model pairs.

An important issue in indexing is to make the extracted features relatively invariant to affine transformation and orthographic/perspective projections [17]. For indexing using three points on the object, an elegant approach based on the *probabilistic peaking effect* can be implemented. It has been observed that the probability density function of certain features in an image tend to peak at the values taken by the features in the model (*probabilistic peaking* effect) [18]. This means there is a large range of viewing directions over which these values change in the image by a small amount. For example, if the joint probability of the angle between segments from one of the three image points and the ratio of the lengths of these segments is determined over a viewing sphere, it peaks at the value of the corresponding features in the model. This information can be used to select only those matches whose feature values are fairly close to the image features and to disregard matches that have a small likelihood of being in actual correspondence.

The probabilistic indexing approach can also be extended to more that three points [19]. As a final note, the probabilistic indexing works well for most object recognition problems except when the objects are considerably foreshortened in the image due to rotation. In such cases they produce angles and distance ratios far from the probability peaks and will be difficult to recognize using probabilistic indexing techniques.

Alignment [20] and geometric hashing [21] [22] are related techniques that are used to recognize 3D object from 2D scenes. Both methods use a small number of points to find a transformation between the model space and the image space. Recognition then consists of finding evidence for instances of the models in the data, either by transforming the image into the model space and voting for an object's pose or by hypothesizing a pose and then transforming it into image space to guide the search.

Alignment: is the process of determining a unique (up to a reflection) affine transformation between an image and the model by matching three image points with three model points. Once the transformation that brings them into alignment is determined, each of these transformations must be tested for correctness (verification). Speedup can be done using the probabilistic indexing scheme described earlier to determine which matches are most likely to be correct. Only these matches are considered and the rest are discarded. Other error measures can be used to further reduce the number of matches that need to be examined. Using this method the speedup obtained is equal to the fraction of the total matches indexed that are used for verification [19].

Geometric Hashing: for object recognition may be summarized as follows. Given m models $M_i, i = 1 \ldots m$, each consisting of a pattern of n points ($p_i, i = 1 \ldots n$), we precompute a hash table for these models where each entry is of the form $M(x, y, M_k, B_i, p_l)$, where a subset of the model points B_i (called a model *basis*) of the model M_k is combined with another point (feature) p_l of M_k not belonging to the set B_i, then it hashes to the location (x, y). The number of basis points is determined based on the application (e.g., 2 points for similarity-invariant recognition and 3 points for recognition under affine transformation and orthographic projection) and a separate entry in the hash table for all possible combinations of the distinct points (B_i, p_l) (called an *model group*) for every model M_k is generated. During the recognition phase, all features (points) in the image are detected and the following process is repeated until a recognition decision can be made: (1) an image basis is picked randomly and all image groups using this image basis is formed. The number of points in an image basis corresponds to the number of points in the model basis. (2) The location that each image group hashes to is located and all entries in the hash table (possible model group matches) register a weighted vote for each model group and the corresponding model basis. The weights depend on how close the image group hash location (u, v) is to the model groups location (x, y). (3) if a model (through the current model basis) receives sufficient votes, an object hypothesis has been found, otherwise the recognition process is repeated with a new image basis.

Geometric hashing is recast in probabilistic terms in [22] and geometric hashing with weighted voting is interpreted as a Bayesian maximum likelihood object recognition system. Using this interpretation, for the case where the basis comprises of two points, it is shown that voting weights (likelihoods) depend on the density functions of the hash values in the hash space under assumptions of a particular model/basis combination and a particular basis set selection in the scene. If it is assumed that the model points after undergoing translation, rotation and scaling are perturbed by a Gaussian-distributed noise before being projected to the scene, then the perturbations lead to an approximately Gaussian distribution of the hash locations, where the covariance matrix is diagonal and depends on both the location of the hash point and the separation between the basis points in the scene. Given a model M_k and its $(n-2)$ hash locations, the expected density function in hash space is simply the superposition of the $(n-2)$ Gaussian distributions, each centered around a distinct hash entry.

2.2 Model representation and matching

The matching process involves matching image features to model features. Matching can be performed either by trying to first get all good matches between the image features and the model features (commonly referred to as *correspondence* matching or *hypotheses* generation) or by trying to estimate the transformation of the image features needed to match the model features (i.e., trying to determine the object pose, commonly referred to as *transformation* matching), or a combination of both. Matching in the *correspondence* space is easily defined using Bayesian statistics. Also, the search in the correspondence space can by cast as an iterative estimation problem using the Bayesian theory.

Both techniques are exemplified by the work done by Wells [23], who uses a two stage statistical formulation for feature based object recognition. In the first stage (*correspondence*), the joint hypotheses of match and pose are evaluated in terms of their a posteriori probabilities for a given image. The model matching is cast as a maximum a posteriori (MAP) estimation problem using Bayesian theory and a Gaussian error model. The parameters that are to be estimated in the matching are the correspondences between the image and the object features, and the pose of the object in the image. The probability densities of image features, conditioned on the parameters of match and pose, are combined with the prior probabilities on the parameters using Bayes' rule to give the a posterior probability density of the parameters. An estimate of the parameters is then obtained by choosing them so as to maximize their a posteriori probability. The probability models for the features are built by assuming that matched features are normally distributed about their predicted positions in the image, and the unmatched features (considered as background features) are uniformly distributed in the image. The prior probabilities for the correspondences between the image and the object are assumed to be uniform for match features and constant for background features. Prior information on the pose is assumed to be supplied as a normal density.

The second stage of the statistical formulation (*transformation*), presents a method that builds on the earlier stage to provide a smooth objective function for evaluating the pose of the object without actually determining the correct match. To obtain the pose of the image, the posterior probability density of the pose is computed from the joint posterior probability on pose and match, by taking the marginal over all possible matches (for a given pose). Limited experiments show that this function is relatively smooth and its maximum is usually close to the correct pose. This maximum is then obtained by iteratively using a variant of the Expectation-Minimization (EM) algorithm to get the correct pose.

Markov random fields (MRFs): provide an efficient tool to incorporate neighborhood/ dependency constraints that can make the matching/recognition process more reliable and effective [24]. During the *hypothesis* generation stage, while trying to determine all possible matches between image features (such as regions or edges) and model features, it is important to include all possible correct matches and exclude as many incorrect ones as possible. Incorrect matches can be excluded by incorporating constraints based on prior knowledge about the problem (such as object models). Dependencies between hypotheses or features can be represented elegantly in the Markov random field framework. MRFs possess characteristics that are particularly suited to solution of spatially oriented vision problems that involve uncertainty (recognition). These include:

- MRFs fit conveniently into a Bayesian framework.
- Prior knowledge about local spatial interactions can be easily expressed.
- Probabilistic constraints based on arbitrary spatial dependencies and neighborhood relationships can be encoded easily.
- Local evidence about hypotheses is easily represented and manipulated.

For a review of MRFs and their applications to computer vision, the reader is referred to [25]. The use of a Markovian framework to improve the efficiency of object recognition is used in [26] where a probabilistic *hypothesis generation* or matching between features in an image and the model features is done. In this work, planar regions (R_i) extracted from range images are used as primitive features that are matched to faces on CAD models (M_j) of each object. Each planar region (feature) is represented by a fixed set of attribute values ($f^k_{R_i}, k = 1, \ldots n_1$) such as region area, second order moments, area diameter etc. Also, relationships between pairs of regions R_i and R_j are described by another set of attributes ($f^l_{R_i,R_j}, l = 1, \ldots, n_2$) such as simultaneous visibility, maximum distances between surfaces, etc. These region-based attributes are then then used to determine a set of model face matches using a Bayesian statistical approach. This is done by obtaining the statistical distribution of the observed attributes of each model face in the form of the conditional probabilities $P(f^k_{R_i}|M_j)$ and $P(f^k_{R_i,R_j})$ and the prior probabilities $P(M_j)$. During the hypothesis generation stage features are extracted from the image, and a set of possible matches of model features to image features is found. To reduce the number of incorrect matches, prior knowledge about the dependencies between different hypotheses is used to reduce the set of possible matches using Markovian framework. Each of the hypotheses can be thought of as being either correct (**ON**) or incorrect (**OFF**). The dependencies between hypotheses are incorporated in a Markovian framework as follows. Let X_{R_i,M_j} represent the hypothesis that the region R_i in the image is assigned to the model face M_j. Since each hypothesis can be either correct or incorrect, it has an associated value $\omega_{R_i,M_j} \in \{\mathbf{ON}, \mathbf{OFF}\}$. The variables X_{R_i,M_j} can be thought of as the MRF random variables (X) and the ω_{R_i,M_j} as the labels that these variables take. To define the dependencies between these MRF variables, two neighborhoods are defined: one for contradictory hypotheses and one for supporting hypotheses. Contradictory hypotheses are defined as those where the same image region is matched with two different model faces, while supporting hypotheses are those that are consistent with the prior constraints, i.e. $P(f_{R_i,R_k}|M_j, M_l) > 0$.

The matching process, which is equivalent to determining the most likely state of the MRF variables given a set of image regions $R_i, i = 1, \ldots, n$, reduces to determining the minimum of the posterior energy function:

$$U(\omega|R_1, R_2, \ldots, R_n) = \sum_{c \in C} V_c(\omega) - \sum_{X_{R_i,M_j} \in X} \log P(R_i|\omega_{R_i,M_j}). \quad (2)$$

The clique potentials $V_c(\omega)$ are easily obtained. For example, if only 1 and 2-cliques are considered, then for 1-cliques the potential at $c = \{X_{R_i,M_j}\}$ equals the prior probability $P(M_j)$ if the hypothesis is true and its complement otherwise. Similarly the 2-cliques energies are obtained by considering two related MRF variables $c = \{X_{Ri,M_j}, X_{R_k,M_l}\}$ in a neighborhood consisting of both contradicting and supporting hypotheses. The clique potentials are chosen according to the amount of supporting evidence that a hypothesis requires to survive the energy minimization process. The probabilities $P(R_i|\omega_{R_i,M_j})$ are found by determining the probability that attributes extracted for the region are similar to the model face M_j if the hypothesis is true ($P(f_{R_i}|M_j)$) and the complementary if the hypothesis is incorrect ($1 - P(f_{R_i}|M_j)$). Given a set of regions extracted from the image, the most likely set of hypotheses is obtained by minimizing the energy function in equation (2). Different techniques such as Highest Confidence First (HCF) [25] and Simulated Annealing [27] procedures can be used to get fairly good results. After the minimization process is completed using either of the procedures mentioned above, all matches that are **ON** are considered for verification. This verification is done by trying to estimate the pose of the object in the image.

View clusters: are often used to to reduce the time complexity involved with matching image features (regions) with all the possible model views. The clusters are obtained by grouping model views into equivalence classes using similarity metrics. Once the clusters are formed, each *view cluster* is represented by a prototype feature vector. This is a viewer-centered approach to object recognition in which matching is reduced to finding the closest view cluster(s) to which the image features may belong. Once the closest match(es) are found, the matching process can be refined to find the exact pose within each *view cluster*. The pose identification process can also be used to verify the initial match(es), as the candidate *view clusters* should have a good pose match with the image region (features). Bayesian statistics provide a complete framework for representing these *view clusters* and for the view class determination problem [8].

In [28] a hierarchical recognition methodology that uses salient object parts as cues for classification and recognition and a hierarchical modular structure (HMS) for parts-based object recognition is proposed. In this system, each level in the hierarchy is made up of modules, each of which is an *expert* on a specific part of a specific object. Each modular *expert* is trained to recognize the part under different viewing angles and transformations (translation, scaling and rotation). When presented with an input object part, each *expert* provides a measure of confidence of that part belonging to the object that the *expert* represents. These confidence estimates are used at the higher levels for refined classification. The modular *experts* (i.e., object parts) are modeled as a mixture density of multivariate Gaussian distributions. For each module, features (Zernike moments in the current implementation) obtained from the object part from all possible viewpoints are used in an *Expectation-Maximization* (EM) approach [29] to determine the module parameters.

When presented with an object part, each module computes the posterior probability of that part belonging to the object the module represents. These posterior probabilities are then pooled using a recursive Bayesian updating rule to compute the final object posterior probabilities given all the input parts. When computing the posterior probabilities of each object, the prior probability for each object part module is determined by its importance in the recognition process.

3 Neural Network Based Methods

Neural networks have been largely used as data driven models for function approximation or classification, or as networks which implicitly optimize a cost function that reflects the goodness of a match. Some promising neural approaches to feature extraction and clustering have also been proposed [36], which are adaptive on-line and may exhibit additional desirable properties such as robustness against outliers [30], as compared to more traditional feature extractors.

Several types of neural networks can serve as adaptive classifiers that learn through examples. Thus, they do not require a good a priori mathematical model for the underlying physical characteristics. These include feed-forward networks such as the Multi-Layer Perceptron (MLP), as well as kernel-based classifiers such as those employing Radial Basis Functions (RBFs). A second group of neural-like schemes such as the Learning Vector Quantization (LVQ) have also received considerable attention. These are adaptive, exemplar-based classifiers that are closer in spirit to the classical K-nearest neighbor method. The strength of both groups of classifiers lies in their applicability to problems involving arbitrary distributions. Most neural network classifiers do not require simultaneous availability of all training data and frequently yield error rates comparable to Bayesian methods without needing a priori information. Techniques such as fuzzy logic can be incorporated into a neural network classifier for applications with little training data. A good review of probabilistic, hyperplane, kernel and exemplar-based classifiers that discusses the relative merit of various schemes within each category, is available in [31].

Although neural networks do not require geometric models, they do do require that the set of examples used for training should come from the same (possibly unknown) distribution as the set used for testing the networks, in order to provide valid generalization and good performance on classifying unknown signals [32]. To obtain valid results, the number of training examples must be adequate and comparable to the number of effective parameters in the neural network. A deeper understanding of the properties of feed-forward neural networks has emerged recently that can relates their properties to Bayesian decision making and to information theoretic results [33]. A survey of neural network approaches to machine inspection can be found in [34].

3.1 Function Approximation for Object Recognition

In this section we describe neural network techniques that have been used for feature-based recognition and for indexing applications. These neural networks are mainly used to approximate certain functions, such as class optimizers (for classifiers) or for interpolation, using some training samples in a supervised fashion.

Feature-based object recognition: is a simplistic, direct approach where neural networks are used in the form of classifiers. Feature-based object recognition system using neural networks generally do the following:

1. Extract and select features from the objects that are *invariant* and/or *salient*. Different types of features (such as shape features, or intensity features, etc.) can be extracted from an image.
2. Features from a set of training images are then used to train a neural network classifier, either using supervised learning or in an unsupervised manner. During the training phase, the neural network can be made to learn different objects, and optionally, the pose of these objects. During the training phase, a set of images, not used for training, can be used to determine how well the neural network is performing.
3. Given a new image, the features extracted from the image are fed into the previously trained neural network, which then classifies the features and recognizes the object. The new features can also be used to further train the neural network.

Based on the above, the main issues in feature-based object recognition are (1) extraction and selection of invariant and salient features and (2) deciding on the type neural network classifier (architecture and size), and type of learning algorithms to use.

A good review of the various issues in feature selection/extraction for pattern recognition is presented in [35], [36]. Feature extraction and selection is used to identify the features which are most important in discriminating among different objects. Also, by retaining a small number of "useful" or "good" features, factors such as computational cost and classifier complexity are reduced. Feature *selection* implies the choosing a subset of the features. Feature *extraction* involves the transformation of the image and selecting a set of features from the transformed space. In object recognition, it is desirable to use features that are invariant to translation (i.e., position in the image), rotation and scale (viewing distance) of the object. A few of the commonly used features are invariant moments, log-polar transforms, shape descriptors such as Fourier descriptors and Zernike moments, and other local features such as curves and corner points, etc [38]. *Saliency* of a feature can be defined as the measure of the feature's ability to impact classification. One way to compare the saliency of features is by using the single probability of error criterion. This technique computes the probability of error separately for each individual feature. These errors are then used to rank the features and select a subset of them.

In [40], object recognition was performed using Zernike moments to represent the shapes of the object. Five different neural network classifiers were tested for this application, with the aim of comparing and evaluating their classification performances. These neural networks were a perceptron, a two-layer perceptron and the three-layer perceptron ART-2; an adaptive resonance theory based network; and a Kohonen associative memory [41]. The data set consisted of images of 6 tactical vehicles viewed from varying distances and angles and under conditions of noise and occlusions. Each object was represented by a vector of 23 features. For this application the multilayer perceptron with 2 hidden layers gave the best results.

In *view-centered* recognition, approaches based on *aspect graphs* are quite common. *Aspect graphs* [7] are created by representing 2D views of a 3D object along the nodes of the graph, with legal view transitions indicated by the arcs among the nodes. Each node represents a characteristic view (CV) of the object in which certain edges and faces are visible, as seen from a contiguous region of viewing angle and distances. Since each CV can be indicated by a binary vector with "1" for observable edges or faces and "0" for hidden ones, an object can be described as a mapping from (view angle, distance) to the CV vectors. In [43], an Radial Basis Function (RBF) network was used to learn this mapping and then predict the CV from a given view angle or to propose a view angle for a given CV.

Clustering techniques: are used in [44] to self-organize aspect graph representations of 3D objects from 2D view sequences. This architecture is based on a neural "cross correlation matrix" which was used to learn both 2D views and 2D view transitions and to associate the 2D views and 2D transitions with the 3D objects that produced them. The characteristic views of the different objects were represented using an Adaptive Resonance Theory (ART2) neural network through unsupervised learning and categorization. These 2D views were then fed into a series of cross-correlation matrices, or view graphs, one for each possible 3D object, so that views and view transitions could be learned by a 3D object categorization layer. The 3D categorization layer incorporated "evidence accumulation" nodes which integrate activations that they receive from learned connections to the correlation matrix. Decay terms in these integrator nodes determine how long they stay active without input support and, hence, determine the amount of evidence that is accumulated from different views of an object. The biggest drawback of this system was in its space (cost) requirements. To reduce the system cost, which is directly related to the complexity of the evidence accumulation parts of the architecture, the VIEWNET architecture proposed in [37] explores the problem of enhancing the preprocessing and categorizing stages in order to generate less ambiguous 2D categories and hence, rely, less on view transitions.

Indexing: as seen earlier, is an efficient method of recovering match hypotheses in model-based recognition. In a number of approaches, the indexing technique is viewed as obtaining indexing functions that associate each index vector from an image to some kind of probability measure with each of the indexed matches. One such method is presented in [45], where *indexing functions* are introduced to estimate these probabilities.

In [45], Radial Basis Functions (RBFs) are used to learn these functions. One advantage of using RBFs is that they smoothly interpolate between training examples (to fill in sections of the viewing sphere where no views exist). Also, a large number of training samples can be represented by a single center, thus reducing the storage requirements of the system. A drawback with using RBFs is that it is often difficult to determine the optimal number and positioning of the centers.

3.2 Matching as an optimization problem

As noted earlier, the main part of a recognition process is to establish the correspondence relationships between the information in the image and the object model. This may be posed as a graph matching problem, which in turn, is often formulated as an optimization problem where an energy function is minimized. In [46], a Hopfield network realizes a constraint satisfaction process to match visible surfaces to 3D objects. In [47], object recognition is posed an inexact graph matching problem and then formulated in terms of constrained optimization. In [48] the problem of constraint satisfaction in computer vision is mapped to a network where the nodes are the hypotheses and the links are the constraints. The network is then employed to select the optimal subset of hypotheses which satisfy the given constraints.

To convey the flavor of such optimization frameworks, we consider here the hierarchical approach of [49], where Hopfield networks are used to recognize objects at two levels: (a) coarse recognition, which is based on the surfaces of the objects, and (b) fine recognition, which is done by determining the correspondences between vertexes in the image and the model. At both levels, object recognition is viewed as a graph matching process differing only in the features being used for matching. At the coarse level, matching is done using surfaces as features, while at the finer level, matching is done using vertex (or corner points) information. The compatibility measures between the features are used to determine the network configuration and as the network iterates to a stable state, the number of active

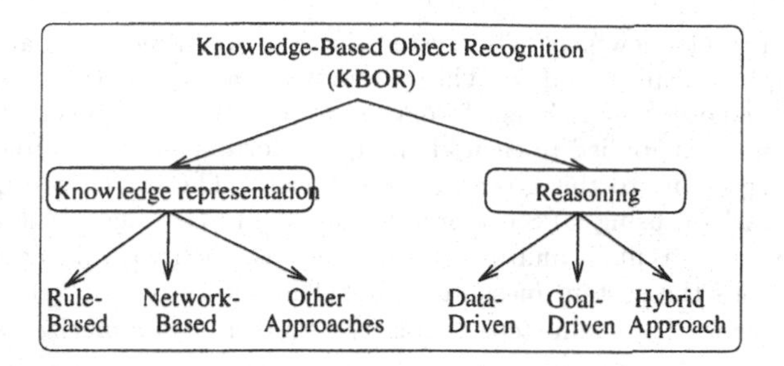

Figure 3. Most common classes of knowledge-based object recognition systems

neurons in the network can be used as a measure of matching between an image and a model. This information can then be used to select a few possible matching models from the model database for further verification. The verification stage (or fine recognition) is done by matching vertex (or corner points) of the image regions to the vertexes of the model in a similar manner.

4 Expert Systems

A Knowledge-Based Object Recognition (KBOR) system may be defined as an object recognition system that uses either a symbolic knowledge format to represent its domain knowledge and/or a knowledge-based inference engine to search its domain knowledge. Several KBOR systems were developed during the 1980s and 1990s [50, 51, 52, 53]. In most of these systems, the knowledge-based paradigm has helped to perform complex and heuristic tasks in a logical and understandable manner. Figure 3 shows some popular approaches taken by KBOR systems, which provide some of the following advantages:

1. *Increased system abstraction level*, due to symbolic representation.
2. *Increased system maintainability*, if the knowledge base and the matching engine can be updated separately.
3. *Better uncertainty handling*, by attaching a measure of belief to output decisions.
4. *Reasoning and explanation capability.*
5. *A built-in control strategy*, via the inference engine, that can be used in a bottom-up, top-down or hybrid top-down and bottom-up fashion. A bottom-up control strategy can be used in a KBOR system when the noise level in the raw data is low or when the search span in the solution space is large and hard to prune. In other cases, when there are many interactions among data in the lower level tasks, a top-bottom or a goal-driven control strategy is more appropriate. However, in both cases, having a built-in control strategy with heuristic search criteria helps to reduce object recognition system complexity and implementation effort.

4.1 Examples

This section summarizes three object recognition systems that use a knowledge-based paradigm, specifically the expert system paradigm, as an integral part.

3D Shape and Orientation Recovery: Shomar et al. have implemented an object recognition system that depends on an expert system to perform 3D shape recovery and orientation from a single view [54]. This system has two main modules: an expert system module and a graphics display module. The system uses some geometric regularity assumptions about perceived objects and image formation to recognize the objects from 2D images. Geometrical reasoning is applied to each 2D image to form a set of possible 3D views and orientations corresponding to this given 2D object view. The search process is done in a forward-chaining fashion using OPS5, a production system language. The outcome of the reasoning process may result in multiple interpretations, each with an attached certainty factor that quantifies the system measure of belief in the recovered 3D object from the given perspective view. The main steps of this system can be summarized as:

1. *Representing geometrical rules:* The system has 35 geometrical heuristic rules that can be divided into five major categories: parallelism, perpendicular, right corners, parallel right corners, focal length, and hidden lines rules. These rules help in recognizing man-made objects since most man-made objects have some geometrical regularities. The first step in the system is to represent these rules in a production format using OPS5. Some supplementary functions were implemented using Pascal and Fortran procedures to facilitate low-level computation.
2. *The 3D reconstruction phase:* Each detected closed region in the given 2D view is assumed to correspond to a planar face in the 3D object. The system then utilizes the stored geometry regularities rules using the inference engine of the OPS5 to reconstruct the 3D object coordinates of the vertices. Each resulting reconstruction may result in different 3D recovery. However, a certainty factor is assigned to each recovered vertex based on the strength of the regularities used to reconstruct it.
3. *The graphical representation phase:* After constructing all possible 3D views from the given scene, the system uses a graphical illustration to display three orthographic views of the reconstructed model after applying some symmetry rules.

The key to the success of this system lies in constraining the domain to geometrical and unoccluded shapes. Regularities in this limited environment are then exploited to limit the search space.

A KB system for Image Interpretation: Chu and Aggarwal [55] developed a knowledge-based multi-sensor image interpretation system using KEE, an expert system shell development package. The AIMS (Automatic Interpretation using Multiple Sensors) system has three main building blocks.

1. *A segmentation module* that integrates segmentation information from thermal, range, intensity, and velocity images and combines them into an integrated segmentation map [56].
2. *A representation module* where the outcome of the segmentation module is represented in a structured knowledge-based format that can be utilized by the KEE package.
3. *An interpretation module* that uses KEE and supplementary LISP procedures in a bottom-up manner to recognize different objects in an image. AIMS' reasoning process depends on knowledge in the form of rules that are based on: (i) knowledge of the imaging geometry and device parameters, which are independent of the imaged scene; (ii) information on the segmented image regions, such as size, average temperature within the region, average distance etc.; (iii) neighborhood relationships between the image regions; (iv) features and models of objects; and (v) other general heuristics, which are derived from known facts about the application domain and common sense.

Using the above knowledge, a forward-chaining reasoning approach is adopted to recognize the objects that appear in an image. Six types of rules are used sequentially: (i) pre-processing rules: to handle the difference between individual segmentation maps and integrated segmentation map and to compute low-level attributes and place them in the corresponding knowledge structure; (ii) coarse recognition rules: to distinguish between Man-Made Objects and Back-Ground (MMO/BG); (iii) grouping rules: to group similar segments (regions) into objects based on neighborhood relationships and other similarity measures; (iv) back-ground classification rules: to classify back-ground (BG) into SKY, TREE, and GROUND types; (v) man-made classification rules: to classify man-made objects (MMO) into different types such as BULLETIN-BOARD, TANK, JEEP, APC, or TRUCK based on shape and size analysis; (vi) consistency check rules: to verify the interpretation of an object and its surrounding objects. For example, a region recognized as a SKY cannot be surrounded by a region classified as GROUND. Any conflicting interpretations lead to reduced certainty factors. One such example is the rule:

IF (Segment A is of type MMO) AND
(Segment A has a cool sub-region located at its lower-half) AND
(Segment A is about 2.0-2.5m high) AND
(Segment A has a trapezoidal contour) AND
THEN (Segment A is an APC with confidence of 0.8)

This system is more versatile in that it allows for occluded objects and integrates knowledge of multisensor characteristics. However, the domain is again specialized, in this case to identify military objects from ground objects.

SIGMA: Matsuyama and Hwang have developed an image understanding system called SIGMA, which is a knowledge-based aerial image understanding system [57]. SIGMA uses expert systems in three different modules.

1. *A geometric reasoning expert* which extracts geometric structures and spatial relations among objects and represents them in a symbolic hierarchical knowledge. Bottom-up and top-down reasoning approaches are integrated into a unified reasoning approach, which is then used to construct a globally consistent description of the scene.
2. *A model selection expert* to reason about specialized objects that match the resulting general description of the geometric reasoning expert module. The model selection expert uses contextual goal reasoning to determined the most plausible object. However, this top-down reasoning approach is not enough to determine the most plausible object in each case. Thus, this module is used to compose objects into specific shape parts.
3. 3)*A low-level vision expert* is used to perform image segmentation and extract specific parts of objects features which the model selection expert has specified as its output. It uses a trial-and-error reasoning approach to find out segmentation features to help the other experts to reason about objects in the image. This module is the only expert in the system that uses domain-independent knowledge.

5 Future Directions in Object Recognition

In the preceding sections, through a review of some existing object recognition systems, we have highlighted the use of Bayesian statistics, neural networks and expert systems for object recognition. This discourse would be incomplete without mentioning that there are similarities among these paradigms and many of the reasoning/modeling abilities of one approach can be mimicked by the others. But, even more importantly, there are features of these approaches that are complementary in nature. To fully exploit these approaches to build robust and comprehensive object recognition systems, they have to be used concurrently in a mutually supportive manner. We now touch upon some areas of research that the authors believe are important and will influence the design of object recognition systems in the future.

5.1 Bayesian Methods and Neural Networks

Bayesian methods and neural networks share several similarities [16] [58] [59]. Both methods generate models that closely fit the data. Many popular artificial neural networks are essentially nonlinear parametric or semi-parametric estimators that are based on general and powerful functional forms such as a linear combination of sigmoidal or radial basis functions. The parameters are the weights which are "learned" or estimated using training data. Due to the specific types of non-linearities used, such functional forms are very flexible and can model complex variations in the data better than simple linear methods. However, with increased flexibility comes the potential problems of over-fitting and poor generalization. Bayesian methods, with their inherent preference for simpler models over complex ones, can complement neural networks by providing information about the amount of flexibility that is warranted by the data. This process can be facilitated by interpreting neural networks as probabilistic models which is possible with several neural networks that are used as regression networks as well as classifiers [16]. For example, the objective function (plus some regularization parameter) that is minimized during the training (weight change) of a neural network can be regarded as the negative log of the probability assigned to the observed data by the model with the current weights. As more data is seen, this objective function is updated to get the most probable weights given the data, using Bayes' theorem.

5.2 Combining Neural Networks and Expert Systems

The main motivation to combine expert systems and neural networks is to revise available domain knowledge and to augment the neural networks' output decision with explanation capabilities. Two popular methods of combining are presented in [60]. In one, an expert system is used to initialize a neural network. In this initialization, antecedents of rules are mapped into input and hidden nodes, certainty factors determine initial weights, while rule consequences are mapped into output nodes. The goal of this mapping is to embed all available prior domain knowledge into the network's initial internal architecture. Due to this embedded prior knowledge, the time required to train such networks is much less than those which are initialized randomly. In the other method, rules are extracted from trained neural networks. In this case, a neural network is mapped into a rule-based system. Rule extraction can provide trained (i.e., adapted) connectionist architectures with explanation power. Extracted rules can also be used to validate the connectionist networks' output decisions.

5.3 Pattern Theory: A Unifying Framework

The statistical framework of pattern theory provides mathematical representations of subject-matter knowledge that can serve as a basis for the algorithmic understanding of images [61] [62]. This powerful theory uses a modified Bayesian approach for hypothesis

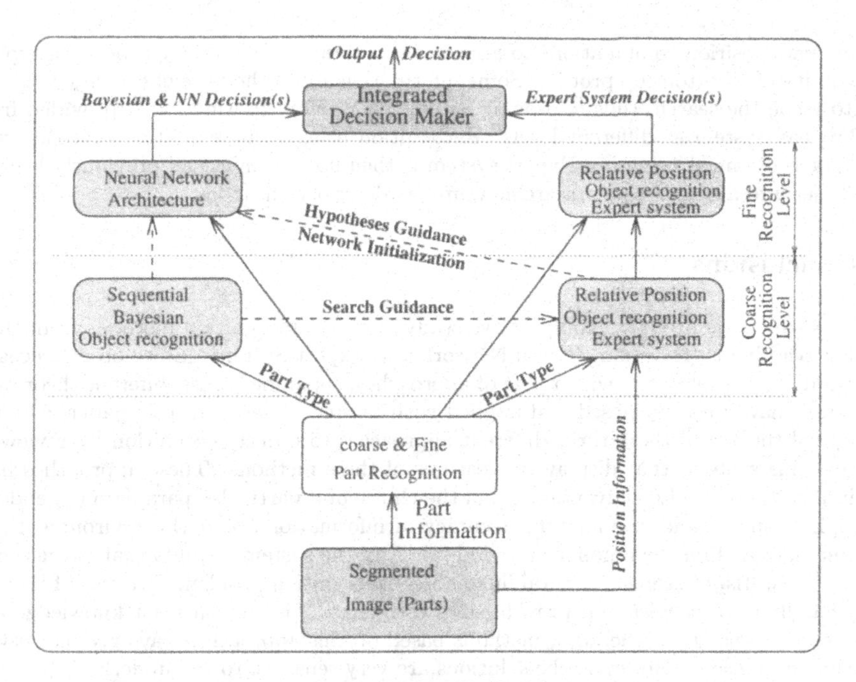

Figure 4. Schematic of a Mutually Guided Hybrid System.

formation that is capable of the creation or annihilation of hypotheses by jumps from one continuum to another in configuration (hypothesis) space. Also, rules can be represented by visual grammars that regulate transformations of or algebraic operations on pattern templates. Thus it provides a common language for both Bayesian and rule-based reasoning. Pattern theoretic approaches have met with success in a number of applications, from describing mitochondria ensembles to multi-target recognition and tracking [62] [63]. Moreover, mixed Markov models inspired by this theory are being suggested as basic tools in object recognition [64].

5.4 Combining all three paradigms: An illustration

To illustrate how all three paradigms can be tightly integrated, we briefly describe an ongoing project in which supplementary information from each sub-system helps to guide the working of the other, as exemplified by figure 4. We are currently implementing this hybrid system for object recognition from second generation Forward Looking Infrared (FLIR) images. This hybrid system uses a methodology based on a hierarchical, modular structure for object recognition by parts. Recognition is performed at different levels in the hierarchy, and the type of recognition performed differs from level to level. Each module is used to represent and recognize parts of objects. In the Bayesian sub-system, the final recognition of the object is based on the evidence accumulated from the sequential presentation of the different parts of the object. However, this does not exploit relative positional information of the different object parts while performing the final recognition. Relative positional information about object parts is readily incorporated into an Expert system, that uses the part evidence from the Bayesian part experts and the relative positional information from the image to recognize the object. However, searching through all possible part position combinations to arrive at a recognition is a costly process; a typical shortcoming of a bottom-up process. Some information about the type of the object can be used to guide the search, thus improving its efficiency. This information is provided from the Bayesian system at different levels. Recognition using relative spatial information of the parts as obtained using the Expert system is then used to inject all previously learned hypotheses into a neural network architecture, as described in [60].

6 Conclusions

In this paper, we present a comparative study of object recognition methods from three different paradigms: Bayesian, Neural Network and Expert Systems. Since object recognition is a difficult problem, a wide range of approaches, spanning across different theoretical paradigms, have been proposed and applied with limited success. In this paper, we have highlighted the use of these three different approaches to object recognition by reviewing some existing systems that display the features of these methods. These approaches have certain advantages and disadvantages, and the choice of a particular paradigm depends on the application at hand, the amount/accuracy of information about the environment, the available data and on the amount of "blind faith" in the system outputs that is tolerable.

Bayesian statistics seems a natural fit to object recognition problems because of its ability to handle uncertainties and provide error estimates. Also, given prior knowledge and some assumptions about the data, methods based on this approach always give consistent and concise solutions. However, the solutions are very sensitive to the underlying assumptions, and are only acceptable if the knowledge about the domain is quite reliable and well understood. Also, this approach can become computationally prohibitive if the size of the parameters needed to reasonably describe the problem becomes large. Neural networks, on the other hand, perform well in complex environments, with their data-driven ability to learn the underlying functionalities and their relationship to the application domain. However, neural networks suffer in their lack of interpretability and in incorporating prior domain knowledge. Another difficultly encountered when using either Bayesian statistics or neural networks is in incorporating high level (symbolic) reasoning capabilities into the system. This kind of reasoning is easily implemented using expert systems. Expert systems also provide an explanation for every decision that is made; which is a desirable property in any system. Based on the above discussion, it is clear that for systems to perform well in complex and dynamic environments, complementary features from each of these paradigms should be incorporated into a system in a mutually supportive manner. In this paper we have briefly addressed issues on how this may be done.

Even with the advance of technology and sophistication of object recognition algorithms, object recognition systems of today are still really limited when compared with human performance. Humans can recognize about 10,000 distinct objects [65] under varying conditions, while a state of the art object recognition system can recognize relatively a few objects and certainly are nowhere near the breadth and depth of the human performance. Automatic target recognition (ATR) is a good example of an exacting situation where the shortcomings of the state of the art systems are evident.

It is evident from the above discussion that object recognition remains an important problem to be resolved. This area has evolved significantly in the past two decades with applications into diverse areas ranging from recognizing targets in a battlefield to recognizing produce at checkout counters. However, we are still not able to design reliable object recognition systems. This may be partly attributed to the absence of theoretical underpinnings for object recognition systems which may enable us to analyze, synthesize and design such systems. It is envisaged that future efforts will be directed at fulfilling this need for theoretical underpinnings for pattern recognition and object recognition.

References

[1] J. W. McKee and J. K. Aggarwal, "Computer recognition of partial views of curved objects," *IEEE Transactions on Computers*, vol. C-26, no. 8, pp. 790–800, 1977.

[2] W. E. L. Grimson, *Object Recognition by Computer: The role of geometric constraints.* MIT Press, Cambridge, 1990.

[3] B. Vemuri, A. Mitiche, and J. K. Aggarwal, "Curvature-based representation of objects from range data," *Image and Vision Computing*, vol. 4, no. 2, pp. 107–114, 1986.

[4] F. Arman and J. K. Aggarwal, "CAD-based vision: Object recognition in cluttered range images using recognition strategies," *Computer Vision, Graphics, and Image Processing*, vol. 58, no. 1, pp. 33–47, 1993.

[5] Y. F. Wang, M. J. Magee, and J. K. Aggarwal, "Matching three-dimensional objects using silhouettes," *IEEE Transactions on Pattern Analysis and Machine Intelligence*, vol. 6, no. 4, pp. 513–518, 1984.

[6] F. Arman and J. K. Aggarwal, "Model-based object recognition in dense depth images - A review," *ACM Computing Surveys*, vol. 25, no. 1, pp. 5–43, 1993.

[7] J. Koenderink and A. van Doorn, "The internal representation of solid shape with respect to vision," *Biological Cybernetics*, vol. 32, pp. 211–216, 1979.

[8] A. Pathak and O. I. Camps, "Bayesian view class determination," *Proceedings of the IEEE Conference on Computer Vision and Pattern Recognition*, pp. 407–412, 1993.

[9] S. Zhang, G. Sullivan, and K. Baker, "The automatic construction of a view-independent relational model for 3D object recognition," *IEEE Transactions on Pattern Analysis and Machine Intelligence*, vol. 15, no. 6, pp. 778–786, 1993.

[10] A. Pope, "Model-based object recognition-a survey of recent research," *Technical Report*, vol. TR-94-04, University of British Columbia, 1994.

[11] J. B. Burns and E. M. Riseman, "Matching complex images to multiple 3D objects using view description networks," *Proceedings IEEE Conference on Computer Vision and Pattern Recognition*, pp. 328–334, 1992.

[12] S. Chen and A. K. Jain, "Strategies of multi-view multi-matching for 3D object recognition," *Computer Vision and Image Processing*, vol. 57, no. 1, pp. 121–130, 1993.

[13] J. Pearl, *Probabilistic Reasoning in Intelligent Systems: Networks of Plausible Inference.* Morgan Kaufmann Publishers, Inc. San Mateo, California, 1988.

[14] R. O. Duda and P. E. Hart, *Pattern Classification and Scene Analysis.* A Wiley-Interscience Publication, 1973.

[15] A. Jain, J. Mao, and K. M. Mohiuddin, "Artificial neural networks: A tutorial," in *Computer*, pp. 31–44, March 1996.

[16] D. J. MacKay, "Probable networks and plausible predictions - a review of practical bayesian methods for supervised neural networks." to appear in *Network.*

[17] Y. Lamdan, Y. Shwartz, and H. Wolfson, "Affine invariant model-based object recognition," *IEEE Transactions on Robotics and Automation*, vol. 6, no. 5, pp. 578–589, 1990.

[18] J. Ben-Arie, "The probabilistic peaking effect of viewed angles and distances with application to 3D object recognition," *IEEE Transactions on Pattern Analysis and Machine Intelligence*, vol. 12, no. 8, pp. 760–774, 1990.

[19] C. F. Olson, "Probabilistic indexing for object recognition," *IEEE Transactions on Pattern Analysis and Machine Intelligence*, vol. 17, no. 5, pp. 518–522, 1995.

[20] D. P. Huttenlocher and S. Ullman, "Recognizing solid objects by alignment with the image," *International Journal on Computer Vision*, vol. 5, no. 2, pp. 195–212, 1990.

[21] D. Gavrila and F. Greon, "3D object recognition from 2D image using geometric hashing," *Pattern Recognition Letters*, vol. 13, no. 4, pp. 263–278, 1992.

[22] I. Rigoutsos and R. Hummel, "Distributed Bayesian object recognition," *Proceedings IEEE Conference on Computer Vision and Pattern Recognition*, pp. 180–186, 1993.

[23] W. M. Wells, *Statistical Object Recognition.* PhD thesis, Cambridge, MIT, November 1993.

[24] P. R. Cooper, *Parallel Object Recognition from Structure The Tinkertoy Project.* PhD thesis, University of Rochester, Rochester, New York, 1989.

[25] P. B. Chou and C. M. Brown, "The theory and practice of Bayesian image labeling," *International Journal on Computer Vision*, vol. 4, pp. 185–210, 1990.

[26] M. Wheeler and K. Ikeuchi, "Sensor modeling, probabilistic hypothesis generation, and robust localization for object recognition," *IEEE Transactions on Pattern Analysis and Machine Intelligence*, vol. 17, no. 3, pp. 252–265, 1995.

[27] S. Geman and D. Geman, "Stochastic relaxation, gibbs distribution, and the bayesian restoration of images.," *IEEE Transactions on Pattern Analysis and Machine Intelligence*, vol. 6, pp. 721–741, 1984.

[28] D. Nair and J. K. Aggarwal, "Hierarchical, modular architectures for object recognition by parts." submitted to *13th International Conference on Pattern Recognition.* November, 1996, Vienna, Austria.

[29] A. P. Dempster, N. M. Laird, and D. B. Rubin, "Maximum likelihood from incomplete data via the EM algorithm.," *Journal of the Royal Statistical Society*, vol. 39-B, pp. 1–38, 1977.

[30] L. Xu and A. L. Yuille, "Robust principal component analysis by self-organizing rules based on statistical physics approach," *IEEE Transactions on Neural Networks*, vol. 6, no. 1, pp. 131–195, 1995.

[31] K. Ng and R. Lippmann, "Practical characteristics of neural network and conventional pattern classifiers," in *Neural Information Processing Systems* (J. M. R.P. Lippmann and D. Touretzky, eds.), pp. 970–976, 1991.

[32] J. Ghosh and K. Tumer, "Structural adaptation and generalization in supervised feed-forward networks," *Journal of Artificial Neural Networks*, vol. 1, no. 4, pp. 431–458, 1994.

[33] C. M. Bishop, *Neural Networks for Pattern Recognition.* New York: Oxford University Press, 1995.

[34] J. Ghosh, "Vision based inspection," in *Artificial Neural Networks for Intelligent Manufacturing* (C. H. Dagli, ed.), pp. 265–297, Chapman and Hall, London, 1994.

[35] A. K. Jain, "Advances in pattern recognition," in *Pattern Recognition Theory and Applications* (F. A. Denijver and J. Kittler, eds.), pp. 1–19, Springer-Verlag, 1986.

[36] J. Mao and A. K. Jain, "Artificial neural networks for feature extraction and multivari-

ate data projection," *IEEE Transactions on Neural Networks*, vol. 6, no. 2, pp. 296–317, 1995.

[37] G. Bradski and S. Grossberg, "Fast-learning VIEWNET architectures for recognizing three-dimensional from multiple two-dimensional views," *IEEE Transactions on Neural Networks*, vol. 8, no. 7/8, pp. 1053–1080, 1995.

[38] J. Wang and F. Cohen, "3D object recognition and shape estimation from image contours using B-splines, shape invariant matching, and neural networks," *IEEE Transactions on Pattern Analysis and Machine Intelligence*, vol. 16, no. 1, pp. 1–23, 1994.

[39] D. P. Casasent and L. M. Neiberg, "Classifier and shift-invariant automatic target recognition neural networks," *IEEE Transactions on Neural Networks*, vol. 8, no. 7/8, pp. 1117–1129, 1995.

[40] D. Nair, A. Mitiche, and J. K. Aggarwal, "On comparing the performance of object recognition systems," in *Proceedings of the Second IEEE International Conference on Image Processing*, (Washington D. C.), pp. 311–315, October 1995.

[41] S. Haykin, *Neural Networks: A Comprehensive Foundation.* MacMillan, 1994.

[42] T. Kohonen, *Self-Organization and Associative Memory.* Kluwer Academic Publishers, 1988.

[43] S. V. Chakravarthy, J. Ghosh, and S. Jaikumar, "Aspect graph construction using a neural network of radial basis functions," in *Proc. ANNIE 91*, pp. 465–472, Nov 1991.

[44] M. Seibert and A. Waxman, "Adaptive 3D object recognition from multiple views," *IEEE Transactions on Pattern Analysis and Machine Intelligence*, vol. 11, no. 3, pp. 107–124, 1987.

[45] J. S. Beis and D. G. Lowe, "Learning indexing functions for 3D model-based object recognition," *Proceedings IEEE Conference on Computer Vision and Pattern Recognition*, pp. 275–280, 1994.

[46] B. Pravin and G. Medioni, "A constraint satisfaction network for matching 3D object," in *International Joint Conference on Artificial Intelligence*, vol. II, pp. 18–22, June 1989.

[47] E. Mjolsness, E. Gindi, and P. Anandan, "Optimization in model matching and perceptual organization," *Neural Computation*, vol. 1, pp. 218–219, 1989.

[48] R. Mohan, "Application of neural constraint satisfaction network to vision," in *International Joint Conference on Artificial Intelligence*, vol. II, pp. 619–620, June 1989.

[49] W. Lin, F. Liao, C. Tsao, and T. Lingutla, "A hierarchical multiple-view approach to three-dimensional object recognition," *IEEE Transactions on Neural Networks*, vol. 2, no. 1, pp. 84–92, 1991.

[50] A. Wong, "Knowledge representation for robot vision and path planning using attributed graphs and hypergraphs," in *Machine Intelligence and Knowledge Engineering for Robotics Applications, Proc. NATO/ASI Workshop* (A. Wong and A. Pugh, eds.), pp. 113–143, Springer Verlag, 1987.

[51] J. T. Tou, "Knowledge-based systems for robotic application," in *Machine Intelligence and Knowledge Engineering for Robotics Applications, Proc. NATO/ASI Workshop* (A. Wong and A. Pugh, eds.), pp. 145–189, Springer Verlag, 1987.

[52] M. De Mathelin, C. Perneel, and M. Acheroy, "IRES: an expert system for automatic target recognition from short-distance infrared images," in *Proceedings of SPIE, Architecture, Hardware, and Forward-Looking Infrared Issues in Automatic Object Recognition* (L. Garn and L. Graceffo, eds.), vol. 1957, pp. 68–84, 1993.

[53] E. Riseman and A. Hanson, "A methodology for the development of general knowledge-based vision system," in *Computer vision: theory and Industrial Applications* (C. Torras, ed.), pp. 293–336, Springer Verlag, 1992.

[54] W. Shomar, G. Seetharaman, and T. Young, "An expert system for recovering 3D shape and orientation from a single view," in *Computer vision and image processing* (L. Shapiro and A. Rosenfeld, eds.), pp. 459–516, Academic press, 1992.

[55] C. Chu and J. K. Aggarwal, "The interpretation of a laser radar images by a knowledge-based system," *Machine Vision and application*, vol. 4, pp. 145–163, 1995.

[56] C. Chu and J. K. Aggarwal, "The integration of image segmentation maps using region and edge information," *IEEE Transactions on Pattern Analysis and Machine Intelligence*, vol. 15, no. 12, pp. 1241–1252, 1993.

[57] T. Matsuyama and V. Hwang, *SIGMA: A knowledge-based aerial image understanding system.* Plenum Press, New York, 1990.

[58] V. Cherkassky, J. Friedman, and H. W. (Eds.), *From Statistics to Neural Networks, Proc. NATO/ASI Workshop.* Springer-Verlag, 1995.

[59] I. Sethi and A. Jain, eds., *Artificial Neural Networks and Statistical Pattern Recognition.* Elsevier Science, Amsterdam, 1991.

[60] I. Taha and J. Ghosh, "A hybrid intelligent architecture for refining input characterization and domain knowledge," in *Proceedings of World Congress on Neural Networks*, vol. II, pp. 284–287, July 1995.

[61] U. Grenander, *General Pattern Theory.* Oxford Univ. Press, 1994.

[62] U. Grenander and M. I. Miller, "Representations of knowledge in complex systems," *Jl. of the Royal Statistical Society Series B*, vol. 56, no. 4, pp. 549–603, 1994.

[63] M. I. Miller, A. Srivastava, and U. Grenander, "Conditional-mean estimation via jump-diffusion processes in multiple target tracking/recognition," *IEEE Trans. Signal Processing*, vol. 43, pp. 1–13, November 1995.

[64] D. B. Mumford, "Pattern theory: a unifying perspective," in *Proc. 1st European Congress of Mathematics*, 1994.

[65] I. Biederman, "Human image understanding: Recent research and a theory," *Computer Vision, Graphics and Image Processing*, vol. 32, pp. 29–73, 1985.

Section 5

Computer Vision Technology in Application

Towards 3-D model-based tracking of humans in action

D.M. Gavrila and L.S. Davis
Computer Vision Laboratory, CfAR,
University of Maryland
College Park, MD 20742, U.S.A.
{gavrila,lsd}@cfar.umd.edu
http://www.umiacs.umd.edu/users/{gavrila,lsd}/

Abstract

We describe a vision system for the 3-D model-based tracking of unconstrained human movement. Using image sequences acquired simultaneously from multiple views, we recover the 3-D body pose at each time instant without the use of markers. The pose-recovery problem is formulated as a search problem and entails finding the pose parameters of a graphical human model whose synthesized appearance is most similar to the actual appearance of the real human in the multi-view images. The models used for this purpose are acquired from the images. We use a decomposition approach and a best-first technique to search through the high dimensional pose parameter space. A robust variant of chamfer matching is used as a fast similarity measure between synthesized and real edge images.

We present initial tracking results from a large new Humans-In-Action (HIA) database containing more than 2500 frames in each of four orthogonal views. They contain subjects involved in a variety of activities, of various degrees of complexity, ranging from the more simple one-person hand waving to the challenging two-person close interaction in the Argentine Tango.

1 Introduction

The ability to recognize humans and their activities by vision is a key feature in the pursuit to design a machine capable of interacting intelligently and effortlessly with a human-inhabited environment. Besides this long-term goal, there are many applications possible in the more near term, e.g. in virtual reality, "smart" surveillance systems, motion analysis in sports, choreography of dance and ballet, sign language translation and gesture-driven user interfaces. In many of these applications a non-intrusive sensory method based on vision is preferable over a (in some cases not even feasible) method that relies on markers attached to the bodies of human subjects.

Our approach to looking at humans and recognizing their activities has two major components:

1. body pose recovery and tracking
2. recognition of movement patterns

Several considerations have to be made regarding body pose determination and tracking, which affect what features can be used for movement matching: the type of model used (stick figure, volumetric model, none), the dimensionality of the space in which tracking takes place (2-D or 3-D), the number of sensors used (single, stereo, multiple), the sensor

 0-8186-7644-2 $5.00

modality (visible light, infrared, range), the sensor placement (centralized vs. distributed) and sensor mobility (stationary vs. moving). We consider the case where we have multiple stationary (visible-light) cameras, previously calibrated, and we observe one or more humans performing some action from multiple viewpoints. The aim of the first component is to reconstruct from the sequence of multi-view frames the (approximate) 3-D body pose of the human(s) at each time instant; this serves as input to the movement recognition component. In an earlier paper [5] we considered movement recognition as a classification problem and we used a Dynamic Time Warping method to match a test sequence with several reference sequences representing prototypical activities. The features used for matching were various 3-D joint angles of the human body. In this paper, we deal only with the pose recovery and tracking component.

The outline of this paper is as follows. First, Section 2 provides a motivation for our choice of a 3-D recovery approach rather than a 2-D approach. In Section 3 we discuss 3-D human modeling issues and the (semi-automatic) model acquisition procedure used by our system. Section 4 deals with the pose recovery and tracking component. Included is a bootstrapping procedure to start the tracking or to re-initialize it if it fails. Section 5 presents new experimental results in which successful unconstrained whole-body movement is demonstrated on two subjects. These are initial results [1]derived from a large Humans-In-Action (HIA) database containing two subjects involved in a variety of activities, of various degree of complexity. We discuss our results and possible improvements in Section 6. Finally, Section 7 contains our conclusions.

2 2-D vs. 3-D

One may question whether it is desirable or feasible to try to recover 3-D body pose from 2-D image sequences for the purpose of recognizing human movement. An alternative approach is to work directly with 2-D features derived from the images, using some form of 2-D model [7] [11] or not [3] [16].

Model-free 2-D features are usually obtained by applying a motion-detection algorithm to the image (assuming a stationary camera) and obtaining the outline of a moving object, presumably human. Frequently, a $K \times N$ spatial grid is superimposed on the motion region, after a possible normalization of its extent. In each of the $K \times N$ tiles a simple feature is computed, and these are combined to form a $K \times N$ feature vector to describe the state of movement at time t. See [3] and [16]. Another possibility is to use 2-D model-based features, where the assumption is that as a result of 2-D segmentation and tracking a sequence of 2-D stick figure poses is available [7].

Recognition systems using 2-D model-free features have been able to claim early successes in matching human movement patterns. For constrained types of human movement (such as walking parallel to the image plane, involving periodic motion), many of these features have been successfully used for classification, as in [16]. This may indeed be the easiest and best solution for several applications. But we find it unlikely that reliable recognition of more unconstrained and complex human movement (e.g. humans wandering around, making different gestures while walking and turning) can be achieved using these types of features exclusively. With respect to using 2-D model-based features, we note that few systems actually derive the features they use for movement matching. Self-occlusion makes the 2-D tracking problem hard for arbitrary movements and thus existing systems assume some a-priori knowledge of the type of movement and/or the viewpoint under which it is

[1]The tracking results described in this paper are also available as video clips from our home pages.

observed. 2-D labeling and tracking under more general conditions is attempted by [11].

We therefore investigate in this paper the more general-purpose approach of recovering 3-D pose through time, in terms of 3-D joint angles defined with respect to a human-centered 3-D motion recovery from 2-D images is often an ill-posed problem. In the case of 3-D pose tracking, however, we can take advantage of the large available a-priori knowledge about the kinematic and shape properties of the human body to make the problem tractable. Tracking also is well supported by the use of a 3-D human model which can predict events such as (self) occlusion and (self) collision. Once 3-D tracking is successfully completed, we have the benefit of being able to use the 3-D joint angles as features for movement matching, which are viewpoint independent and directly linked to the body pose. Compared with 3-D joint coordinates, they are less sensitive to variations in the size of humans.

The techniques described in this paper lead to tracking on a fine scale, with the obtained joint angles being within a few degrees of their true values. Besides providing meaningful generic features for a movement matching component, such techniques are of independent interest for their use in virtual reality applications. In other applications, such as surveillance, continuous fine-scale 3-D tracking will not always be necessary, and can be combined with tracking on a more coarse level (for example, considering the human body as a single unit), changing the mode of operation from one to another depending on context.

3 3-D body modeling and model acquisition

3-D graphical models for the human body generally consist of two components: a representation for the skeletal structure (the "stick figure") and a representation for the flesh surrounding it. The stick figure is simply a collection of segments and joint angles with various degree of freedom at the articulation sites. The representation for the flesh can either be surface-based (using polygons, for example) or volumetric (using cylinders, for example). There is a trade-off between the accuracy of representation and the number of parameters used in the model. Many highly accurate surface models have been used in the field of graphics [1] to model the human body, often containing thousands of polygons obtained from actual body scans. In vision, where the inverse problem of recovering the 3-D model from the images is much harder and less accurate, the use of volumetric primitives has been preferred to "flesh out" the segments because of the lower number of model parameters involved.

For our purposes of tracking 3-D whole-body motion, we currently use a 22-DOF model (3 DOF for the positioning of the root of the articulated structure, 3 DOF for the torso and 4 DOF for each arm and each leg), without modeling the palm of the hand or the foot, and using a rigid head-torso approximation. See [1] for more sophisticated modeling. Regarding shape, we felt that simple cylindrical primitives (possibly with elliptic XY-cross-sections) [4] [8] [18] would not represent body parts such as the head and torso accurately enough. Therefore, we employ the class of *tapered super-quadrics* [12]; these include such diverse shapes as cylinders, spheres, ellipsoids and hyper-rectangles. So far, we have obtained satisfactory modeling results with these primitives alone (see experiments); a more general approach also allows deformations of the shape primitives [12] [14].

We derive the shape parameters from the projections of occluding contours in two orthogonal views, parallel to the zx- and zy-planes. This involves the human subject facing the camera frontally and sideways. We assume 2-D segmentation in the two orthogonal views; a way to obtain such a segmentation is proposed in [10]. Back-projecting the 2-D projected contours of a quadric gives the 3-D occluding contours, after which a coarse-to-fine search procedure is used over a reasonable range of parameter space to determine the best-fitting quadric. Fitting uses chamfer matching (see the next section) as a similarity measure between the fitted and back-projected occluding 3-D contours. Figure 4 shows frontal and side views of the recovered torso and head for two persons: DARIU and ELLEN. Figure 5 shows their complete recovered models in a graphics rendering. These models are used in the tracking experiments of Section 5.

4 Pose recovery and tracking

The general framework for our tracking component is inspired by the early work of O'Rourke and Badler [19] and is illustrated in Figure 1. Four main components are involved: prediction, synthesis, image analysis and state estimation. The prediction component takes into account previous states up to time t to make a prediction for time $t+1$. It is deemed more stable to do the prediction at a high level (in state space) than at a low level (in image space), allowing an easier way to incorporate semantic knowledge into the tracking process. The synthesis component translates the prediction from the state level to the measurement (image) level, which allows the image analysis component to selectively focus on a subset of regions and look for a subset of features. Finally, the state-estimation component computes the new state using the segmented image.

The above framework is general and can also be applied to other model-based tracking problems. In the remainder of this section, we discuss how the components are implemented in our system for the case of tracking humans, and how this relates to existing work. In the first subsection we cover the pose estimation component, the second subsection briefly covers the other components.

4.1 Pose estimation

Our approach to pose recovery is based on a generate-and-test strategy. The problem is formulated as a search problem and entails finding the pose parameters of a graphical human model whose synthesized appearance is most similar to the actual appearance of the real human (see Figure 2). This approach has the advantage that the measure of similarity between synthesized appearance and actual appearance can now be based on whole contours and/or regions rather than on a few points. So far, existing systems which work on real images using this strategy have had limitations: Perales and Torres [15] describe a system which involves input from a human operator. Hogg [8] and Rohr [18] deal with the restricted movement of walking parallel to image plane, for which the search space is essentially one-dimensional. Downton and Drouet [4] attempt to track unconstrained upper-body motion, but must conclude that the tracking gets lost due to propagation of errors. Goncalves *et al.* [6] use a Kalman-filtering approach to track arm movement from single-view images where the shoulder remains fixed. Finally, work by Rehg and Kanade [17] is geared towards finger tracking. We aim to improve the previous approaches, where applicable, along the following lines.

- Similarity measure

In our approach the similarity measure between model view and actual scene is based on arbitrary edge contours rather than on straight line approximations (as in [18], for example); we use a robust variant of *chamfer matching* [2]. The *directed* chamfer distance $DD(T, R)$ between a test point set T and a reference point set R is obtained by summing the distances between each point in set T to its nearest point in R

$$DD(T,R) = \sum_{t \in T} dd(t,R) = \sum_{t \in T} min_{r \in R} \| t - r \| \quad (1)$$

and its normalized version is

$$\overline{DD}(T,R) = DD(T,R)/|T| \quad (2)$$

$DD(T,R)$ can be efficiently obtained in a two-pass process by pre-computing the chamfer distance on a grid to the reference set. The resulting distance map is the so-called "chamfer image" (see Figures 7b and 7c). It would be efficient if we could use only $DD(M,S)$ during pose search (as done in [2]), where M and S are the projected model edges and scene edges, respectively. In that case, the scene chamfer image would have to be computed only once, followed by fast access for different model projections. However, using this measure alone has the disadvantage (which becomes apparent in experiments) that it does not contain information about how close the reference set is to the test set. For example, a single point can be really close to a large straight line, but we may not want to consider the two entities very similar. We therefore use the *undirected* normalized chamfer distance $\overline{D}(T,R)$

$$\overline{D}(T,R) = (\overline{DD}(T,R) + \overline{DD}(R,T))/2 \quad (3)$$

A further modification is to perform outlier rejection on the distribution $dd(t,R)$. Points t for which $dd(t,R) > \theta$ are rejected outright; the mean μ and standard deviation σ of the resulting distribution is used to reject points t for which $dd(t,R) > \mu + 2\sigma$.

We note that other measures could (and) have been used to evaluate a hypothesized model pose, which work directly on the scene image: correlation (see [6] and [17]) and average contrast value along the model edges (a measure commonly used in the snake literature). The reason we opted for preprocessing the scene image (i.e. applying an edge detector) and chamfer matching is that it provides a gradual measure of similarity between two contours while having a long-range effect in image space. It is gradual since it is based on distance contributions of many points along both model and scene contours; as two identically contours are moved apart in image space the average closest distance between points increases gradually. This effect is noticeable over a range up to threshold θ, in the absence of noise. The two factors, graduality and long-range, make (chamfer) distance mapping a suitable evaluation measure to guide a search process. Correlation and average contrast along a contour, on the other hand, typically provide strong peak responses but rapidly declining off-peak responses.

- Multiview approach

By using a multi-view approach we achieve tighter 3-D pose recovery and tracking of the human body than from using one view only; body poses and movements that are ambiguous from one view can be disambiguated from another view. We synthesize appearances of the human model for all the available views, and evaluate the appropriateness of a 3-D pose based on the similarity measures for the individual views (see Figure 2).

- Search

Search techniques are used to prune the high dimensional pose parameter space (see also [13]). We currently use *best-first* search; we do this because a reasonable initial state can be provided by a prediction component during tracking or by a bootstrapping method at start-up. The use of a well-behaved similarity measure derived from multiple views, as discussed before, is likely to lead to a search landscape with fairly wide and pronounced maxima around the correct parameter values; this can be well detected by a local search technique such as best-first. Nevertheless, the fact remains that the search-space is very large and high-dimensional (22 dimensions per human, in our case); this makes "straight-on" search daunting. The proposed solution to this is *search space decomposition*. Define the original N-dimensional search space Σ at time t as

$$\Sigma = \{\{p_1\} \times .. \times \{p_N\}\},$$
$$\{p_i\} = \{\hat{p}_i - \Delta_{1i}, .., \hat{p}_i + \Delta_{2i}\}, \;\; step\ \Delta_{3i} \tag{4}$$

where $\hat{\mathbf{P}} = (\hat{p}_1, ..., \hat{p}_N)$ is the state prediction for time t. We define the decomposed search space Σ^* as

$$\Sigma^* = (\Sigma_1, \Sigma_2) \tag{5}$$
$$\Sigma_1 = \{\{p_{i_1}\} \times .. \times \{p_{i_M}\} \times \{\hat{p}_{i_{M+1}}\} \times .. \times \{\hat{p}_{i_N}\}\} \tag{6}$$
$$\Sigma_2 = \{\{\tilde{p}_{i_1}\} \times .. \times \{\tilde{p}_{i_M}\} \times \{p_{i_{M+1}}\} \times ... \times \{p_{i_N}\}\} \tag{7}$$

where $(\tilde{p}_{i_1}, .., \tilde{p}_{i_M})$ is derived from the best solution to searching Σ_1. The above search space decomposition can be applied recursively and can be represented by a tree in which non-leaf nodes represent search spaces to be further decomposed and leaf nodes are search spaces to be actually processed. The recursive scheme we propose for the pose recovery of K humans is illustrated in Figure 3. In order to search for the pose of the i-th human in the scene we synthesize humans 1, ..., $i-1$ with the best pose parameters found earlier, and synthesize humans $i+1$, ..., K with their predicted pose parameters. We search for the best torso/head configuration of the i-th human while keeping the limbs at their predicted values, etc.

We have found in practice that it is more stable to include the torso-twist parameter in the arms (or legs) search space, instead of in the torso/head search space. This is because the observed contours of the torso alone are not very sensitive to twist. Given that we keep the root of the articulated figure fixed at the torso center, the dimensionalities of the search spaces we actually search are 5, 9, and 8, respectively.

- Initialization

Our bootstrapping procedure for starting the tracking currently handles the case where moving objects (i.e. humans) do not overlap and are positioned against a stationary background. The procedure starts with background subtraction, followed by a thresholding operation to determine the region of interest; see Figure 6. This operation can be quite noisy, as shown in the figure. The aim is to determine from this binary image the major axis of the region of interest; in practice this is the axis of the prevalent torso-head configuration. Together with the major axis of another view, this allows the determination of the major 3-D axis of the torso. Additional constraints regarding the position of the head along the axis (currently, implemented as a simple histogram technique) allow a fairly precise estimation of all torso parameters, with the exception of the torso twist which is searched for, together with the arms/legs parameters, in a coarse to fine fashion.

The determination of the major axis can be achieved robustly by iteratively applying a principal component analysis (PCA) [9] on data points sampled from the region of interest. At each iteration the "best" major axis is computed using PCA and the distribution of the distances from data points to this axis is computed. Data points whose distances to the current major axis are more than the mean plus twice the standard deviation are considered outliers and removed from the data set. This process results in the removal of the data points corresponding to the hands if they are located lateral to the torso, and also of other types of noise. The iterations are halted if the parameters of the major axis vary by less than a user defined fraction from one iteration to another. In Figure 6 the successive approximations to the major axis are shown by straight lines in increasingly light colors.

4.2 The other components

Our prediction component works in batch mode and uses a constant acceleration model for the pose parameters. In other words, a second degree polynomial is fitted at times $t, ..., t-T+1$, and its extrapolated value at time $t+1$ is used for prediction. The synthesis component uses a standard graphics renderer to give the model projections for the various camera views. Finally, the image analysis component applies an edge detector to the real images, performs linking, and groups the edges into constant curvature segments. These segments are each considered as a unit and either accepted or rejected into the filtered scene edge map, a decision which is based on their directed chamfer distances to the projected model edges; see Figure 7. This process facilitates the removal of unwanted contours which could disturb the scene chamfer image (in Figure 7, for example, background edges around the head area in the original edge image are absent in the filtered edge image).

5 Experiments

We compiled a large data base containing multi-view images of human subjects involved in a variety of activities. These activities are of various degrees of complexity, ranging from single-person hand waving to the challenging two-person close interaction of the Argentine Tango. The data was taken from four (near-) orthogonal views (FRONT, RIGHT, BACK and LEFT) with the cameras placed wide apart in the corners of a room for maximum coverage; see Figure 8. The background is fairly complex; many regions contain bar-like structures and some regions are highly textured (observe the two VCR racks in lower-right image of Figure 8). The subjects wear tight-fitting clothes. Their sleeves are of contrasting colors, simplifying edge detection somewhat in cases where one body part occludes another.

Because of disk space and speed limitations, the more than one hour's worth of image data was first stored on (SVHS) video tape. A subset of this data was digitized (properly aligned by its time code (TC)) and makes up the HIA database, which currently contains more than 2500 frames in each of the four views.

The cameras were calibrated using an iterative, non-linear least squares method developed by Szeliski and Kang [20] and kindly made available to us. Figure 8 illustrates the outcome; the epipolar lines shown in the RIGHT, BACK and LEFT views correspond to the selected points in the FRONT view. One can see that corresponding points lie very close to or on top of the epipolar lines. Observe how all the epipolar lines emanate from one single point in the BACK view: the FRONT camera center lies within its view.

Our system is implemented under A.V.S. (Advanced Visualization System). Following its data flow network model, it consists of independently running modules, receiving and passing data through their interconnections. The implemented A.V.S. network bears a close resemblance to Figure 2. The parameter space was bounded in each angular dimension by $\pm$ 15 degrees, and in each spatial dimension by $\pm$ 10 cm around the predicted parameter values. The discretization was 5 degrees and 5 cm, respectively. We kept these values constant during tracking.

Figure 10 illustrates tracking for persons DARIU and ELLEN. The movement performed can be described as raising the arms sideways to a 90 degree extension, followed by rotating both elbows forward. Moderate opposite torso movement takes place for balancing as arms are moved forward and backwards. The current recovered 3-D pose is illustrated by the projection of the model in the four views, shown in white. The displayed model projections include for visual purposes the edges at the intersections of body parts; these were not included in the chamfer matching process. It can be seen that tracking is quite successful, with a good fit for the recovered 3-D pose of the model for the four views. Figure 9 shows some of the recovered pose parameters for the DARIU sequence.

6 Discussion

As we process more sequences of our HIA database our aim is to be able to process the more complex sequences, involving fast-varying poses, multiple bodies and close interaction (see for example Figure 11). We consider several improvements to our system. On the image processing level, we are interested in a tighter coupling between prediction and segmentation. Currently, the image processing component applies a general- purpose edge-detector and uses prediction only for filtering purposes. We are interested in more actively using the prediction information through the use of deformable templates. On the algorithmic level, we are interested in methods of further constraining the search space, based on either image flow or stereo correspondence.

7 Conclusions

We have presented a new vision system for the 3-D model-based tracking of unconstrained human movement from multiple views. A large Humans In Action database has been compiled for which initial tracking results were shown. We draw the following two conclusions from these initial experimental results. First, our calibration and human modeling procedures support a (perhaps surprisingly) good 3-D localization of the model such that its projection matches the all-around camera views. This is good news for the feasibility of *any* multi-view 3-D model-based tracking method, not just ours. Second, the proposed pose recovery and tracking method based on, among others, the chamfer distance as similarity measure, is indeed able to maintain a good fit over time. This is encouraging as we turn to the more complex sequences.

8 Acknowledgements

We would like to thank Ellen Koopmans, P.J. Narayanan and Pete Rander for their help in acquiring the Humans-In-Action database at CMU's 3-D Studio. This work was supported by the Advanced Research Projects Agency (Order No. C635) and by the Office of Naval Research (Grant N00014-95-1-0521).

References

[1] N.I. Badler, C.B. Phillips, and B.L. Webber, "Simulating Humans," Oxford University Press, Oxford, UK, 1993.

[2] H.G. Barrow *et al.*, "Parametric Correspondence and Chamfer Matching: Two New Techniques For Image Matching," *Proc. IJCAI*, vol.2, pp.659-663, 1977.

[3] T. Darrell and A. Pentland, "Space-Time Gestures," *Looking at people, Proc. IJCAI*, Chambery, France, 1993.

[4] A.C. Downton and H. Drouet, "Model-Based Image Analysis for Unconstrained Upper-Body Motion," *Proc. Int. IEE Conf. on Image Processing and its Applications*, pp. 274-277, 1992.

[5] D. M. Gavrila and L.S. Davis, "Towards 3-D Model-based Tracking and Recognition of Human Movement," *Proc. Int. Work. on Face and Gesture Recognition*, Zurich, Switzerland, 1995.

[6] L. Goncalves *et al.*, "Monocular Tracking of the Human Arm in 3D," *Proc. ICCV*, pp.764-770, 1995.

[7] Y. Guo, G. Xu and S. Tsuji, "Understanding Human Motion Patterns," *Proc. ICPR*, 1994.

[8] D. Hogg, "Model Based Vision: A Program to See a Walking Person," *Image and Vision Computing*, vol.1, nr.1, pp.5-20, 1983.

[9] I.T. Jolliffe, *Principal Component Analysis*, Springer Verlag, New York, 1986.

[10] I. Kakadiaris and D. Metaxas, "3D Human Body Model Acquisition from Multiple Views," *Proc. ICCV*, 1995.

[11] M.K. Leung and Y.H. Yang, "First Sight: A Human Body Outline Labeling System," *IEEE Trans. on PAMI*, vol.17, no.4, pp.359-377, 1995.

[12] D. Metaxas and D. Terzopoulos, "Shape and Nonrigid Motion Estimation through Physics-Based Synthesis," *IEEE Trans. on PAMI*, vol.15, no.6, pp.580-591, 1993.

[13] J. Ohya and F. Kishino, "Human Posture Estimation from Multiple Images Using Genetic Algorithm," *Proc. ICPR*, 1994.

[14] A. Pentland, "Automatic Extraction of Deformable Models," *Int. J. Computer Vision*, vol.4, pp.107-126, 1990.

[15] F.J. Perales and J. Torres, "A System for Human Motion Matching between Synthetic and Real Images," *IEEE Work. on Motion of Non-Rig. and Art. Objects*, Austin, TX, 1994.

[16] R. Polana and R. Nelson, "Low Level Recognition of Human Motion," *IEEE Workshop on Motion of Non-Rigid and Articulated Objects*, Austin, TX, 1994.

[17] J. Rehg and T. Kanade, "Model-Based Tracking of Self-Occluding Articulated Objects," *Proc. ICCV*, pp.612-617, 1995.

[18] K. Rohr, "Towards Model-Based Recognition of Human Movements in Image Sequences," *CVGIP: Image Understanding*, Vol.59, No.1, pp.94-115, 1994.

[19] J. O'Rourke and N.I. Badler, "Model-based image analysis of human motion using constraint propagation," *IEEE Trans. on PAMI*, vol.2, pp.522-536, 1980.

[20] R. Szeliski and S.B. Kang, "Recovering 3D Shape and Motion from Image Streams Using Nonlinear Least Squares," *J. Vis. Comm. and Im. Rep.*, vol.5, no.1, pp.10-28, 1994.

[21] J. Yamato, J. Ohya and K. Ishii, "Recognizing Human Action in Time-Sequential Images using Hidden Markov Model," *Proc. IEEE CVPR*, pp 379-385, 1992.

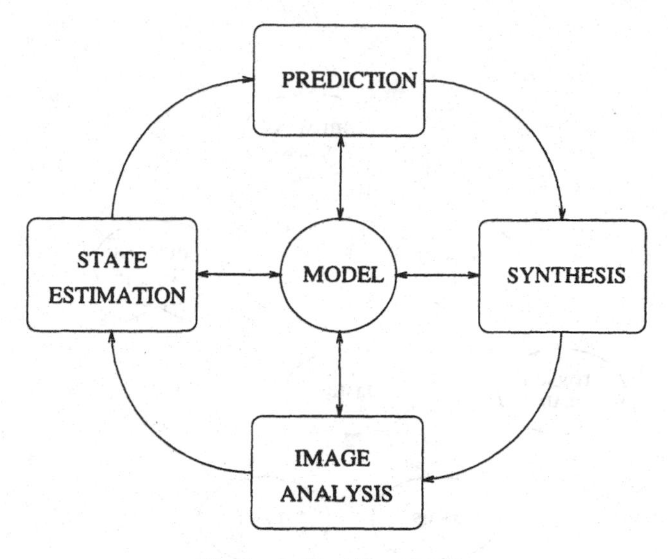

Figure 1. Tracking cycle

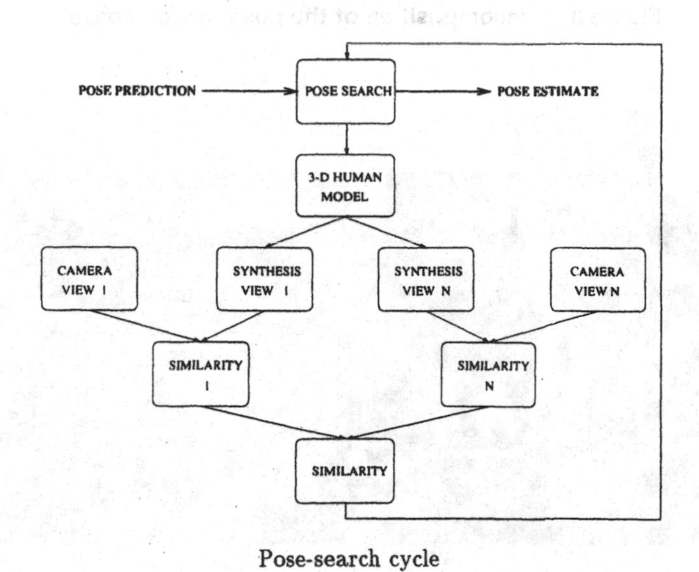

Figure 2. Pose-search cycle

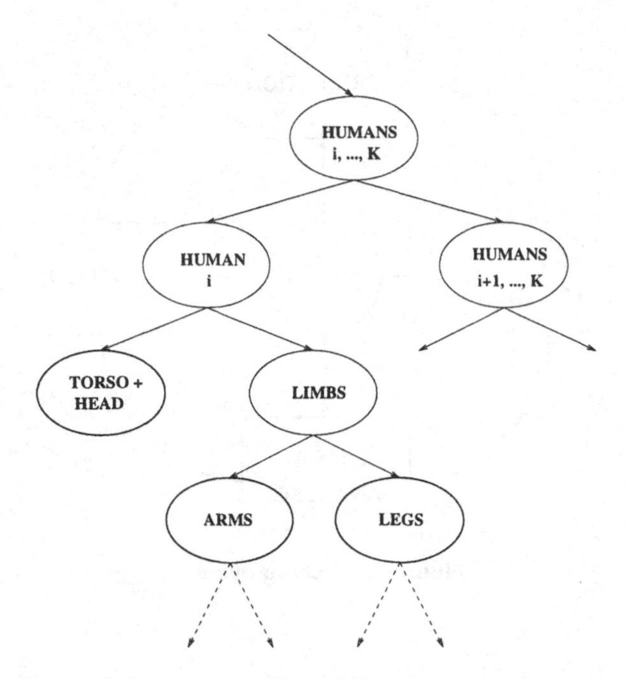

Figure 3. A decomposition of the pose-search space

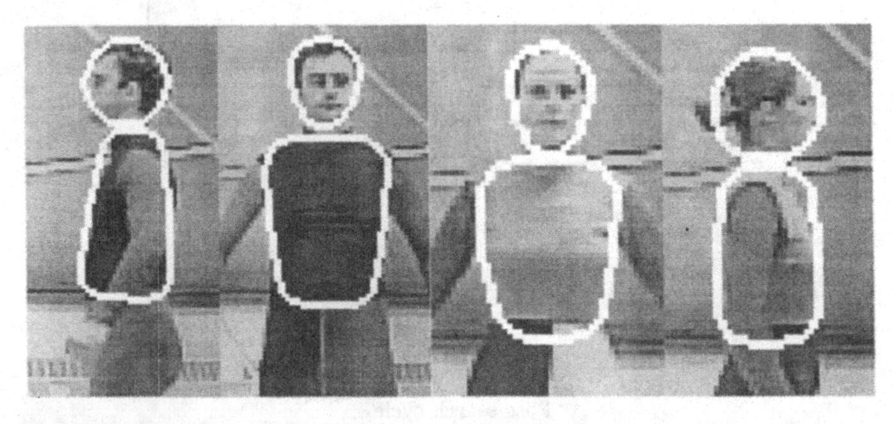

Figure 4. Frontal and side views of the recovered torso and head for the DARIU and ELLEN model

Figure 5. The recovered 3-D models ELLEN and DARIU say "hi!"

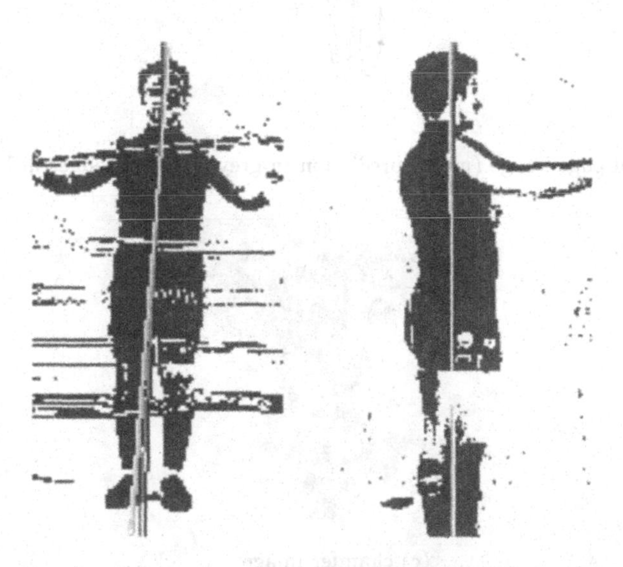

Figure 6. Robust major axis estimation using iterative PCA (cameras FRONT and RIGHT). Successive approximations to the major axis are shown in lighter colors.

(a) Scene edge image (after preprocessing)

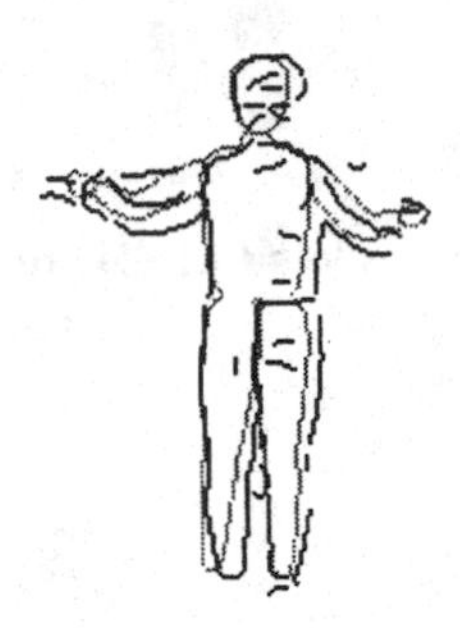

(b) filtered edge image (model prediction in grey, accepted edges in black)

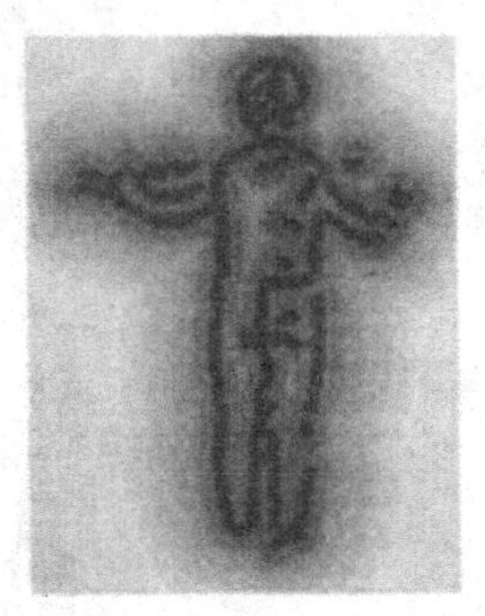

(c) chamfer image

Figure 7. Image processing

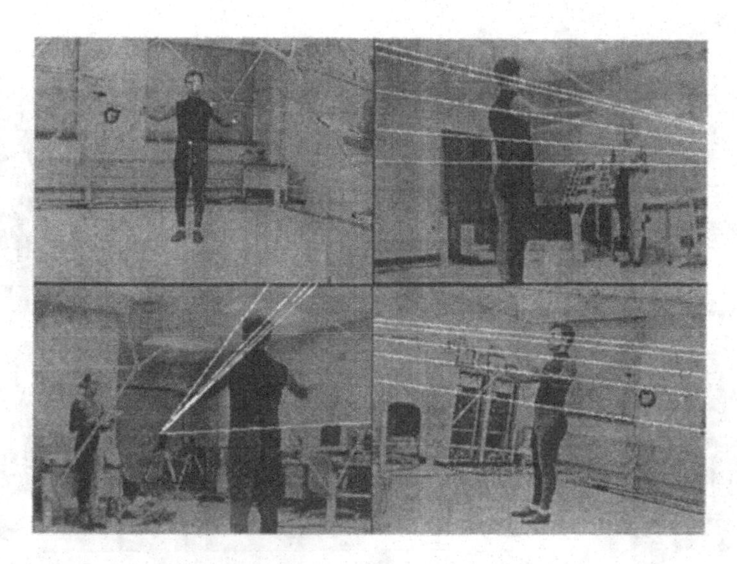

Figure 8. Epipolar geometry of cameras FRONT (upper-left), RIGHT (upper-right), BACK (lower-left) and LEFT (lower-right): epipolar lines are shown corresponding to the selected points from the view of camera FRONT

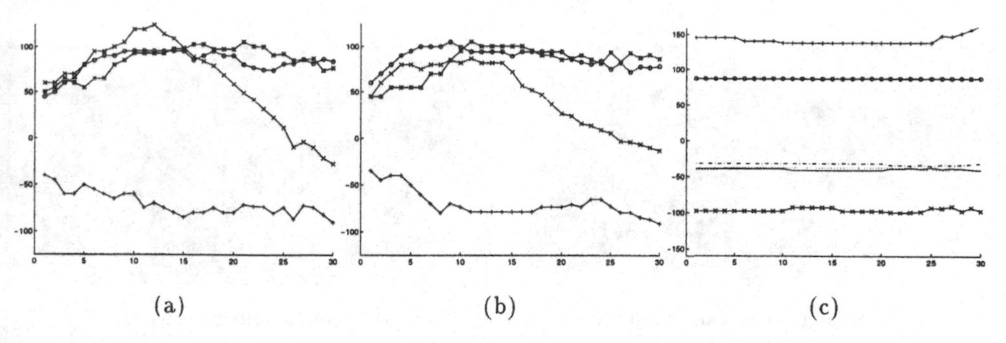

Figure 9. Recovered 3-D pose parameters vs. frame number, D-TwoElbRot; (a) **and** (b)**: LEFT and RIGHT ARM, abduction- (x), elevation- (o), twist- (+) and extension-angle (*)** (c)**: TORSO, abduction- (x), elevation- (o), twist-angle (+) and x- (dot), y- (dashdot) and z-coordinate (solid)**

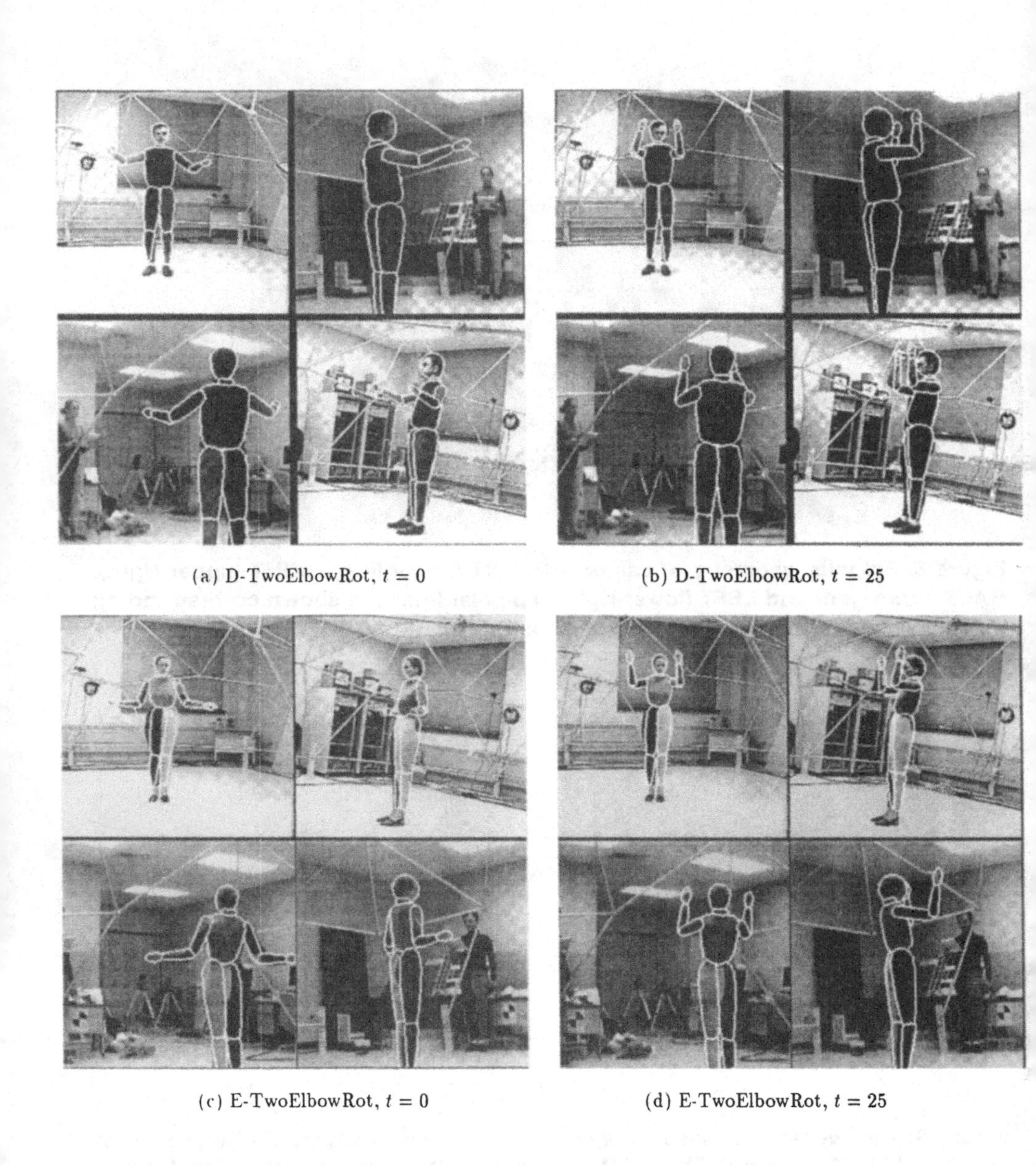

(a) D-TwoElbowRot, $t = 0$

(b) D-TwoElbowRot, $t = 25$

(c) E-TwoElbowRot, $t = 0$

(d) E-TwoElbowRot, $t = 25$

Figure 10. (a)-(b) Tracking sequence D-TwoElbowRot, (c)-(d) Tracking sequence E-TwoElbowRot, cameras FRONT, RIGHT, BACK and LEFT.

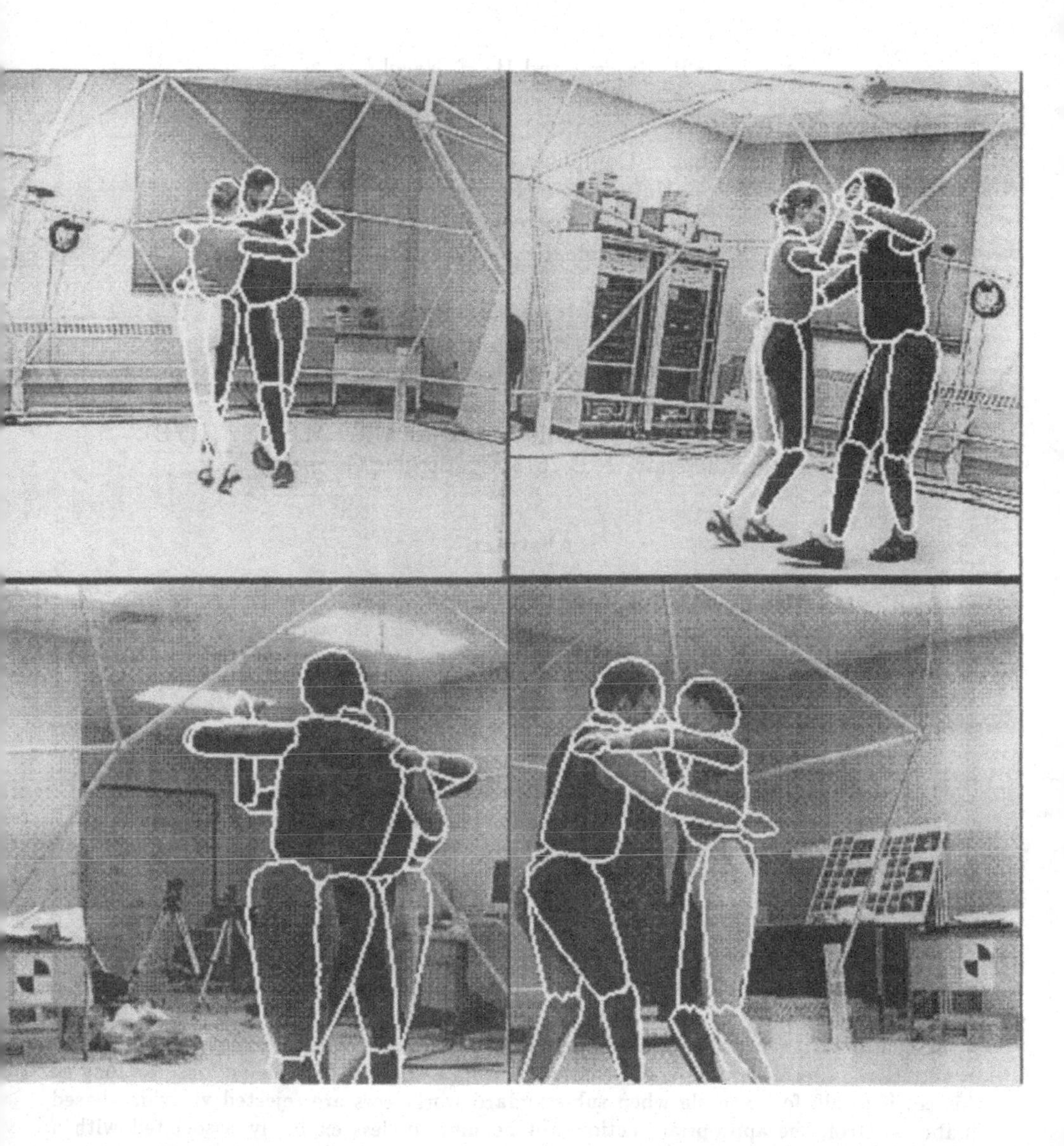

Figure 11. "Manual" 3-D pose recovery for two persons dancing the Argentine Tango (cameras FRONT, RIGHT, BACK and LEFT)

Descriptive and Prescriptive Languages for Mobility Tasks: Are They Different?

R. Bajcsy[†] and H.-H. Nagel [‡]

[†] GRASP Laboratory
Department of Computer and Information Science
University of Pennsylvania
3401 Walnut Street, Room 301C
Philadelphia, PA 19104, USA

[‡] Institut für Algorithmen und Kognitive Systeme
Fakultät für Informatik der Universität Karlsruhe (TH)
Postfach 6980, D-76128 Karlsruhe, Germany
and
Fraunhofer-Institut für Informations- und Datenverarbeitung (IITB)
Fraunhoferstr. 1, D-76131 Karlsruhe, Germany

Abstract

We start from the observation that non-trivial mobility tasks can be recognized, described, and performed automatically, based on machine vision. We compare several approaches to representations for tasks at the conceptual level used in this context. All the approaches taken into account have been implemented and used for extensive experiments. It is thus of particular interest to study the relations between descriptive *and* prescriptive *versions of the formulation of a mobility task. In addition, analogies and differences between applications for (indoor) mobile robots and for (outdoor) vision-based automatic road vehicles are compared to better understand the general properties of the required representations.*

1: Introduction and motivation

Imagine a system which evaluates signals from video-sensors to initiate, control, and terminate the execution of operations selected from a set of admissible actions. As long as this set is small, for example when sub-standard workpieces are rejected via video-based quality control, the appropriate action will be more or less explicitly associated with a suitable combination of "features" extracted from the input video signal. Under these conditions, a straightforward pattern recognition approach (Sense, Preprocess, Extract features, Classify, Act) may link signal and (re-)action. For more complex systems, such as video-based control of assembly robots, automatic vehicles, or automatic road traffic surveillance, longer sequences of elementary operations must be performed, in particular

0-8186-7644-2 $5.00 © 1996 IEEE

within a non-stationary environment. The "short-cut" link provided by pattern recognition no longer remains feasible. Not only must the set of admissible actions be greatly expanded for these systems; the selection of the appropriate action will depend on both the *actual* signal values and the *internal state* of the system. The set of system states and the signal-dependent transitions between them propagate a condensed representation of past signals and related actions up to the current point in time. The current system state may restrict the selection of the next action to one from a subset of the principally admissible actions, thereby influencing the order in which actions can be concatenated into action sequences.

The designer of such a system is thus confronted with the necessity of specifying a complex, signal-dependent system behavior. To facilitate a design process which minimizes the inadvertent omission of a part of the specification or the occurrence of design errors, specification tools must be developed. In this context, properties of the conceptual vocabulary used to formulate the desired system behavior must studied.

Compared to human abilities, video-based (re-)active systems are still in an exploratory state. It is thus no surprise that almost every group active in this area has developed its own conceptual vocabulary and representational tools. While discussing our experience with the design, implementation, experimentation, and assessment of various such systems, we became interested in comparing the representational tools used in our groups. Could we associate commonalities and differences in our approaches with particular properties of the tasks to be performed or with the boundary conditions under which these tasks had to be performed? Would it be possible to generalize the range of validity of our observations? What is the necessary and sufficient *vocabulary* that will *describe* scenes with moving agents and at the same time will serve as a *task description language* to command, for example, driving tasks for mobile indoor robots or autonomous land vehicles? We view this problem as fundamental to the *representation* of the necessary dynamical knowledge/information about the agent (i.e., a mobile robot or a car and its driver), the environment (e.g., the road and all the objects on the road, including other cars), the task and the interactions between all these different aspects of the task domain.

The issue of representation, in our interpretation, is a matter of how one partitions the continuous dynamical process of the agent/environment/task interaction into discrete (symbolic) states, places, events and behaviors. The *descriptive* aspect is the classical `signal-to-symbol` transformation, in our case going from (video) images to symbolic descriptions; the *prescriptive* aspect is the reverse, going from `symbol-to-signal`, i.e., translating symbolic expressions into continuous behaviors.

2: A look at the background

It might be easier to follow the subsequent discussion if we first outline the background of experience on which our deliberations have been based. We will then state our problem formulation in more specific terms.

2.1: From satellite images to cooperating mobile indoor robots

For well over a decade, the research of one of us (RB) has addressed the question of how to transform the information given by a picture into a conceptual description — see, e.g., [2]. Initially, these investigations centered on the extraction of object descriptions from single image frames and on the determination of planar spatial relations between objects detected in an image. During this research phase, the definition of a vocabulary for the desired conceptual descriptions was less of a problem than the design of computational processes which linked the image signal to instantiations of the given vocabulary. The extension of research toward the study of cooperation between several autonomously mobile vision-based indoor robots (see, e.g., [3, 11]) required incorporation of conceptual descriptions of *actions* into the representational vocabulary of the system. This necessitates more than merely adding object nouns, adjectives, or other types of spatial relations to the system vocabulary: the specification of an action needs to characterize spatial attributes of the agent with respect to its environment as well as the *temporal development* of these attributes.

Actions of individual mobile robots should be *described* as well as *performed*. The specification must therefore incorporate computational processes for the recognition *and* execution of actions. In experiments with several cooperating robots, many different *combinations* of actions have to be specified, and thus it pays if each action is represented by a symbol: relations between different actions can be specified as those between symbols. This facilitates the inspection of the specification, possibly even a *formal* check for internal consistency.

Such an approach, however, necessitates investigating the transition from signals to symbols (recognition of an action specification) and the inverse transition from symbol to signal (for execution of an action specification). Pushing this line of thought a bit further, the question arises whether there are *elementary* actions. If so, more complicated compound actions could then be designed by appropriate combinations of elementary actions. An immediate consequence is the question of what constitutes an elementary action (see [1]).

2.2: From video sequence evaluation to autonomously navigating road vehicles

The recording and subsequent digitization of video sequences, if necessary followed by storage of each digitized video image on a regular computer disk, offers the potential to develop computer programs which transform the video signal into a description of temporal developments within the scene recorded by a video camera. A quarter century ago, one of us (HHN) set out to study this potential by linking a minicomputer, which facilitated the design and operation of an interface, to "home-grown" experimental video digitization equipment. In the mid-seventies, first experiences with the automatic detection of the images of moving vehicles from a road scene demonstrated that at least an initial step in the desired direction was feasible. This motivated a long-range research program — of longer range than anticipated at that time — to derive conceptual descriptions from image sequences (see [16]). About a decade later, the ability to detect, initialize, and track images of road vehicles in video sequences had progressed to the point where it became realistic to start investigations into the extraction of conceptual descriptions for elementary road vehicle movements from video image sequences (see [17]). Continuous efforts in this direction have resulted in a system which has begun to stabilize (see, e.g., [9]), with current efforts being directed at making it more robust (see [21]).

When progress in this direction became discernible, it motivated attempts to advance from the extraction of descriptions of a movement corresponding to a *single maneuver* of a road vehicle to descriptions of *entire sequences* of vehicle maneuvers. System internal representations for admissible sequences of maneuvers can be interpreted as representations for different goals of an agent: since an observable execution of the entire maneuver sequence will bring the vehicle to a certain location, it appears reasonable to postulate that reaching this location constitutes the goal of the agent (see [18]). The instantiation of an initial subsequence from such a representation can then be associated with a hypothesis that the remaining maneuvers to be observed should correspond to those which constitute the remaining, not yet instantiated subsequence of the entire maneuver sequence representation. Preliminary results for this approach have been recently published in [23].

The idea to infer the intentions of an agent based on its observed sequence of actions is pursued for two different purposes within the context of an autonomously mobile road vehicle [19] :

- One potential application consists in inference processes by which an automatic driver assistance system, a *copilot system*, should attempt to infer the (short-term) intentions of the vehicle's driver, thus obviating the necessity of the driver to explain to the system in detail what he intends to do.
- Provided that the ability to automatically infer the intentions of drivers of *other* vehicles — visible from the one equipped with the envisaged assistance system — can be advanced to the degree that it meets certain reliability requirements, such a system could restrict warnings about potentially dangerous traffic situations to those for which it cannot infer on the basis of the driver's maneuvers that he or she has already noticed them and takes them into account.

As a preparatory step in this direction, an attempt has been made to formally specify a subset of admissible sequences of driving maneuvers on highways [20]. In parallel with the investigations mentioned in this section, two generations of vision-based, autonomously mobile road vehicles have been specified, acquired, commissioned, and operated (see [22]).

2.3: Intermediate resumé

The results referred to in the two preceding sections should support our claim that enough experience has been accumulated to justify the abstraction step outlined in the introduction. We are looking for a common framework which enables us to position the various results relative to each other, to check for conflicts and mutual confirmation among them, and to watch out for "white patches" on our map of this problem area.

In all three domains of discourse — cooperating mobile robots, description of road traffic scenes, and vision-based copilots for road vehicles — we conceive of at least two agent categories, namely automatic agents that communicate with human counterparts, and agents that perform activities. To the latter category belong mobile robots, vehicle(s) observed on the road, or the driver to be assisted. In all three cases, an agent must perform a task for which a schema is known to the system: basically, the agent must execute a *sequence* of maneuvers according to some general rules. One of the rules stipulates, for example, that the agent recognizes and avoids obstacles. Additional rules might restrict the movements of an agent, for example to stay within a (subset of) lane(s) in the case of road traffic.

Although the description of a task may be given as a concatenation of individual actions, the actual performance of each task usually implies a smooth transition between consecutive actions. We are thus confronted with the necessity of developing a representational vocabulary which is discrete, making provisions to establish a mapping between this discrete vocabulary and an agent's continuous activity. It is important to note here that:

- This formulation implies the *postulate* that the vocabulary should be the same, irrespective of whether we want to employ it for *prescriptive* or *descriptive* purposes.
- We *postulate* that both types of vision-based (re-)active systems, a command-oriented one (corresponding to the prescriptive use of the vocabulary) and an observer-oriented one (corresponding to the descriptive use of the vocabulary), could be generated from essentially the *same schematic* mold.

In case this latter postulate appears as an unnecessary complication, one may think of a copilot system which can not only inform or warn, but which is in principle capable of actually performing what it recommends: if the driver demands, the copilot system should be able to guide the vehicle according to the actual traffic situation surrounding it and in accordance with traffic laws. This could allow a driver, for example, to concentrate on a (mobile) phone conversation while moving in stop-and-go traffic. A different example of the advantage of switching easily between prescriptive and descriptive use of a vocabulary for the representation of action sequences is provided by a video-based, semi-autonomous robot which detects a hindrance that prevents it from performing its subtask. By switching between different uses of the representational vocabulary, the semi-autonomous robot could more easily and efficiently communicate the cause for its inability to perform as instructed.

2.4: Reactive systems

We recognize that an agent is a special case of a more general class of *reactive systems.* Reactive systems are autonomous systems embodied with sensors, actuators, and controllers that can accomplish a task in partially unknown and changing environments. Systems like these occur, for example, in manufacturing and military applications. Recently, there have been efforts to formulate and formalize requirements for such systems. For example, [13] is a study of 18 different methods that are applied to the problem of developing automation software to control typical industrial production cells used in a metal processing factory. Applications such as these belong to the category of *safety-critical systems,* as a number of properties must be enforced by control software to avoid injury to people. It is a *reactive system,* as the control software must react permanently to changes in the environment, i.e., the production cell. The methods surveyed are evaluated on the basis of the richness of the language they use to capture the agent's environment, constraints and behavior and on how well the language can *prove/verify* the performance of the reactive system.

While the study cited is focused on production cells in a manufacturing environment, we believe that the formal requirements can be easily adapted to more general domains, such as road traffic. These requirements comprise:

- Safety, meaning that the controller must avoid collisions and damage to agents of any kind;
- Liveness, meaning that the agent is sufficiently robust to successfully reach its goals under reasonable conditions;
- Real-time constraints, meaning that the agent reaches the goal within a certain interval of time.

The mathematical tools used to express the reactive system's language of behaviors under the above requirements are temporal logic, dynamic logic, first order predicate logic (see, e.g., [4]), declarative or imperative synchronous and/or asynchronous programming languages and their derivatives. The goal of these formalisms is to provide specification, synthesis, and verification of agent behaviors.

We mention these studies only to show the reader the scientific community's wide interest in formal analysis of reactive systems. What is missing in these and other related studies is how one partitions the behavior space and places for the given task, i.e., how granular the language should be to have the flexibility for proper specification of a driver's or robot's task (see Section 4 for an elaboration).

3: Driving road vehicles: description and prescription

The discourse domain of road traffic offers an opportunity to investigate the evolution of descriptive and prescriptive language fragments in the context of surveillance as well as of autonomous driving tasks based on computer vision. The structuring of a driving task can be thought about on different levels. In [22] the authors speak about a four level hierarchy for the description of a driving mission: a Mission Master (MM) Level, a Navigation (N) Level, a Maneuver (M) Level, and the bottom one, a Control (C) Level, corresponding to continuous control. Here we are particularly interested in the connection between the Maneuver and Control Levels.

3.1: Elementary maneuvers for driving task specifications

To assess a set of maneuver definitions, one must indicate the range of applications within which this set should be used. The set of "elementary" driving maneuvers discussed in [22] was intended to facilitate the specification of experiments with driver assistance (copilot) systems for road vehicles. This set comprises the following maneuvers:

1. `Start_and_continue`
2. `Follow_lane`
3. `Approach_obstacle_ahead`
4. `Overtake`
5. `Stop_in_front_of_obstacle`
6. `Pass_obstacle_to_the_left/right`
7. `Start_after_preceding_car`
8. `Follow_preceding_car`
9. `Cross_intersection`
10. `Merge_to_left/right_lane`
11. `Turn_left/right`
12. `Slowdown_to_right_road_edge_and_stop`
13. `Back_up`
14. `U-turn_to_the_left/right`
15. `Reverse_direction`
16. `Enter_parking_slot`
17. `Leave_parking_slot`
18. `Standing`

Apart from the parking maneuvers, most of the other maneuvers have already been performed by automatic vision-based vehicles in some form or another (see, e.g., [24, 15, 22, 12]). Thus, it is justified to consider them as examples for prescriptive specifications of road vehicle driving tasks.

It turns out to be advantageous to slightly modify this set of "elementary" maneuvers. Maneuver 2 (`Follow_lane`) actually implies that the vehicle *speed* will be continuously modified such that the vehicle remains controllable in curved sections of a lane, i.e., the speed is reduced sufficiently to prevent the vehicle from being carried out of the lane by centrifugal forces. In addition, its *direction* has to be modified to follow a curved lane. It is understood, however, that these modifications can be derived from a *steady state* longitudinal and lateral *control* regime based on machine vision, set up to keep the vehicle within its lane. To emphasize the steady-state character of this maneuver, we consider `Follow_lane` to be a synonym for "Cruise," i.e., stay within the current lane with as constant a speed as possible. For completeness' sake, we add a trivial 18th "maneuver" `Standing`, i.e., the steady-state of zero-motion.

As Tölle [25] has pointed out, this set can be decomposed into three subsets. One comprises the parking-slot related maneuvers which — depending on the geometric conditions of the parking slot and the maneuverability of the agent vehicle — can become too complex to be discussed here in detail and thus will be excluded from further consideration. Among the remaining 16 maneuvers, a subset comprises those maneuvers which relate solely to the agent vehicle and the road, namely maneuvers 1, 2, 9 through 15 and, in addition, the added trivial maneuver 18 (`Standing`). Let us denote this subset as the Agent-related or A-set. The other subset, comprising the maneuvers 3 through 8, refers to the relation between the agent vehicle and another object (different from the road) and will be denoted as the Object-related or O-set.

A closer look at the A-set reveals that maneuver 12 (`Slowdown_to_right_road_edge-_and_ stop`) actually consists of a combination of three maneuvers, namely maneuver 12a

(Merge_to_the_right_road_edge), maneuver 12b (Slowdown), and maneuver 12c (Stop). For symmetry's sake, we generalize maneuver 12b to (Slowdown / Accelerate), thus capturing both positive and negative longitudinal accelerations. Following this slight modification, we may subdivide the A-set further into three subsets:

1. A set of maneuvers specifying the longitudinal motion characteristics. This subset will be referred to as the Agent-Velocity set or A-V-set. It comprises first of all two complementary maneuvers which change either from no motion at all to non-zero motion (maneuver 1), or its inverse maneuver, namely changing from non-zero motion to a standstill (maneuver 12c). The generalized maneuver 12b (Slowdown/Accelerate) accounts for deliberately intended, significant changes in speed to be distinguished from those which are due to the desire to keep cruising on a geometrically non-stationary (i.e., laterally and/or vertically curved) lane. In addition to these maneuvers which characterize instationarities of vehicle motion, we have the complementary set of maneuvers describing steady-state vehicle motion, namely maneuver 2 (Follow_lane) and the trivial maneuver 18 (Standing).
2. A subset referring to a change in direction, denoted as the Agent-Direction or A-D-set, which comprises maneuver 11 (Turn_left/right), maneuver 13 (Back_up, i.e., drive backwards without essential change in the orientation of the vehicle), maneuver 14 (U-turn_to_the_left/right), and maneuver 15 (Reverse_direction). This latter maneuver implies a sequence of alternating backward/forward movements to reverse the orientation of the vehicle even in cases where this can not be effected, due to limitations of maneuvering space, by a U-turn.
3. The remaining maneuvers from the A-set refer, in addition to the agent itself, explicitly to the road in some form or other. We shall refer to this subset as the Agent-Road-set or A-R-set. It comprises maneuver 9 (Cross_intersection), maneuver 10 (Merge_to_left/right_lane), and maneuver 12a (Merge_to_the_right-road_edge).

Each maneuver within the first two subsets of the A-set is characterized by either a significant change or by exactly the absence of a significant change in the longitudinal (A-V-set) or lateral (A-D-set) control regime for the vehicle. In addition, we obtained the A-R-set where the relation between the vehicle and particular sections of the road is emphasized, implying essentially steady-state conditions for the vehicle's velocity parameters.

The O-set may be similarly subdivided into two subsets:

1. One subset of object-related maneuvers implies a concatenation of *directional* changes in order to avoid a collision with the object, namely maneuvers 4 (Overtake) and 6 (Pass_obstacle_to_the_left/right). The difference between these two maneuvers consists in the fact that in the case of maneuver 4, the object moves in the same direction as the agent vehicle whereas in the case of maneuver 6, the object is either stationary or moves in a direction significantly different from that of the agent vehicle, in particular in the opposite direction (evading an obstacle approaching in the vehicle's lane). We shall denote this subset as the Object-Directional or O-D-set.
2. The other subset specifies characteristics of the longitudinal component of the relative distance and/or velocity between the agent vehicle and the object, namely either a decrease (maneuver 3, Approach_obstacle_ahead), a stop (maneuver 5, Stop_in_front_of_obstacle), an increase (maneuver 7, Start_after_preceding_car), or the steady-state (maneuver 8, Follow_preceding_car). We shall denote this latter subset as the Object-Velocity or O-V-set.

This refinement of the set of maneuvers discussed in [22] thus yields a symmetric and intuitively plausible vocabulary of basic maneuvers for road vehicles.

3.2: A terminal symbol set for the representation of road vehicle maneuver sequences on highways

Given the experience of automatically guiding road vehicles by computer vision, [20] studies the representation of concatenations of maneuvers. This corresponds to a level of abstraction above the interface between the Control Level and the Maneuver Level, reaching up to the Navigation Level in the hierarchical decomposition discussed in [22]. In an attempt to completely cover a restricted discourse world, opposed to covering a larger discourse world in an incomplete and difficult-to-judge manner, the study in [20] was limited to a restricted discourse world of "driving on highways." A context free grammar approach was used to define an admissible set of maneuver sequences in an intuitively accessible manner by recursive transition diagrams. Based on the results of the preceding Subsection, we shall now discuss and refine the set of 29 "terminal" maneuver symbols from [20]:

1. `Accelerate_in_lane`: Corresponds to the combination of a maneuver from the A-V-set regarding the action of acceleration and from the A-R-set regarding the explicit property of staying within a lane.
2. `Approach_vehicle_preceding_in_same_lane`: Corresponds to the O-V-set.
3. `Continue_in_deceleration_lane_to_junction_or_exit`: Represents the combination of an elementary maneuver from the A-V-set (`Continue`) and from the A-R-set, due to its explicit reference to a road location (junction or exit).
4. `Cross_virtual_entrance_of_highway_system`: A-R-set.
5. `Decelerate_in_lane`: Inverse maneuver to `Accelerate_in_lane`, i.e., belongs to a combination of A-V-set and A-R-set.
6. `Drive_with_constant_speed`: A-V-set.
7. `Evade_obstacle_without_leaving_lane`: Corresponds to the combination of a maneuver from the O-D-set (`Evade_obstacle`) and from the A-R-set (`withoutleaving_lane`).
8. `Follow_lane_deviation`: A-V-set, corresponding to `Follow_lane`.
9. `Follow_vehicle_beginning_to_move`: Corresponds to O-V-set (`Start_after_preceding_car`).
10. `Follow_vehicle_preceding_in_same_lane`: Corresponds to the combination of O-V-set and A-R-set.
11. `Leave_deceleration_lane`: A-R-set.
12. `Leave_virtual_exit_of_highway_system`: A-R-set.
13. `Merge_into_deceleration_lane`: A-R-set.
14. `Merge_into_left_lane`: A-R-set.
15. `Merge_into_right_lane`: A-R-set.
16. `Merge_left_into_main_lane`: A-R-set.
17. `Merge_left_into_overtaking_lane`: A-R-set.
18. `Merge_onto_directional_lanes`: A-R-set.
19. `Merge_to_left`: A-D-set.
20. `Merge_to_right`: A-D-set.

21. `Overtake`: O-D-set.
22. `Pull_onto_emergency_lane`: A-R-set.
23. `Pull_up_to_left_curb`: A-R-set.
24. `Pull_up_to_right_curb`: A-R-set.
25. `Start_driving`: A-V-set.
26. `Start_driving_on_emergency_lane`: Corresponds to the combination of a maneuver from the A-V-set (`Start_driving`) and from the A-R-set (`on_emergency_lane`).
27. `Stop`: A-V-set.
28. `Stop_in_front_of_vehicle`: O-V-set.
29. `Stop_on_emergency_lane`: Corresponds to the combination of a maneuver from the A-V-set (`Stop`) and from the A-R-set (`on_emergency_lane`).

As a result of this discussion, we may conclude that the refined set of elementary maneuvers given in the preceding Subsection 3.1 suffices completely to categorize all terminal maneuver symbols introduced in [20]. This significantly enlarges the range of validity of our current tentative considerations up to quite complex maneuver sequences describing a task like *"driving from a highway entrance via several highway junctions to a distant highway exit."*

3.3: Descriptive approach to road traffic reporting

Whereas the two preceding subsections addressed the specification of road vehicle movements from a prescriptive point of view, we shall now study a descriptive one, based on the developments reported in [8] for German vehicle motion verbs and extended to English vehicle motion verbs in [9]. These latter authors enumerate 67 road vehicle motion verbs extracted in a systematic manner from a practically complete set of over 9000 German verbs. Recognition automata have been implemented and are routinely used to assess a degree of confidence with which each verb can be associated with every point of a vehicle trajectory, which in turn has been extracted automatically from a digitized video image sequence. These 67 motion verbs have been subdivided into four categories, depending on the dominant reference type, namely Agent Reference, Location Reference, Road Reference, and Object Reference. Apart from the Location Reference type, the three other categories have already been encountered in the categorization discussed in Subsection 3.1.

Since the authors of [8, 9] use synthetic concepts comprising occasionally spatial or other adverbs, they speak of occurrences rather than verbs if they refer to items of the vocabulary. We shall now treat each occurrence in turn to further corroborate the evidence for our line of argumentation.

Descriptive vocabulary for "agent reference" motion verbs The occurrences discussed in this Subsection do not explicitly or implicitly require reference to anything beyond the agent vehicle. The occurrences are treated in a kind of natural order encountered when entering a road vehicle for a trip.

1. `Be_standing`: A-V-set.
2. `Drive_off`: A-V-set.
3. `Accelerate`: A-V-set.
4. `Drive_slowly`: A-V-set, with additional specification of speed range.
5. `Drive_at_regular_speed`: Analogous to preceding occurrence.
6. `Run_fast`: A-V-set, with additional specification of speed range.

7. `Run_very_fast`: A-V-set, with additional specification of speed range.
8. `Drive_at_constant_speed`: Analogous to preceding occurrence.
9. `Brake`: A-V-set, synonym for `Decelerate` or `Slow_down`; remember that from outside a vehicle it can not be decided in general — based on an evaluation of a digitized image sequence — whether the driver just has disengaged his foot from the gas pedal or whether he has activated the brake.
10. `Stop`: A-V-set.
11. `Run_straight_ahead`: A-D-set.
12. `Turn_left`: A-D-set.
13. `Turn_right`: A-D-set.
14. `Revolve_around_a_vertical_axis`: A-D-set.
15. `Slide`: A-D-set; the assignment of this occurrence to the A-D-set is based on the consideration that a sliding motion *without change in vehicle orientation* can not be differentiated from a deceleration maneuver, given only video input signals.
16. `Skid`: A-D-set.
17. `Reverse`: A-D-set.
18. `Run_forward`: A-D-set.

Descriptive vocabulary for "location reference" motion verbs The introduction of this category actually suggests extending the categorization introduced in Subsection 3.1. The location in question will be referred to explicitly by a spatial relation which has been incorporated into the occurrence definition. We shall denote this new subset of vehicle maneuvers as Agent-Location or A-L-set. The order will again be chosen analogously to that which one may encounter while driving.

1. `Drive_to_location`: Combination of A-D-set and A-L-set.
2. `Arrive_at_location`: Combination of A-V-set and A-L-set.
3. `Run_over_location`: Combination of A-V-set and A-L-set.
4. `Pass_location`: Combination of A-V-set and A-L-set.
5. `Stop_at_location`: Combination of A-V-set and A-L-set.
6. `Park_at_location`: Combination of A-V-set and A-L-set.
7. `Depart_from_location`: Combination of A-V-set and A-L-set.
8. `Leave_location`: Combination of A-V-set and A-L-set.
9. `Leave_location_behind`: Combination of A-V-set and A-L-set.

Descriptive vocabulary for "road reference" motion verbs The descriptive vocabulary employed by [9] with respect to road reference corresponds — with one exception — to the prescriptive vocabulary discussed in Subsection 3.1.

1. `Leave_driving_lane`: A-R-set.
2. `Enter_lane`: A-R-set.
3. `Turn`: A-D-set.
4. `Change_section`: A-R-set.
5. `Drive_on_lane`: A-R-set.
6. `Cross_a_lane`: A-R-set.

The exception (Turn) should probably be transferred from the Road Reference to the Agent Reference category above.

Descriptive vocabulary for "object reference" motion verbs This category is much more populated than the corresponding Object-set (O-set) introduced in Subsection 3.1, due to the fact that experiments with the evaluation of image sequences are simpler regarding the extraction of different object types than experiments with vision-based automatic driving tasks referring to particular objects.

1. `Catch_up_with_object`: O-V-set.
2. `Fall_behind_object`: O-V-set.
3. `Follow`: O-V-set.
4. `Follow_object_closely`: Corresponds to a maneuver from the O-V-set, but incorporates an additional spatial relation.
5. `Run_into_object`: O-V-set.
6. `Pull_out_from_behind_object`: Corresponds to a maneuver from the O-V-set, but incorporates an additional spatial relation.
7. `Get_out_of_the_way_of_object`: O-D-set.
8. `Cut_in_in_front_of_object`: O-D-set.
9. `Slip_in_in_front_of_object`: O-D-set.
10. `Pull_up_to_object`: O-V-set.
11. `Flank_object`: Corresponds to the combination of a maneuver from the O-V-set (approximately same speed) and the O-D-set (approximately same direction), but incorporates an additional spatial relation.
12. `Move_past_object`: Corresponds to a maneuver from the O-V-set, but incorporates an additional spatial relation.
13. `Let_run_into`: O-V-set.
14. `Pass`: Corresponds to the combination of a maneuver from the O-V-set (higher speed) and the O-D-set (approximately same direction), but incorporates an additional spatial relation.
15. `Drive_in_front_of_object`: O-V-set.
16. `Loose_a_lead_on_object`: O-V-set.
17. `Draw_ahead_of_object`: O-V-set.
18. `Approach_oncoming_object`: Corresponds to the combination of a maneuver from the O-V-set (Approach) and the O-D-set (implying more or less a directional alignment with the oncoming object).
19. `Make_way_for_oncoming_object`: Analogous to the preceding occurrence.
20. `Leave_an_object_driving_off_in_opposite_direction`: O-D-set.
21. `Approach_crossing_object`: O-D-set.
22. `Leave_crossing_object`: O-D-set.
23. `Close_up_to_object`: O-V-set.
24. `Merge_in_front_of_object`: O-D-set (merging into the lane on which the reference object is driving is implied, e.g., in the terminal phase of an overtaking maneuver).

25. **Move_towards_stationary_object**: O-V-set; analogous to **Approach_an_object**, but for the special case of a stationary object.
26. **Stop_behind_stationary_object**: Analogous to the preceding occurrence, but with an explicit additional spatial relation.
27. **Be_standing_near_stationary_object**: Analogous to the preceding occurrence.
28. **Start_in_front_of_stationary_object**: Analogous to the preceding occurrence.
29. **Pull_out_behind_stationary_object**: Analogous to the preceding occurrence.
30. **Drive_around_stationary_object**: Analogous to the preceding occurrence.
31. **Pass_stationary_object**: O-D-set.
32. **Merge_in_front_of_stationary_object**: O-D-set, but with an explicit additional spatial relation.
33. **Move_away_from_stationary_object**: O-D-set.
34. **Collide_with_object**: Corresponds to the combination of a maneuver from the O-V-set and one from the O-D-set.

It can be seen that no new categories need to be introduced to integrate this considerable vocabulary into the basic schema outlined in Subsection 3.1.

Discussion of the descriptive vocabulary for road traffic motion verbs At first, it might appear surprising that even such a rich descriptive vocabulary can be assigned to a relatively small number of categories which have been defined in the context of a prescriptive vocabulary. After some consideration, though, this should no longer appear so: the differences between the entries can be accounted for by specifying parameters like velocity, direction, and spatial relations between the agent vehicle and particular objects or locations. The large number of occurrences collected in this Subsection 3.3 essentially represents a shorthand notation for a small number of basic maneuvers, supplemented by adverbs and spatiotemporal relations.

Each occurrence could be read as a task prescription with some more or less implicit default option for additional specifications which can be derived from the occurrence identifier. This hypothesis appears to be supported by additional experiments conceived and conducted after the descriptive vocabulary discussed above had been essentially fixed.

3.4: Descriptive approach for gas station traffic

The idea to represent an intention of an agent in road traffic as a sequence of maneuvers [17] eventually resulted in an example of a discourse world specifically selected to simplify the demonstration of this idea, namely road vehicle traffic at a filling station (see Figures 1, 2, 3, and [18]). The selection of this discourse world enabled use of a stationary video camera for recording complex sequences of maneuvers taking place within a relatively small area.

The geometry of a filling station is fairly standard nowadays, facilitating the association of generic vehicle positions with specific goals of a driver: for example, pulling up to a petrol pump implies in general that the driver intends to obtain gas for his vehicle. Different petrol pumps can be associated with different types of gas (low or high octane gas, diesel), thus facilitating an additional differentiation of goals.

Several years after this "Gedanken experiment" was suggested, it became possible to actually record vehicles at a filling station. This provided input which exercised the capabilities of the Xtrack system (see [7, 9]) to the extreme and still presents problems not

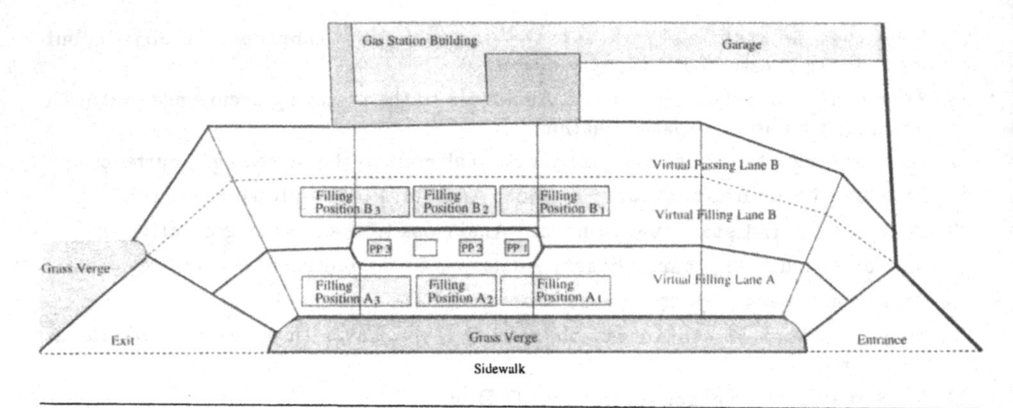

Figure 1. The gas station model representing three petrol pumps (marked as PP 1, PP 2, and PP 3), two times three virtual filling positions (marked as A_1, A_3, A_3, B_1, B_2, and B_3) as well as virtual filling and passing lanes (from [23]).

yet fully mastered, in particular regarding the tracking of heavily occluded vehicles in low-contrast shadowed areas, such as under a roof. Preliminary results from such experiments, however, enabled the algorithmic derivation of a rich set of additional descriptions (see, e.g., [23]).

To do this, the set of occurrences was expanded beyond the basic ones discussed so far by the inclusion of synthetic ones which incorporate additional adverbs, location and object references. These occurrences could be arranged into maneuver sequences representing a considerable number of variants of how to move around on the premises of a filling station. We shall discuss here only those occurrences which have not yet been mentioned.

1. `Start_driving_backwards`: Combination of A-V-set (`Start_driving`) and A-D-set (`Back_up`).
2. `Slow_down_on_public_road`: Combination of a maneuver from the A-V-set with one from the A-R-set, comprising an additional spatial relation (`on_public_road`).
3. `Turn_right_to_gas_station_entrance`: Combination of a maneuver from the A-D-set with an additional spatial relation (`to_gas_station_entrance`) — corresponding to a maneuver from the A-L-set.
4. `Enter_gas_station`: Combination of a maneuver from the A-R-set (`Enter`) with an additional spatial relation (`gas_station`).
5. `Drive_on_virtual_lane`: A-R-set (a virtual lane is a section of the filling station premise intended for driving a car, but without particular lane markings).
6. `Drive_on_passing_lane`: A-R-set (a passing lane is a special kind of virtual lane).
7. `Drive_on_a_filling_lane`: A-R-set (a filling lane is a lane next to the petrol pumps).
8. `Drive_on_filling_position`: A-L-set (a filling position is the space on a filling lane in the immediate vicinity of a petrol pump).
9. `Drive_on_last_filling_position`: A-L-set.
10. `Change_to_filling_lane`: A-D-set (with an additional predicate specifying a particular lane, i.e., a combination of maneuvers from the A-D-set and the A-R-set).

Figure 2. First and last frame of an image subsequence: a vehicle is pulling up to filling position A_1 on Lane_A and continues to filling position A_3 (from [23]).

Figure 3. Trajectories of vehicles stopping at filling positions A_2 and B_2 (from [23]).

11. `Change_to_passing_lane`: Analogous to the preceding occurrence.
12. `Stop_on_passing_lane`: A combination of maneuvers from the A-V-set and the A-R-set with an additional predicate specifying a particular lane.
13. `Stop_on_filling_lane`: Analogous to the preceding occurrence.
14. `Stop_on_filling_position`: A combination of maneuvers from the A-V-set and the A-L-set with an additional predicate specifying a particular position.
15. `Stop_on_last_filling_position`: Analogous to the preceding occurrence.
16. `Advance_to_exit`: A-L-set.
17. `Leave_filling_station`: Combination of a maneuver (`leave`) from the A-D-Set with one from the A-R-set (`filling_station`).
18. `Merge_into_through_traffic_on_public_road`: Combination of a maneuver from the A-D-Set (`Merge`) with one from the A-R-set (`on_public_road`).
19. `Approach_vehicle_which_gets_gas`: Combination of a maneuver from the O-D-set (`Approach`) with an additional spatial relation (vehicle on a filling position).
20. `Pull_up_to_vehicle_on_filling_position`: Combination of a maneuver from the O-V-set (`Pull_up`) with an additional spatial relation (vehicle on a filling position).
21. `Pass_vehicle_on_filling_position`: Combination of a maneuver from the O-D-set

(`Pass`) with an additional spatial relation (vehicle on a filling position).

As in the preceding subsections on the descriptive task vocabulary for road vehicle motion, no new categories for occurrences need to be introduced. The simplest way of proceeding with the experiments turned out to be an adaptation of the available formalism for the extraction of occurrence descriptions: the introduction of additional road and location references. This set of experiments can thus be considered to corroborate the working hypothesis that the elementary maneuvers for road traffic introduced above are sufficient. The discussion in this Subsection illustrates nicely, however, which objects and spatial relations have to supplied in addition to the elementary maneuvers to obtain a satisfactory description. It becomes evident that essentially the same additional pieces of information will either have to be supplied by a task description or have to be made available as *a priori* knowledge to convert an abstract ("high level") prescription into a sequence of executable maneuvers. The knowledge alluded to consists of a generic gas station map: specific locations will then have to be determined by machine vision based instantiation of (part of) this schematic knowledge, which requires the recognition of objects and of their locations as well as the evaluation of various kinds of spatial relations.

4: Prescriptive (task specification) language for autonomously driving agents

The basis of this approach implies a formulation of the geometry of the road model, and of the dynamical model of the vehicle in terms of state and control variables which describe *continuous* states. The mathematical tools used here are the system of linear or linearized nonlinear partial differential equations with well established solutions.

An alternative line of thinking came from the field of control engineering in the late 1970s and early 1980s in an effort to address the problem of the representations needed for autonomous driving (e.g., Dickmanns [5], other similar efforts documented in [15, 24], and conferences on intelligent vehicles [6]). While it is clear that this later formulation provides an appropriate control mechanism on a straight or continuously curved road, and even enables smooth obstacle avoidance, it is still necessary to give (*prescribe*) instructions about the task to the agent-driver if the path to the goal is not straightforward and some intermediate decisions must be made during task performance. Hence, there is a need for discretizing and structuring the task.

The question, then, is this: what is the desiderata for the vocabulary of the descriptive-task for a driving specification language? The elementary maneuvers as listed in Section 3.1 are sufficiently rich to prescribe the necessary vehicle maneuvers in the present context. As stated in [20], these maneuvers are considered to be generic in the sense that the distance to the reference line for lateral control, the speed, and other entities are parameters which must be either properly chosen by some higher control level or estimated based on the current state of the vehicle and road.

A driving task, however, is more than just a list of maneuvers! We need to specify places and locations, and thus encounter the interesting problem of how finely or coarsely one must specify the route for the agent. Consider the following three cases:

Case (1) The agent has no knowledge about its environment and limited sensory capabilities to verify places from the environment. In this case the task specification must give very detailed instructions in terms of the elementary maneuvers and places on its path, for example, `You_are_in_place_A`, `Go_straight_10m`, `Cross_intersection`, `Turn_right`, `Follow_road_for_20m`, `Turn_left`, and so on.

Case (2) The agent has a full knowledge of the environment and sensory capabilities for detecting places/objects. In this case, the task specification can be high level, i.e., `GoTo_place_X`.

Case (3) The agent has some knowledge of the environment and sensory capabilities for detecting places/objects, but for whatever reasons, the task of getting to the goal is specified in more detail than in the preceding case. In [22], the authors refer to the first case as the "beginner driver" and the second case as the "experienced driver."

In the first case, the agent has no autonomy. The positive aspect of this situation is that the symbols can be directly translated into continuous behaviors. If the environment does not change between the time the behavior is specified and the time the task is to be performed, the system can guarantee the attainability of the goal. On the negative side, the system is not adaptive and flexible. Case (2) gives the agent full flexibility: it can optimize for the shortest path and can adapt to conditions unforeseen at the time of task specification. The agent has two options:

1. Since it has a full knowledge of the environment/map, the agent can generate a sequence of elementary subtasks. The problem is then converted to that of Case (1).
2. The agent invokes elementary maneuvers — such as `Start_and_continue` — until the sensory apparatus indicates a discontinuity of an event (here we define an event as a coherence in some perceptual feature, such as in visual depth or optical flow), which in turn causes a different elementary maneuver to be invoked.

Case (3) is between Case (1) and (2) and needs no further explanation. Case (2), on the other hand, is at the heart of our problem; hence, we shall elaborate on this further.

Consider an agent given the task `Drive_to_my_house/office`. The agent begins with elementary maneuver `Start_and_continue`. If there is an obstacle or anything else that necessitates changing lanes, the agent can do so in a continuous manner *without* invoking another elementary maneuver. Hence, it does not need to generate all the symbolic subtasks as described in Case (1). Only when the agent comes to a *decision place* where a new control regime (new elementary maneuver) must be invoked, will this place be given a new symbol or landmark. In summary, while for descriptive purposes the fine symbolic granularity of elementary maneuvers and places is necessary, for task-prescriptive purposes the level of symbolic granularity depends upon the complexity of the environment, the assumed knowledge of the agent regarding the environment, and the perceptual capabilities of the agent to maneuver in a continuous fashion (i.e., the experience of the agent!). The *complexity of the environment* will determine how often the agent will need to switch its behavioral regime. Here the agent determines the granularity of the subtasks dependent on the dynamics of the environment and, of course, on its own perceptual capabilities. This is different from Case (1) where the task specifies explicitly and symbolically the granularity of the subtasks.

To substantiate our claims, Košecká designed a task description language for the task of *navigation* [10]. This language is a derivative of the Robot Schema ($\mathcal{RS}$) model introduced by Lyons [14]. The language designed by Košecká, however, is far richer than the original language, namely, it includes a composition of elementary processes, modeled as Finite State Machines (FMS). These FMS's are linked with continuous control strategies very much in spirit of Discrete Event Systems. In addition to systematic modeling of action control strategies, we have also modeled the environment, namely places, as a network of perceptual strategies which are yet another FMS. The details are far too complex to do them justice in this paper; however, we shall present two examples.

Task description for Case (1) `Init; GoTo 10m; GoToHead 90° ; GOTO 20 m; Stop.` The event `GoTo` initiates the control law described as

$$\dot{x}_d = -\nabla(U_a(x) + U_r(x))$$

where x is a vector $(x, y)^T$, $\dot{x}_d$ is the derived velocity at each time, and U_a and U_r represent an attractor/goal and repellor/obstacle respectively in the artificial potential field. Similarly, the event `GoToHead` initiates the strategy for pure rotation controlled by the equation

$$\dot{\omega}_M = k_\theta(\theta - \theta_d)$$

where $\dot{\omega}_M$ is the computed desired angular velocity, θ_d is the desired heading, θ is the current heading, and k_θ is a constant.

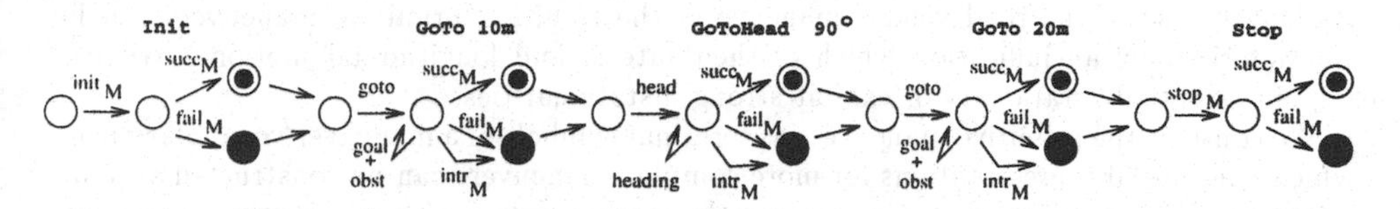

Figure 4. Case (1) task description.

Task description for Case (2) `GoTo_place_X` This assumes that the agent has a place graph (see Figure 5) in its head where words are associated with some perceptual strategies, such as how to find a doorway, how to find a corner, and so on. `GoTo` has the same interpretation as in Case (1).

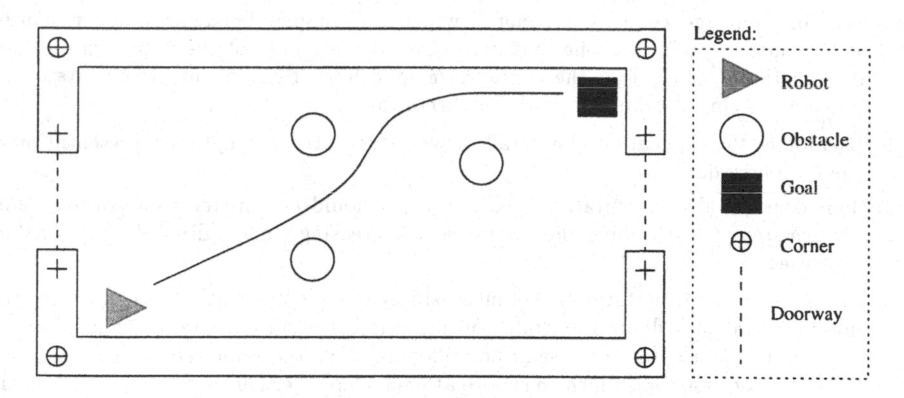

Figure 5. Case (2) place graph.

5: Conclusions

We generalized a set of previously reported road vehicle maneuvers such that we now can describe the movement of a road vehicle (the "agent") on the premises of a simple gas station as well as on intersection-free highways. Based on preceding experimental work by the Karlsruhe group, we claim that this set is sufficient, given rather generous boundary conditions. We then systematically reduced this set to a smaller one of more "*primitive* (road vehicle) *maneuvers*" by explicating objects and locations, together with spatial relations between these and the agent vehicle. A lane of a road is treated here as a special type of object which just happens to be more often involved in spatial relations with the agent vehicle than other objects. Once objects, locations, and spatial relations have been factored out as parameterized attributes of the resulting set of primitive maneuvers, we can differentiate between these primitive maneuvers on the basis of significant longitudinal or lateral control instationarities.

In a similar procedure, we have collected a set of maneuvers from experiments with multiple, cooperative mobile robots in the GRASP Laboratory. Again, we could show that the resulting set of "*primitive* (vision-controlled mobile robot) *maneuvers*" can be decomposed into a smaller set by factoring out objects, locations, and spatial relations. As in the case of the road vehicle maneuvers, the resulting primitive maneuvers can be characterized by an indication which of their lateral and longitudinal motion attributes remain essentially stationary or exhibit strong instationarities.

We consider the primitive maneuvers as terminal symbols of a maneuver vocabulary from which conceptual representations for more complex maneuvers can be constructed systematically by composition rules. Considering the representation of non-primitive maneuvers as words of a formal language allows us to cast the (de-)composition rules into a grammar of this formal language. It almost comes naturally to associate higher level abstractions for the description of complex combinations of maneuvers with the non-terminal part of the action vocabulary.

Returning to primitive maneuvers, the association of their instationarities with "events" offers a way to study in a systematic manner the transition between a signal representation of an action and its symbolic representation. The two discourse domains of road traffic and of mobile cooperating indoor robots are broad enough to justify the hypothesis that the observable convergence in the manner in which the complex behavior of agents can be formalized is not an accidental one, but indicative of the merits of the approach outlined here. It is gratifying to see that the *appropriate* modeling of significant ingredients of the discourse domain almost naturally works in three ways:

1. It enforces the explication of a detailed, well-structured conceptual representation of the discourse domain.
2. This conceptual representation specifies what should be omitted as a *relative detail* at each step while climbing the "staircase of increasingly generalized descriptions" of activities.
3. When descending the "staircase of increasingly specific descriptions," i.e., transforming a general task description into the parameters of an executable system control routine, this conceptual representation likewise offers the *same relative details* in the form of options among which to choose at each step — either implicitly, based on the *a priori* knowledge of the agent, or explicitly, by giving additional hints regarding how the task must be performed.

Thus, the approach discussed in this contribution is:

- *Model based.* There are geometric expectations of vehicles and road structures.
- *Behavior based.* The perception/action connection is embedded in the elementary maneuvers.
- *Situated.* There are expectations to find vehicles on roads, lanes, etc.

What are the essential characteristics of concepts to be represented by the set of terminal symbols, i.e., by the terminological axioms (see, e.g., [4, Kapitel 2.7: Terminologische Systeme]) of a vocabulary? The extension of the discourse domain from describing the information content of a single image frame to describing activities of cooperating mobile robots forces us to consider the interaction between hitherto essentially separate disciplines:

- Signal Processing and Pattern Recognition for the transition from video signals to symbols.
- Artificial Intelligence for the representation and manipulation of complex conceptual knowledge by symbol systems.
- Control Theory for the transition from a specification to the actual performance of an action by a mechanical system.

It is important to note that while today's AI field has accepted the above approaches, it was not that common at the time when this work was developing. It came as a necessity when dealing with real applications of dynamic scenes.

6: Acknowledgment

We thank M. Haag, IAKS Karlsruhe, and C. Reynolds, GRASP Lab Administrator, for their help in the preparation of this manuscript during our transatlantic cooperation. We thank P. Deussen and the Deutsche Forschungsgemeinschaft (DFG) for support through the 'Sonderforschungsbereich 314 Künstliche Intelligenz – Wissensbasierte Systeme' during the recent stay of RB at the Universität Karlsruhe which facilitated the research presented here. RB would also like to acknowledge the support of the Army Research Office grant DAAH04-96-1-0007, Advanced Research Projects Agency grant N00014-92-J-1647 and National Science Foundation grant IRI93-07126.

References

[1] R. Bajcsy: *Signal-to-Symbol Transformation and Vice Versa: From Fundamental Processes to Representation.* ACM Computing Surveys **27**:2 (1995) 310–313

[2] R. Bajcsy, A. Joshi, E. Krotkov, and A. Zwarico: *LandScan: a Natural Language and Computer Vision System for Analyzing Aerial Images.* Proc. 9th Intl. Joint Conf. on Artificial Intelligence IJCAI '85, August 1985, Los Angeles/CA, pp. 919–921

[3] R. Bajcsy and J. Košecká: *The Problem of Signal and Symbol Integration: A Study of Cooperative Mobile Autonomous Agent Behaviors.* In: Mustererkennung 1995, G. Sagerer, S. Posch, F. Kummert (eds.), 17. DAGM-Symposium, Bielefeld/Germany, 13–15 September 1995, Informatik aktuell, Springer-Verlag Berlin Heidelberg New York/NY 1995, pp. 618–633

[4] W. Bibel: *Wissensrepräsentation und Inferenz, eine grundlegende Einführung.* Friedrich Vieweg & Sohn Verlagsgesellschaft mbH, Braunschweig und Wiesbaden, Germany 1993 (ISBN 3-528-05374-7)

[5] E.D. Dickmanns and V. Graefe: *Applications of Dynamic Monocular Machine Vision.* Machine Vision and Applications **1** (1988) 241–261

[6] I. Masaki (Ed.): Proc. Intelligent Vehicles '95 Symposium, 25-26 September 1995, Detroit/MI

[7] D. Koller, K. Daniilidis, and H.-H. Nagel: *Model-Based Object Tracking in Monocular Image Sequences of Road Traffic Scenes.* Intl. Journal of Computer Vision **10** (1993) 257–281

[8] H. Kollnig and H.-H. Nagel: *Ermittlung von begrifflichen Beschreibungen von Geschehen in Straßenverkehrsszenen mit Hilfe unscharfer Mengen.* Informatik – Forschung und Entwicklung **8** (1993) 186–196 (in German)

[9] H. Kollnig, H.-H. Nagel, and M. Otte: *Association of Motion Verbs with Vehicle Movements Extracted from Dense Optical Flow Fields.* Proc. Third European Conf. on Computer Vision ECCV '94, Vol. II, Stockholm/Sweden, 2-6 May 1994, J.-O. Eklundh (ed.), Lecture Notes in Computer Science 801, Springer-Verlag Berlin Heidelberg New York/NY 1994, pp. 338-347

[10] J. Košecká: *A Framework for Modeling and Verifying Visually Guided Agents: Design, Analysis and Experiments.* Ph.D. thesis. Department of Computer and Information Science, University of Pennsylvania, Philadelphia, PA, USA, 1996.

[11] J. Košecká, H.I. Christensen, and R. Bajcsy: *Discrete Event Modeling of Visually Guided Behaviors.* Intl. Journal of Computer Vision **14** (1995) 179–191

[12] K. Kluge and C.W. Thorpe: *The YARF System for Vision-Based Road Following.* Mathematical and Computer Modelling **22** (1995) 213–233

[13] C. Lewerentz and T. Lindner (Eds.): *Formal Development of Reactive Systems.* Lecture Notes in Computer Science 891, Springer-Verlag Berlin Heidelberg New York/NY 1995

[14] D.M. Lyons: *A Process-based Approach to Task Representation.* Proc. IEEE Intl. Conf. on Robotics and Automation, 1990, pp. 2142–2150

[15] I. Masaki (ed.): *Vision-Based Vehicle Guidance.* Springer-Verlag Berlin Heidelberg New York/NY 1992

[16] H.-H. Nagel: *Analysing Sequences of TV-Frames: System Design Considerations.* Proc. 5th Intl. Joint Conf. on Artificial Intelligence IJCAI '77, August 1977, Cambridge/MA, p. 626

[17] H.-H. Nagel: *From Image Sequences Towards Conceptual Descriptions.* Image and Vision Computing **6** (1988) 59–74

[18] H.-H. Nagel: *The Representation of Situations and Their Recognition from Image Sequences.* Invited presentation, Proc. AFCET 8e Congrès Reconnaissance des Formes et Intelligence Artificielle, Lyon-Villeurbanne/France, 25-29 November 1991, pp. 1221–1229

[19] H.-H. Nagel: *AI Approaches towards Sensor-Based Driver Support in Road Vehicles.* In B. Nebel and L. Dreschler-Fischer (Eds.): 'Proc. KI-94: Advances in Artificial Intelligence', 18th German Annual Conf. on Artificial Intelligence, Saarbrücken/Germany, 18–23 September 1994, Springer-Verlag Berlin Heidelberg New York/NY etc. 1994, pp. 1–15

[20] H.-H. Nagel: *A Vision of 'Vision and Language' Comprises Action: An Example from Road Traffic.* Artificial Intelligence Review **8** (1994) 189–214

[21] H.-H. Nagel and H. Kollnig: *Evaluation of Image Sequences from Outdoor Scenes: Selected Problems and Solutions.* Proc. Second Asian Conf. on Computer Vision ACCV '95, 5–8 December 1995, Singapore, Vol. III, pp. 16–20

[22] H.-H. Nagel, W. Enkelmann, and G. Struck: *FhG-Co-Driver: From Map-Guided Automatic Driving by Machine Vision to a Cooperative Driver Support.* Mathematical and Computer Modelling **22** (1995) 185–212

[23] H.-H. Nagel, H. Kollnig, M. Haag, and H. Damm: *The Association of Situation Graphs with Temporal Variations in Image Sequences.* In: Working Notes AAAI-95 Fall Symposium Series 'Computational Models for Integrating Language and Vision', 10–12 November 1995, Cambridge/MA, pp. 1–8

[24] C.W. Thorpe (ed.): *Vision and Navigation - The Carnegie Mellon Navlab.* Kluwer Academic Publishers, Boston/MA Dordrecht/NL London/UK 1990

[25] W. Tölle: *Hierarchisches Regelungskonzept für die Fahrzeugführung.* Fraunhofer-Institut für Informations- und Datenverarbeitung IITB, Karlsruhe/Germany, Interner Bericht (15. März 1993, in German)

Computer Vision and Visual Information Retrieval

Ramesh Jain
Electrical and Computer Engineering
University of California at San Diego
La Jolla, CA 92093-0407
and
Amarnath Gupta
Virage, Inc.
177 Bovet Road, Suite 520
San Mateo, CA 94402

Abstract

The discipline of Computer vision has gone through many different doctrines and applications. At the same time, advances in computing and communication technologies have made visual information much more easily accessible to the general user. Consequently, the realm of information is fast expanding from alphanumeric to multimedia. In this new computing world, images and videos are becoming a major source of information. Techniques to manage these Visual Information Management Systems (VIMS) require techniques that are very different compared to the conventional information management systems. We contend that Computer Vision, simultaneously with database systems, will play a central role in defining the functional capabilities of these systems. But we also believe that in order to fulfill this role, many new challenging research issues are posed to Computer Vision research. In this paper, we briefly discuss VIMS and the role of Computer Vision in these systems.

1: Introduction

Over the last three decades the field of Computer Vision has witnessed several major phases of evolution. These have ranged from basic digital picture processing techniques to understanding the physics of image appearance, automatic recognition of objects in a natural scene, and modeling the processes of biological vision. In course of this evolution, computer vision has benefited from many diverse research disciplines like signal processing, optics, pattern recognition, psychophysics, neuroscience, computer graphics, and even hardware technologies. Each of these disciplines has provided Computer Vision with a new way of looking at the machine vision problem and often, a new set of problems that has enriched the scope of the discipline. Perhaps equally important as the contribution of these fundamental sciences, Computer Vision has also benefited significantly from applications, because they have fostered the growth of a large body of techniques that has put the science of Computer Vision to practical use in engineering, medicine and commerce.

Another major driving force that has helped shape the course of computer vision is the evolution of computing technology itself. In the early days of Computer Vision, a major problem used to be how to acquire images and how to store and process them on the computers available during those days. With the current state of the art in real time signal processing hardware, powerful image

0-8186-7644-2 $5.00 © 1996 IEEE

rendering mechanisms, and high bandwidth image communication facilities, Computer Vision practitioners and researchers think about developing algorithms very differently from what they used to a decade ago. A recent phenomenon in the world of computing is the multimedia revolution. For early computers, information processing always implied number or at best symbol crunching. Now it is expected that computers will be equally adept in dealing with graphics, images, video, and audio as pieces of information that any user may want to acquire, organize, store, search, manipulate, transmit and distribute as easily as alphanumeric information. This revolution is resulting in a marriage of computing with communications making George Gilder's teleputer a possibility in just a few more years. The information highway is becoming more and more visual.

These advances in computing are forcing computer scientists to look at images in the same capacity as alphanumeric information. Techniques to deal with images and videos are being addressed by researchers in all aspects of computer science and technology. A new field of Visual Information Management Systems is receiving increasing attention from researchers and practitioners [1]. We believe that Computer Vision, with its large storehouse of image and video processing techniques, will play a key role in this emerging field. We show that many of the techniques pioneered by Prof. Rosenfeld in his long and continuing association with Computer Vision, have direct impact on VIMS research. But simultaneously, this emerging field will have a profound influence on Computer Vision, not only as yet another application but as a completely new set of processing requirements posed on Computer Vision research. As we see it, Visual information management poses many interesting challenges to Computer Vision. In this paper, we briefly discuss this emerging field and discuss the interplay between Computer Vision and visual information retrieval.

2: Visual Information Management Systems

Suppose we have a digital collection containing a large number of images. It could be one million trademark images maintained by a government agency, or several hundred thousand images collected by a stock photography company, or a terrabyte-order picture archive of radiological images shared by a network of hospitals. We would like to build a system which will store these images and retrieve them upon the user request. However the user request may, unlike regular databases, be posed in terms of image properties, such as, "Find me images with a texture like this and color distribution like this". The challenge is: the system should know how to compute, represent and search image properties such that it produces correct results, and should do it in a reasonable turnaround time (3 seconds is a standard guideline for database systems, but the acceptable time depends on applications). If this can be done, the utility is immense and immediately obvious. A doctor has access to a world-wide archive of ultrasonic images of the heart. He takes a new ultrasound image and wants to retrieve images with a *comparable degree of left ventricular hypertrophy*. The trademark office wants to know if a new company's design is sufficiently unique and will not infringe on any other registered trademark. A football coach wants to know how to stop the star running back from his opponents in the coming match. For this, he wants to see the sequence of all runs in the last 25 games. An auto designer wants to see the parts of simulations in which temperature at a point crosses limits set a priori. Many other examples can be cited.

As the examples suggest, the key issue in VIMS is visual information retrieval in response to a user request. We can therefore describe the functionality of the system by an exposition of the types of queries[1] that the user can make.

[1] Please note that here we loosely refer to all interactions initiated by a user as queries.

2.1: Query Classes for a Visual Information Management System

In the following query classes, we mention many issues which show the role of Computer Vision techniques in Visual Information Management Systems. In doing so, we intentionally ask more questions than answered by current research because we believe these questions will help to put the research need in perspective.

1. **Search**: The most important task to be supported by a generic VIM system is to allow the user to search for information. Two classes of search are recognized in literature: search by specification, and search by example. One could also consider search modes which are hybrids of the two.

   (a) *Search by Specification* may be expressed as, in the ultimate case, a combination of keywords that describe semantic information (such as date and time of imaging), image attributes (such as background color), image object properties (such as a circular shape), domain defined objects (such as the heart in an X-ray image), measurements (such as the area) of domain or image objects, and their absolute or relative positions (on the lower right corner of the heart). An example of such a query can be formulated say by a X-ray technician trying to find an artifact produced the X-ray machine: *Retrieve all chest X-ray images taken after June 19, 1995, which have a poor contrast, and show a circular object with about 1 cm diameter, about 2 cm below the apex of the heart and to its right.* This query, albeit artificial, illustrates many key concerns of a VIM system.

   First, how does the system know what the heart is and how to locate it in the front view of an X-ray image? Then, given that it can identify the heart, how does it know where its apex is? This is the issue of domain object modeling in a VIM system, and we are interested in both how the model is specified to the system (the modeling language) and in what goes in the model (the model content). An important research question is to decide what kind of base-level features should be used in defining the domain model. Also crucial to the domain definition is the concept of constraints, which can span from a restriction on the range of values to complex constraint equations that must be satisfied among several domain objects.

   Second, how does it interpret the term "poor contrast" and select all images satisfying this qualification? Is it defined based on the global contrast, or worst local contrast given some definition of locality? Is it defined on absolute values or with respect to a reference population? Is the classification of images on the basis of their contrast crisp or fuzzy? Distinct from domain object modeling, the problem here is to provide the system a model of images and their attributes. Thus MRI images will have model different from X-ray images, and contrast for fluorescent images will be different from that for images obtained from back-lit optical microscopes. We believe explicit image modeling is a very significant aspect of VIM system for images and has been largely ignored so far.

   Consider the criterion of "circular object of about 1 cm. diameter". How well-segmented must the circular object be? How good should be the fit of a candidate object to a circle? What is the error limit on the diameter? For all these questions, the basic issue is how much of the accuracy parameters should be hardwired in the system, how much should the user specify at application definition time and how much control the user can have in tweaking the accuracy parameters on rough-cut search results to suit his or her specific needs *at query time*?

The basic point in the above discussion was to emphasize that most working application-oriented Computer Vision systems answer these issues in very specific, deeply implementation dependent ways. For example, an aerial image analysis system may use the assumption that buildings typically have vertical edges. But this information, is embedded inside the code. However, to build VIM systems, we would need to support the use of such domain specific knowledge so explicitly that the user of the system can define and modify application requirements without having to redesign features and matching metrics.

(b) *Search by Example* is a different paradigm, where the system *constructs* the query parameters from an example image object such as a colored region or a hand-sketched shape, or, alternatively, a complete image. An example of this situation is the trademark search mentioned above. This paradigm shares all concerns raised in the previous case, and has additional issues to address. As the user no longer mentions the matching criteria, the system has to have a set of features as comparison axes like color, texture etc. against which two images can be "matched". As the query is not precisely specified, there is no "perfect match" for the query image in this case. Therefore, the "matching" is on the basis of a overall similarity ranking made up of similarity measures for each comparison axis.

The very nature of this query paradigm ushers in a set of significant issues. What are the axes of comparison? Is texture a valid axis of comparison by itself, or do we have to break it up into individual measures of randomness, orientation, periodicity etc. to give the correct level of granularity? Second, for any specific search are the axes all equally important? Do they depend on the domain, or on the individual query image (if an image has very strong edge components but poorly identifiable texture segments, should the system automatically increase the weights of a edge-dependent shape measures to formulate the query?) Third, as different axes of comparison may need different matching criteria or similarity distance functions, what is a meaningful way to combine them to produce a master rank list? For example, is a Euclidean measure good for RGB color space? Again, is a weighted Euclidean measure theoretically justified and practically effective to combine an RGB color distance and a texture based similarity distance into one single metric? Fourth, how is similarity defined in terms of parts of interest in an image? For example, if the query image shows a chair and the user intends to find all living room pictures with similar chairs in them, what notion of similarity should be used?

Common to both these modes of search is the issue of processing. Should one cull the set of images by color first or texture first? Does this depend on the image class, the image population or both? What kind of data structure should one construct for efficient processing? Given the high-dimensionality of visual features, where is the trade-off between accuracy and retrieval efficiency? Not many of these questions have been satisfactorily addressed let alone answered in the current state of the art, and needs strong research attention of the community.

2. **Browsing**: Browsing is a navigational mode of search in which the next query is based upon the results of the previous query. While the task of *searching* is to "locate" an image from its partial specification, a browser starts with little user specification. The browser assumes that exploring user has an unexpressed mental model of the images of interest. Consider the stock photography example cited above, and suppose the user is looking for an image that "best suits the theme of a product to be advertised". Stated as such, the query is far too ill-specified and given a very large image collection, the user is likely to lose interest in his search endeavor

soon, unless the system can navigate him to his interest subspace within a small finite number of steps. Thus initially a browser behaves as a random sampler of the database. The browser mode is most effective if it can smoothly guide the user interaction to a potential search-set without the user having to make a conscious choice of query parameters until he or she is in a position to start querying in the locator mode. In other words, an important aspect of the browser design is to make an *informed* sampling based on an *incrementally constructed model of the user's needs*. Once the user reaches an identifiable superset of the images of interest, he or she can instantly switch to the locator mode and continue with more specific query specification.

A somewhat different notion of a browser is that of an incremental *query refinement* mechanism. In this case, the browser starts with the results of a search, and lets the user change query parameters by adding (or removing) restrictions on the search conditions, changing relative weights on a "search by example" query, using relevance feedback (a method by which a user specifies how much a specific result item suits the information need). The system then partially reevaluates the query given the previous results and the new criteria.

The common thread between these two views of a browser is the incremental modification of feature space starting from an initial subspace. Are all visual features amenable to quick incremental recomputation? What kind of population statistics should be maintained to decide which of several dimensions of the feature space should be recomputed first? How should one use negative results of a search (say by relevance feedback) to tune the process of search?

3. **Analysis**:In many scientific, engineering and medical applications, the visual information would be used for analysis. Combining visual information retrieval with support for analysis places some design requirements on the retrieval process.

   (a) **Aggregate Queries**: An aggregate query wants to retrieve either a value (number of sunset images where there is no human figure in the picture), or a set of images, each satisfying a condition which needs the computation of some aggregate function (all images of the fundus containing at least three distinct hemorrhage spots having an average diameter of 2 mm or more), or a set of images together satisfying some aggregate conditions (all images having contrast values within 10% of the population average). The two extreme methods for handling aggregate queries are to precompute all domain objects into an alphanumeric meat-database which is then subjected to standard statistical database techniques, or to perform on-demand computation at query time. An important question is whether it is meaningful to determine an intermediate representation between image objects and domain objects for aggregate queries. The speed-space-accuracy tradeoff for image retrieval in scientific applications has not been explored rigorously.

   (b) **Classification Queries**: In many scientific and cultural applications, retrieval of actual data is somewhat less important than determination of data classes. These applications, as evinced by the growing research interest in data mining, need classification functionalities to be integrated with retrieval. In the domain of images, classification queries can appear in several forms. Given a set of images, which are known to belong to the same class by an application (manifestations of the same disease in radiological images) which image properties make them so similar? Again, Given a set of images (200 sculptures from a single layer of recent archaeological excavation), and a given database with known classification (a nationwide collection of 500,000 sculptures), what archaeological category (e.g., stylistic class of sculpting) would the query sculptures belong to? What is the minimal set of categories that cover all the query sculptures? Third, given a set of painting images, is there a pattern in their treatment of light and shade and formation of human figure? Are some techniques applied to knowledge and data mining adaptable to the domain of images?

(c) **Sampling Queries**: In addition to random sampling as in browsing, sampling a large database can itself be an important class of queries in many applications. Given that images never "match" in all their features, a certain degree of variability is inevitable among the results of a single query. If the result set is very large, it may be more meaningful to retrieve a sample of size n representative of the images that satisfy a certain condition. A complementary class of queries can be to retrieve a sample of size n representative of outliers. A more involved query is to retrieve a sample of images where the user provides a sampling scheme on the image features. An intriguing question for this class of queries is to decide what image features would be amenable to sampling, and what kind of data structure for the features in the database would facilitate traversal and search.

(d) **Feature Space Navigation and Query**: In many applications, it is worthwhile to formulate the query by "looking" at the feature space instead of images. Many scientific users prefer to consider image features such as color or *eigendescriptors* as points in a suitable multidimensional space. Meaningful spatial queries can be formulated by locating clusters in this space, or by using feature densities, or by selecting images related to specific subregions of this space or its projections (e.g., on the plane defined by the first two principal components).

(e) **Metadata Generation and Post-Association**: A very important requirement of scientific applications is that scientists analyze information and would prefer to store the results of analysis as further data into the database itself. Thus images may be classified or annotated, subregions of images may be identified as significant, images may be grouped together by a causal network, a rank order may be imposed on images by some domain-specific criterion. Regardless of how this metadata is generated, or its structural form, the user may subsequently wish to query on this manually associated metadata, as a first class data object. It is therefore the task of the VIM system to establish and maintain the connections from image features onto these associations and vice versa.

2.2: Characteristics of Visual Information Management Systems

Based upon the foregoing discussion, let us highlight some key issues that characterize VIM systems.

1. **Uncontrolled Collection**: The system designer of the VIM system have little control over the images or videos that go inside a collection. If the collection is domain specific, like a radiological image archive, there can be one collection for chest X-rays, another for mammograms and so forth. If on the other hand, it is general, like stock photography, then the content may range from nature photographs to computer generated fractal art. This implies that *VIM systems in general cannot make assumptions about an entire collection having one underlying image model*. Thus assumptions like all textures arise from Markov Random Fields, or all surfaces are Lambertian cannot be made. Even when a VIM collection is domain specific, the variability of the images and in the imaging conditions would be significant enough to preclude such assumptions.

2. **Abstraction Support**: A VIM typically needs multiple levels of abstraction, although the notion of abstraction levels differ somewhat from the low-level, mid-level, and high-level vision familiar to Computer Vision researchers. A VIM needs to support queries specified at the image level (e.g., similar color distribution) or at a segmented image level (e.g., containing a circular region) or at a predefined object level (e.g., a face image with large eyes). The role of the abstraction is to maintain explicit mappings between the information of different levels so that a search initiated at a higher level can be either directly answered because the system had precomputed and indexed the information, or in can be computed at query time following an explicit definitional chain of progressively lower level information. Hence, when defined adequately, looking for a sunset image may translate to a search for an image containing a strong horizon line (an edge like information) and a dominant golden yellow color above and darker color below it (a combination of spatial and chromatic information). It is also possible to have alternate feature-level definitions of a higher level object. The implication here is that *VIM systems will embed multiple image models within a data model of images or videos*, a data model being a collection of well defined objects, their attributes and their inter-relationships.
3. **Flexible Data Model**: A VIM system is successful only to the extent it produces reasonable search results for the user. It is less stringent than object recognition in the sense that no one expects a zero false positive result. But is more important that the user finds search results by each individual feature perceptually meaningful. Typically, most object recognition systems depend only on one (or at best a couple) of features to characterize an object. A query in a VIM system is often data-directed and can easily span across more than one feature in a given level of abstraction, and across different levels of abstractions. So, the notion of "matching" is different from Computer Vision systems. *It is essential that all features in a VIM system have explicit comparison functions and that there are ways to combine different features into a perceptually meaningful results.* The combination functions are interesting research areas.
4. **Suitable Feature Set**: VIM systems are meant for database-like use, requiring feature computation to be fast. The features can be computed either at insertion time or at query time. In either case, features that require long recursive or iterative computation are unlikely to qualify the time constraint. The endeavor is not so much to capture the physics of the image or object by rigorously analyzed features, but address the issues of computation, access and comparison costs as equally important feature design criteria as description and discrimination capacity. Visual Information Management would continue to need more Computer Vision algorithms which are sensitive to these additional requirements. More research should also be devoted to systems that use a very narrow domain with enough prior knowledge so that training session can be used to facilitate more efficient feature computation, and to systems that use approximate and human assisted computations. Especially necessary are mechanisms to allow interactive feature definition at query time, and yet do not degrade the retrieval performance of system, because they make optimal use of existing features that have been precomputed and stored.
5. **Speed of Query Processing**: Even more than the speed of feature computation, query response should be as fast as database systems. This puts strict constraints on the admissibility of features in VIM systems containing large collections, and imposes the need to organize and index the features to facilitate searching and browsing. This means dense point-sets used by many Computer Vision algorithms, are not very good candidate features for visual information retrieval, because they increase the database size and total comparison cost can be high. It also means that techniques such as computing image correlation may have to be used as a

secondary matching criteria, after the candidate result set has been initially pruned by some faster technique. Other techniques, such as those used for pose matching by solving complex differential equations are not likely to be relevant in VIM systems, at least in the near term. In addition to speed of comparison, access cost of features should be small. So features need to be "orderable" in some sense, so that they can be organized in a search structure. Attempts have been made in Computer Vision to compute geometric invariants and organize them in "hash tables", a technique referred to as geometric hashing [2, 3]. While in principle the approach is very logical, the features used for such retrieval has been low level like corners, curvatures and bi-tangents, which are difficult to recover faithfully for an uncontrolled collection of images, and, for a large collection, search using geometric hashing have never been tested to be reliable.

6. **Data Definition Support**: Data definition support is an important component of VIM systems, largely ignored by the Computer Vision community. Hence most Computer Vision systems tend to be inflexible, where the user has no means to define the requirements of his or her collections to the system. This does not refer to tweaking the convolution window and noise suppression threshold in an edge detection algorithm. Rather, it refers to a declarative mechanism to impart the user's knowledge ranging from object definitions to segmentation guidance.

2.3: Information Value of Features in Visual Information Retrieval

In this section we shall touch upon a significantly underexplored research area which, we believe, will be a good way for evaluating a feature for the purpose of visual information retrieval and management. The idea is to use principles from Information Retrieval research to estimate the "goodness" of a visual feature. It has been demonstrated by several authors in Information Retrieval [4, 5], database research [6], and Visual Information Management [7] that a histogram of the number of information sources (e.g., images) plotted against their relevance toward a specific query tends to have a J-like shape (Figure 1), often called Bradford's law [8]. This means if in a randomly chosen collection, we perform an "all-pairs" (i.e., a query asking "Find all pairs of images that are within a distance d of each other using a certain feature f_i"), and plot the number of images for a scale of d ranging from 0 to 100, the curve would have a shape showing three different regions. The first region usually small, denotes very similar images; the second region, a valley denotes images which are similar but not very significantly, and the third region shows a monotonically increasing series of progressively dissimilar images. Sometimes there is a fourth region, where the monotonic curve levels off or even droops down. Let us denote the areas under the three regions as R_1, R_2 and R_3 respectively. In order to qualify as an acceptable feature, it needs to preserve the shape of the curve over random collections, possibly with application imposed limits on the proportions of R_1, R_2 and R_3. However, the real measure of effectiveness is the average number of *relevant* images in each of these zones. Since a well-ordered ranking of relevance of images can be hard to determine, loosely defined ordinal scales like "top n" should be subjectively assessed in these collections by user experiments. An analysis of how the relevance ranking is distributed in these three zones, may determine the general effectiveness of the feature. Faloutsos *et al* [9] suggested a measure called AVRR that gives the average rank of relevant items that can be used for getting a more quantitative assessment of feature performance. The next level of test should assert that the feature is not only good in finding relevant information, but that it is stable and robust. It is common experience that tweaking parameters can be produce very different results for the same test scenario. This stability should be established by measuring the "sensitivity" of the ranking with respect to each free parameter of the feature function

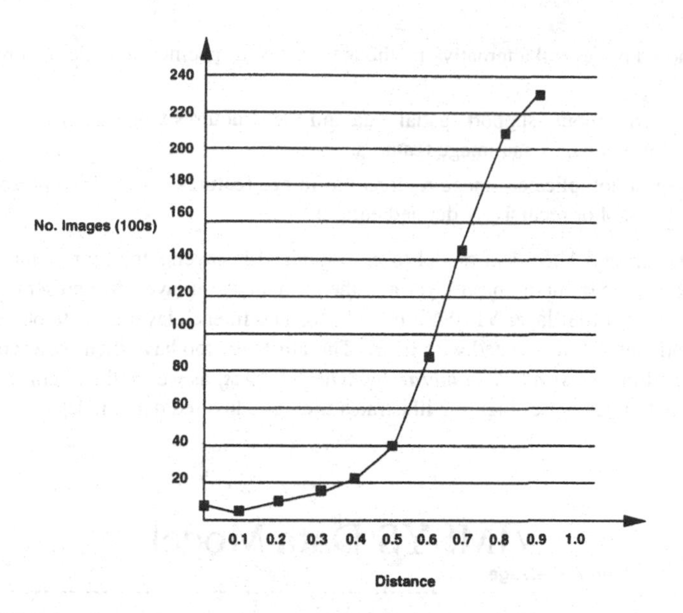

Figure 1. The Bradford curve

near its operating range. The robustness of the feature will be established if perturbing none of the parameters can cause the ranking to jump zones R_1, R_2 and R_3 appreciably, or change affect the AVRR too much. At the third level, when two or more features are used for similarity ranking, their combined performance should not degrade their individual performance. Experimental research in this area by Computer Vision scientists should provide us more insight into the problem of feature performance.

3: Data Model and Semantics in Visual Information Management Systems

In the foregoing sections we have emphasized that VIM systems have much in common with databases, and need to be designed through a data model. In this section we discuss the characteristics of a such a data model and, the relationship between Computer Vision and the data model.

The role of a data model in database systems is to provide the user a textual or visual language to express the properties of the data objects that are to be stored and retrieved using the system. The database language should allow users to define, update (insert, delete, modify) and search objects and properties. For Visual Information Management systems the data model assumes the additional role of specifying and computing different levels of abstraction form images and videos. Accordingly, the data model needs to be satisfy the following properties:

- We should be able to access an image matrix completely or in partitions.
- The image features should be considered as both independent entities, and as related to the image.
- The image features should be ordered in a hierarchy such that more complex features can be constructed out of simpler ones.

- There should be several alternative methods to derive a specific semantic feature from image features.
- The data model should support spatial data and file structures which infer spatial parameters associated with images and image features.
- In the case of complicated image regions the image features should be represented as a sequence of nested or recursively defined entities.

We perceive the general VIM data model to be organized in layers: the representation layer, the image object layer, the domain object layer and the domain event layer. We present here a refined version of our original four layer VIMSYS model [10, 11]. In each layer all data objects have a set of *attributes* and *methods* associated with them. The attributes too have their own representations, and are connected in a *class attribute hierarchy*. The *relations*, as we shall explain shortly, may be spatial, functional or semantic. Figure 2 illustrates the basic layered data model.

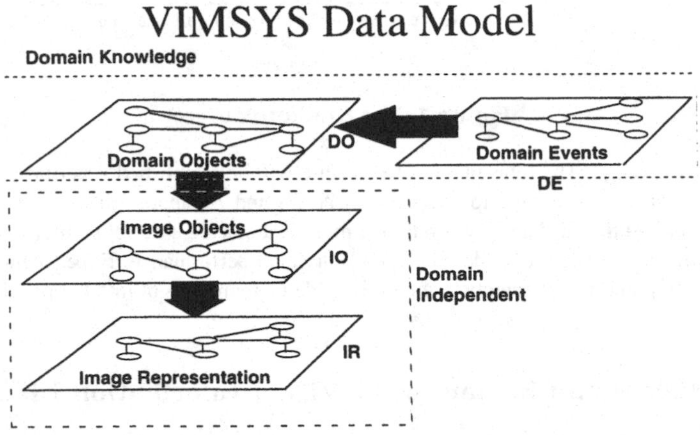

Figure 2. The Layered VIMSYS Data Model

1. **Representation Layer**: The representation layer contains the image matrix and any transformation that results in an alternate but complete representation of the image. For example, an image originally received as an RGB matrix and its LUV conversion required for color processing are both members of the representation layer. Similarly, if a raster scanned image of a line drawing is converted to a vector format, the latter belongs to this layer. Obviously this layer is itself not very rich in processing user queries. Only queries that requests pixel based image information can be provided by this layer. However the value of the layer is giving the system designer an explicit handle to define, maintain and interconvert between representations which are used by other layers of the data model. Since image transformation is part of the layer's intended functionality, the system designer has the option to exercise his knowledge about the image model by meaningful transformations. For example a class of transformations can be defined to enhance the input image under user or designer selected noise

models. Implemented this way, the transformation becomes part of the data model, and upon insertion to the database, every image goes through the enhancement routine and only the enhanced image is used for all computations downstream. Similarly corrections for the Gamma Factor or for specular reflection, if known a priori can be made in this layer. In case of videos, the representation layer functionality can get more complex. Key frame composting, which converts a temporal sequence of frames covering a large spatial area into a large single image constructed to show the entire spatial coverage can be placed in this layer. In case key frames are extracted from the sequence, the input stream is first directed to the segmentation layer, and the extracted key frame(s) is transmitted to the representation layer for storage and usual static content analysis.

2. **Image Object Layer**: This layer has two sublayers – the segmentation sublayer and the feature sublayer.

    - **Segmentation Sublayer**: If the user of a collection would search the images in a collection only by their global properties, or by the global properties of direct image partitions, then the segmentation layer can be omitted. In this case, the designer may directly use the feature layer. If however, the user would search by computed local properties, the system must additionally maintain and allow searching on the segmentation information. We see the process of segmentation as information reduction: – performed both by asserting that not all parts of the image or video are equally informative, and by condensing information about an entire image or video into spatial or temporal clusters of summarized properties. Hence the output of segmentation is the spatial (and temporal) localization of image properties, implemented by using suitable spatial and temporal data structures. While the properties themselves can be searched for in the feature layer, their localization can be searched for in the segmentation layer. Since an image or video can be segmented in several different ways (e.g., based on dominant edges, color, motion), the localization information is indexed by the associated feature category. The separation of the existence of a property from its locality of occurrence serves another important purpose. It allows a query processing system to optimize whether a locality based search or a property based search is likely to produce faster results for a given collection of images or videos. A complexity of the segmentation layer arises from the options that a segmentation can be automatic and based on an image model (e.g., region growing with an assumption on the limit of gray level variance within a segment), or it can be automatic and guided by designer-specified knowledge of the content (e.g. a procedural knowledge of how to find the optic nerve in registered images of the retina or how to find a cut in a video), or interactive (e.g. using a user-initiated deformable contour). However, our data model allows image segmentation models and knowledge modules to be registered with the system during schema [2] declaration. Thus if an image model is registered, a segmentation scheme using it can be declared in the system. This way, the designer can choose libraries of rich segmentation methods developed as part of mainstream Computer Vision.

    - **Feature Sublayer**: As mentioned before, a feature to us is information which is efficient to compute, organizable in a data structure and is amenable to a numeric distance computation to produce a ranking score. Since the goal is not to reconstruct the original image, the features need not be invertible. Most of the literature [12, 13, 14] and products in this area [15, 16] use rather simple macro-level features. Methods derived from basic

[2] A schema is one instance of the data model for a specific database.

digital image processing techniques developed and compiled by Prof. Rosenfeld [17] turn out to be very effective for retrieval. For example, histogram-based features (color histogram, orientation histogram, gray-level co-occurrence matrix) and their properties (e.g., their moments) which have been used from the very early days of digital image processing, produce convincing results. An example of the results obtained by using a

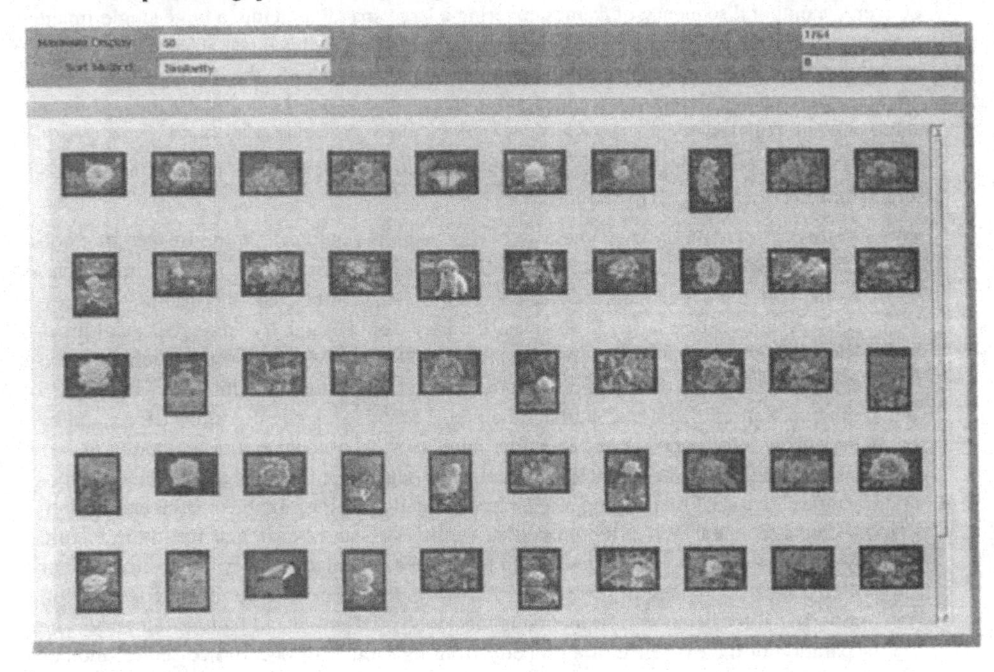

Figure 3. Results of local color and structure query in the Virage System.

set of simple features like local color histogram and local edge properties for image retrieval can be seen in Figure 3. Here, in a database of more than 3000 images, the first image was used as the query image, and the rest are a ranked list of 50 most similar images. Interestingly, although object recognition has not been attempted, 42 of the 50 are flowers, and 41 of them are roses.

The feature layer poses some interesting problems. We mention two such problems relevant to Computer Vision and pattern recognition.

The first problem deals with the correspondence between perceptual similarity and computable feature distance. Most of the feature distances used in literature are convex functions (see [19, 20]) which also satisfy the metric conditions. This is a limitation in many cases, because in the feature space, the shape of the cluster(s) that correspond to perceptual similarity can be far from convex blocks. An obvious alternative is to use unsupervised clustering resulting in nonlinear regions in the feature space. However, it can be very expensive on the one hand, and not amenable to dynamic updates on the other. The problem of dynamic update becomes important because maintaining dynamic statistics of the data population introduces a significant overhead in a multi-user fast transaction environment. Research is needed to find algorithms that find a suitable tradeoff between the perceptual and computational criteria. Moreover, suppose perceptually meaningful clusters are determined by some method and have arbitrary shapes in the feature space. The questions are: what families of distance functions should be used to rank visual objects belonging to a given cluster? Would such a function need to adapt itself as the cluster evolves and changes shape and topology?

The second problem relates to the issue of containment search. Consider the query "Show all images that contain a table and a chair like this", where the system is provided with a reference image with the desired objects. The correct result is that the system will retrieve all images that embed the objects preserving their spatial relationship. The first problem is to find a segmentation scheme which, with the specification of "what" it is expected to segment, produces a reasonable segmentation given all inaccuracies that may occur due to unpredictable contrast and occlusions that may occur due to other parts of the query and candidate images. The second problem is to find a set of shape based features, which are robust to segmentation inaccuracies inevitable in an uncontrolled collection. Also, although complete invariance of affine transformations may not be required, the features should be tolerant to translation and some perturbations of rotation and scale. One approach adopted by the QBIC group [15] is to have the user segment all the important objects in every image which gets inserted in a collection. While this is a very practical, and possibly the surest method to ensure correctness of results, more automatic methods from Computer Vision are warranted.

3. **Domain Object Layer**: A domain object is a user defined entity representing a physical object or a concept that can be translated in terms of one or more features in the lower layers. Thus, a concept like "sunset" or an object like "heart" in a medical image are both domain objects. The domain object layer is analogous to a *conceptual schema* as defined in database systems. It consists of three components. It is a graph that relates an object with its attributes and other objects though different relationships. Many of these relationships are semantic, meaning that they cannot be inferred by any computation on the image or video, but have to be told to the system by the designer. An important category of relationship is *classification*: B is a subclass of A when from all instances of A in a database, a subset satisfying some condition is labeled as B. For visual information, the condition can be tested and labeled automatically if sufficient domain knowledge is built into the system. For example, a heart in systole and a heart in diastole have widely different shapes and occupy different spatial extents, but if their semantic similarity is by a classification hierarchy, the system should be able to correctly search for either the heart or any of its individual subclasses. For the heart, and other objects that go through a finite range of variations, the domain knowledge can be encoded in terms of a basis templates. A domain object can be expressed as a direct mapping to one of these template categories (the *visual thesaurus* approach). Alternately, the templates can be treated as basis vectors and an instance can be expressed as their linear combination (the *eigenimage* approach [18]). Yet another way to specify domain knowledge, is by a set of rules that relate domain objects to image objects. For example, in a database of MRI images of the brain, the rules can be like: a segmented object with shape like *this*, and situated in a bounding box of (x_1, y_1, x_2, y_2) of a normalized T2 image, having a positive local contrast of about δ_1, and a segmented object in the same location of a T1 image, having a similar shape, and a negative local contrast of about δ_2, can be mapped to the domain object "gray matter". Such specification of domain objects in terms of their image properties is not new to Computer Vision. But this model provides generic and explicit methods to specify such knowledge to a retrieval system, and users can use it for his or her application specific data definition.

4. **Domain Event Layer**: The purpose of the domain event layer is to allow "events" computed from image sequences or videos to be queriable entities.These events can be result from pure motion (e.g., when the velocity of the centroid of a segmented object exceeds 20 pixels/frame), from spatial interactions (e.g, when two object centroids come to about 5 pixels from each other), spatio-temporal interaction (e.g., when the object is approaching *this* region in space and it less than 10 pixels away from it), appearance, disappearance or morphing (e.g., when a ballet dancer transitions from one move to another). Domain events also include events that are not instanteneous but occur over a period of time. (e.g, a tumor that has grown beyond 15 pixels in diameter over a sequence of six images acquired monthly). In order to manage domain events,a VIM system not only needs an event detection mechanism but also an event organization mechanism,such as a temporal data structure that allows to maintain and search through detected events of different types and time granularities. We are currently working on effective methods of mapping domain events to motion segmentation results from Computer Vision.

4: Conclusion

In this paper we have discussed the interrelationship between Computer Vision and the emerging field of Visual Information Management. With the growing number of VIM application areas and the advent of commercial products, we believe that VIM systems will require increasingly powerful computer vision algorithms. However, to be applicable in the VIM systems, these algorithms must satisfy several performance and competence requirements. Some of these requirements are outlined in this paper, and many of them are still reearch issues. We expect that closer interaction between the two communities will lead to deeper understanding and more useful results for both.

References

[1] Jain, R., Pentland, A.P., and Petkovic, D., *Workshop Report: NSF-ARPA Workshop on Visual Information Management Systems*.held before ICCV 1995, June 1995.

[2] Grimson, W.E.L.*Object recognition by computer: the role of geometric constraints*. Cambridge, MA, USA: MIT Press, 1990.

[3] Sengupta, K.and Boyer, K.L. "Using geometric hashing with information theoretic clustering for fast recognition from a large CAD modelbase".IN: *Proc. Int. Symp. on Computer Vision*. Los Alamitos, CA, USA: IEEE Comput. Soc. Press, pp. 151-6, 1995.

[4] Vickery B. C., "Bradfords's law of scattering", *J. Documentation*, Vol. 4, No. 3, pp. 198-203, 1948.

[5] Chen, Y-S.Chong, P.P., and Tong, M.Y., "Dynamic behavior of Bradford's law", *J. Americal Soc. for Info. Science*, vol. 46, No. 5, pp. 370-83, 1995.

[6] Brin, S, "Near neighbor search in large metric spaces", IN: *VLDB '95. Proceedings of the 21st International Conference on Very Large Data Bases*, Dayal, U., Gray, P.M.D., Nishio, S. Eds, San Francisco, CA, USA: Morgan Kaufmann, pp. 574-84, 1995.

[7] Stricker, M.; Swain, M., "The capacity of color histogram indexing", IN: *Proc.Conf. on Computer Vision and Pattern Recognition* Seattle, WA, IEEE Comput. Soc. Press, pp. 704-8, 1994.

[8] Bradford, B.C., "Sources of information on specific subjects", *Engineering*, vol 137, pp. 85-86, 1934.

[9] Faloutsos, C., Barber, R., Flickner, M., Hafner, J. and others. "Efficient and effective querying by image content", *J.Int. Info. Syst.: Integrating Artificial Intelligence and Database Technologies*, vol.3, no.3-4, pp.231-62, 1994.

[10] Gupta, A., Weymouth, T.and Jain, R., "Semantic queries with pictures: the VIMSYS model", IN: *Proc. Int. Conf. Very Large Data Bases* Barcelona, Spain, Lohman, G.M.; Sernadas, A.; Camps, R. (Eds.) San Mateo, CA, USA: Morgan Kaufmann, pp. 69-79, 1991.

[11] Bach, J.R., Paul, S.and Jain, R, "A visual information management system for the interactive retrieval of faces", *IEEE Trans. Knowledge and Data Engineering*, vol.5, No.4, pp. 619-28, 1993.

[12] *Proc. IFIP 2.6 3rd Working Conference on Visual Database Systems (VDB-3)*, Lausanne, Switzerland, Eds: Spaccapietra, S.; Jain, R. London, UK: Chapman and Hall, 1995.

[13] Niblack, W. and Jain, R.(eds.), *Storage and Retrieval for Image and Video Databases III*, San Jose, CA, Proc. SPIE, vol.2185, 1994.

[14] Jain, R. and Niblack, W. (eds.), *Storage and Retrieval for Image and Video Databases III*, San Jose, CA, Proc. SPIE, vol.2420, 1995.

[15] Flickner, M., Sawhney, H., Niblack, W., Ashley, J. and others, "Query by image and video content: the QBIC system", *IEEE Computer*, vol.28, No.9, pp.23-32, 1995.

[16] Gupta, A., "Visual Information Retrieval: A VIRAGE perspective", Tech. Report TR95-01, Virage, Inc. 1995

[17] Rosenfeld,A.and Kak, A.C.,*Digital picture processing*, New York : Academic Press, 1976.

[18] Pentland, A.; Moghaddam, B.; Starner, T. " View-based and modular eigenspaces for face recognition", IN: Proc. Conf. Computer Vision and Pattern Recognition, Seattle, WA, pp. 84-91, 1994.

[19] Jain, R.; Murthy, S.N.J.; Chen, P.L.-J.; Chatterjee, S. "Similarity measures for image databases", IN: *Storage and Retrieval for Image and Video Databases III*, San Jose, CA, Proc. of the SPIE, vol.2420:58-65, 1995.

[20] Santini, S. and Jain, R., "Similarity Matching", submitted to *IEEE Trans. PAMI*, 1995.

Object-Based and Image-Based Representations of Objects by their Interiors*

Hanan Samet

Abstract

An overview is presented of object-based and image-based representations of objects by their interiors that can be used to answer two fundamental queries in image understanding: (1) Given an object, determine its constituent cells (i.e., their locations in space). (2) Given a cell (i.e., a location in space), determine the identity of the object (or objects) of which it is a member as well as the remaining constituent cells of the object (or objects). Representations based on collections of unit-size cells and blocks are described that are designed to answer just one of these queries, while representations based on the use of hierarchies of space and objects are presented enabling the efficient response to both queries.

1: Introduction

The representation of spatial objects and their environment is an important issue in applications of computer graphics, computer vision, image processing, robotics, and pattern recognition as well as in building databases to support them (e.g., [25, 26]). We assume that the objects are connected[1] although their environment need not be. The objects and their environment are usually decomposed into collections of more primitive elements (termed *cells*) each of which has a location in space, a size, and a shape. These elements can either be subobjects of varying shape (e.g., a table consists of a flat top in the form of a rectangle and four legs in the form of rods or similar shapes whose lengths dominate their cross-sectional areas), or can have a uniform shape. The former yields an *object-based* decomposition while the latter yields an *image-based* or *cell-based* decomposition. Another way of characterizing these two decompositions is that the former decomposes the objects while the latter decomposes the environment in which the objects lie. This distinction is commonly used to characterize algorithms in computer graphics (e.g., [9]). Regardless of the characterization, the objects (as well as their constituent cells) can be represented either by their interiors or by their boundaries. In this article our focus is on interior-based representations.

Each of the decompositions has its advantages and disadvantages. They depend primarily on the nature of the queries that are posed to the database. The most general queries ask *where*, *what*, *who*, *why*, and *how*. The ones that are relevant to our application are *where* and *what*. They are stated more formally as follows:

1. Given an object, determine its constituent cells (i.e., their locations in space).

*This work was supported in part by the National Science Foundation under Grant IRI-92-16970.

[1] Intuitively, this means that a d-dimensional object cannot be decomposed into disjoint subobjects so that the subobjects are not adjacent in a $(d-1)$-dimensional sense.

 0-8186-7644-2 $5.00

2. Given a cell (i.e., a location in space), determine the identity of the object (or objects) of which it is a member as well as the remaining constituent cells of the object (or objects).

Not surprisingly, the queries can be classified using the same terminology that we used in the characterization of the decomposition. In particular, we can either try to find the cells (i.e., their locations in space) occupied by an object or find the objects that overlap a cell (i.e., a location in space). If objects are associated with cells so that a cell contains the identity of the relevant object (or objects), then query 1 is analogous to retrieval by contents while query 2 is analogous to retrieval by location.

Queries 1 and 2 are the basis of two more general classes of queries. Query 1 is known as a *feature-based* (also *object-based*) query, while query 2 is known as a *location-based* (also *image-based* or *cell-based*) query. Query 2 is a special case of a wider range of queries known as *window queries* which retrieve the objects that cover an arbitrary region (often rectangular). These queries are used in a number of applications including geographic information systems (e.g., [2, 26]).

The generation of responses to these queries is facilitated by building an index (i.e., the result of a sort) either on the objects or on their locations in space. Ideally, we wish to be able to answer both types of queries with one representation. At times, more compact representations are desired in which case we make use of techniques that aggregate identically-valued contiguous cells, or even objects which, ideally, are in proximity. These queries and a number of different representations and indexes that facilitate responding to them are discussed in this article which is organized as follows. Section 2 examines representations based on collections of unit-size cells while Section 3 treats the case of blocks. Section 4 looks at the use of hierarchies of space and objects which enable efficient responses to both queries 1 and 2, while Section 5 contains concluding remarks.

2: Unit-Size Cells

The most common representation of the objects and their environment is as a collection of cells of uniform size and shape (termed *pixels* and *voxels* in two and three dimensions, respectively) all of whose boundaries (with dimensionality one less than that of the cells) are of unit size. Since the cells are uniform, there exists a way of referring to their locations in space relative to a fixed reference point (e.g., the origin of the coordinate system). An example of a specification of a location of a cell in space is the set of coordinate values that enable us to find it in the d-dimensional space of the environment in which it lies. It should be clear that the concept of a *location* of a cell in space is quite different from that of an *address* of a cell which is the physical location (e.g., in memory, on disk, etc.), if any, where some of the information associated with the cell is stored.

In most applications (including the ones that we consider here), the boundaries (i.e., edges and faces in two and three dimensions, respectively) of the cells are orthogonal to each other and parallel to the coordinate axes. In our discussion, we assume that the cells comprising a particular object are contiguous, or equivalently continuous (i.e., adjacent), and that a different unique value is associated with each distinct object thereby enabling us to distinguish between the objects. Depending on the underlying representation, this value may be stored with the cells. For example, Figure 1 contains three two-dimensional objects A, B, and C and their corresponding cells.

Interior-based methods represent an object o by using the locations in space of the cells that comprise o. Operations on the cells (i.e., locations in space) comprising o are facilitated

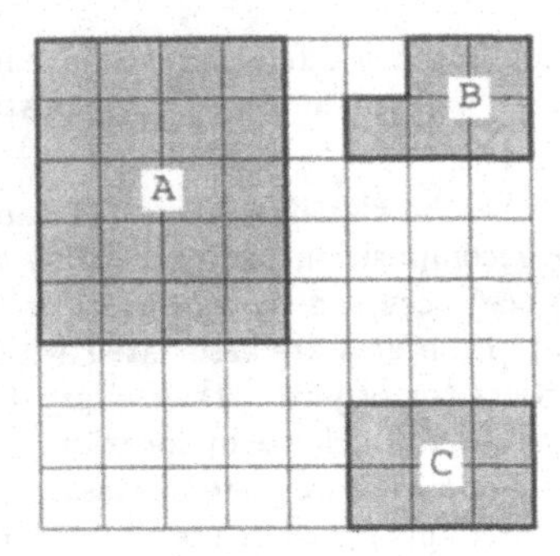

Figure 1: Example collection of three objects and the cells that they occupy.

by aggregating the cells of the objects into subcollections of contiguous identically-valued cells. The aggregation may be implicit or explicit depending on how the contiguity of the cells that make up the aggregated subcollection is expressed.

An aggregation is *explicit* if the identity of the contiguous cells that form the object is hardwired into the representation. An example of an explicit aggregation is one that associates a set with each object o that contains the location in space of each cell that comprises o. In this case, no identifying information (e.g., the object identifier corresponding to o) is stored in the cells. Thus there is no need to allocate storage for the cells (i.e., no addresses are associated with them). One possible implementation of this set is a list. For example, consider Figure 1 and assume that the origin `(0,0)` is at the upper-left corner. Assume further that this is also the location of the pixel that abuts this corner. Therefore, the explicit representation of object `B` is the set of locations {`(5,1) (6,0) (6,1) (7,0) (7,1)`}. It should be clear that using the explicit representation, given an object o, it is easy to determine the cells (i.e., locations in space) that comprise it (query 1).

Of course, even when using an explicit representation, we must still be able to access object o from a possibly large collection of objects which may require an additional data structure such as an index on the objects (e.g., a table of object-value pairs where *value* indicates the entry in the explicit representation corresponding to *object*). This index does not make use of the spatial coverage of the objects and thus may be implemented using conventional searching techniques such as hashing [17]. In this case, we will need $O(n)$ additional space for the index, where n is the number of different objects. We do not discuss such indexes here.

The fact that no identifying information as to the nature of the object is stored in the cell means that the explicit representation is not suited for answering the inverse query of determining the object associated with a particular cell at location l in space (i.e., query 2). Using the explicit representation, query 2 can only be answered by checking for the presence of location l in space in the various sets associated with the different objects. This will be time-consuming as it may require that we examine all cells in each set. It should be clear that the explicit representation could also be classified as being *object-based* as it clearly lends itself only to retrieval on the basis of knowledge of the objects rather than of the locations of the cells in space. We shall make use of this characterization in Section 4.

Note that since the explicit representation consists of sets, there is no particular order for the cells within each set although an ordering could be imposed based on spatial proximity of the locations of the cells in space, etc. For example, the list representation of a set already presupposes the existence of an ordering. Such an ordering could be used to obtain a small, but not insignificant, decrease in the time (in an expected sense) needed to answer query 2. In particular, now whenever cell c is not associated with object o, we will be able to cease searching the list associated with o after having inspected half of the cells associated with o instead of all of them, which is the case when no ordering exists.

An important shortcoming of the use of the explicit representation, which has an effect somewhat related to the absence of an ordering, is the inability to distinguish between occupied and unoccupied cells. In particular, in order to detect that a cell c is not occupied by any object we must examine the sets associated with each object, which is quite time-consuming. Of course, we could avoid this problem by forming an additional set which contains all of the unoccupied cells, and examine this set first whenever processing query 2. The drawback of such a solution is that it slows down all instances of query 2 that involve cells that are occupied by objects.

We can avoid examining every cell in each object set, thereby speeding up query 2 in certain cases, by storing a simple approximation of the object with each object set o. This approximation should be of a nature that makes it easy to check if it is impossible for a location l in space to be in o. One such approximation is a minimum bounding box whose sides are parallel to the coordinate axes of the space in which the object is embedded. For example, for object `B` in Figure 1 such a box is anchored at the lower-left corner of cell `(5,1)` and the right corner of cell `(7,0)`. The existence of a box b for object o means that if b does not contain l, then o does not contain l either, and we can proceed with checking the other objects. This bounding box is usually a part of the explicit representation

Query 2 can be answered more directly if we allocate an address a in storage for each cell c where an identifier is stored that indicates the identity of the object (or objects) of which c is a member. Such a representation is said to be *implicit* because in order to determine the rest of the cells that comprise the object associated with c (and thus complete the response to query 2), we must examine the identifiers stored in the addresses associated with the contiguous cells and then aggregate the cells whose associated identifiers are the same. However, in order to be able to use the implicit representation, we must have a way of finding the right address a corresponding to c, taking into account that there is possibly a very large number of cells, and then retrieving a.

Finding the right address requires an additional data structure, termed an *access structure*, like an index on the locations in space. An example of such an index is a table of cell-address pairs where *address* indicates the physical location where the information about the object associated with the location in space corresponding to *cell* is stored. The table is indexed by the location in space corresponding to *cell*. The index is really an ordering and hence its range is usually the integers (i.e., one-dimensional). When the data is multidimensional (i.e., cells in d-dimensional space where $d > 0$), it may not be convenient to use the location in space corresponding to the cell as an index since its range spans data in several dimensions. Instead, we employ techniques such as laying out the addresses corresponding to the locations in space of the cells in some particular order and then making use of an access structure in the form of a mapping function to enable the quick association of addresses with the locations in space corresponding to the cells. Retrieving the address is more complex in the sense that it can be a simple memory access or may involve an access to secondary or tertiary storage if virtual memory is being used. In most of our discussion, we assume that all data is in main memory, although, as we will see, a number of the representations do not rely on this assumption.

Such an access structure enables us to obtain the contiguous cells (as we know their locations in space) without having to examine all of the cells. Therefore, we will know the identities of the cells that comprise an object thereby enabling us to complete the response to query 2 with an implicit representation. In the rest of this section, we discuss a number of such access structures. However, before proceeding further, we wish to point out that the implicit representation could also be classified as being *image-based* as it clearly lends itself to retrieval on the basis of knowledge only of the cells rather than the objects. We shall make use of this characterization in Section 4.

The existence of an access structure also enables us to answer query 1 with the implicit representation although it is quite inefficient. In particular, given an object o, we must exhaustively examine every cell (i.e., location l in space) and check if the address where the information about the object associated with l is stored contains o as its value. This will be time-consuming as it may require that we examine all the cells.

There are many ways of laying out the addresses corresponding to the locations in space of the cells each having its own mapping function. Some of the most important ones for a two-dimensional space are illustrated in Figure 2 for an 8×8 portion of the space. To repeat, in essence, what we are doing is providing a mapping from the d-dimensional space containing the locations of the cells to the one-dimensional space of the range of index values (i.e., integers) to a table whose entries contain the addresses where information about the contents of the cells is stored. The result is an ordering of the space, and the curves shown in Figure 2 are termed *space-filling curves*.

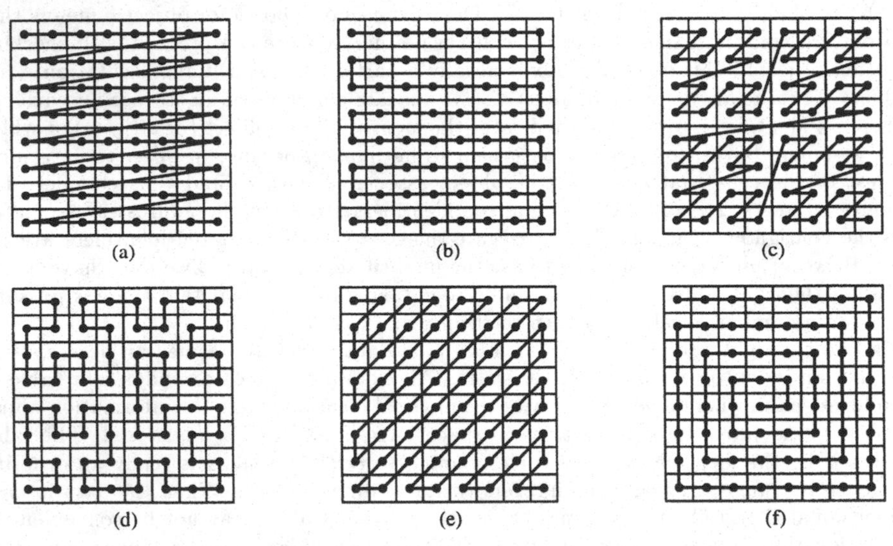

Figure 2: The result of applying a number of different space-ordering methods to an 8×8 collection of cells whose first element is in the upper-left corner: (a) row order, (b) row-prime order, (c) Morton order, (d) Peano-Hilbert order, (e) Cantor-diagonal order, (f) spiral order.

The multidimensional array (i.e., having a dimension equal to the dimensionality of the space in which the objects and the environment are embedded) is an access structure which given a cell c at a location l in space enables us to calculate the address a containing the identifier of the object associated with c. The array is only a conceptual multidimensional structure (it is not a multidimensional physical entity in memory) in the sense that it is a mapping of the locations in space of the cells into sequential addresses in memory. The actual addresses are obtained by the array access function which is based on the extents of the various dimensions (i.e., coordinate axes). The array access function is usually the mapping function for the row order illustrated in in Figure 2a (although, at times, the column order is also used). Thus the array enables us to implement the implicit representation with no additional storage except for what is needed for the array's descriptor. The descriptor contains the bounds and extents of each of the dimensions which are used to define the mapping function (i.e., they determine the values of its coefficients) so that the appropriate address can be calculated given the cell's location in space.

The array is called a *random access structure* as the address associated with a location in space can be retrieved in constant time independent of the number of elements in the array and doesn't require search. Note that we could store the object identifier o in the array element itself instead of allocating a separate address a for o thereby saving some space.

The array is an implicit representation because we have not explicitly aggregated all the contiguous cells that comprise a particular object. They can be obtained given a particular cell c at a location l in space belonging to object o by recursively accessing the array elements corresponding to the locations in space that are adjacent to l and checking if they are associated with object o. This process is known as depth-first connected component labeling (e.g., [21]).

Interestingly, depth-first connected component labeling could also be used to answer query 1 efficiently with an implicit representation if we add a data structure such as an index on the objects (e.g., a table of object-location pairs where *location* is one of the locations in space that comprise *object*). Thus given an object o we use the index to find a location in space that is part of o and then proceed with the depth-first connected component labeling as before. This index does not make use of the spatial coverage of the objects and thus it can be implemented using conventional searching techniques such as hashing [17]. In this case, we will need $O(n)$ additional space for the index where n is the number of different objects. We do not discuss such indexes here.

Of course, we could also answer query 2 with an explicit representation by adding an index which associates objects with locations in space (i.e., having the form location-objects). However, this would require $O(s)$ additional space for the index where s is the number of cells. The $O(s)$ bound assumes that only one object is associated with each cell. If we take into account that a cell could be associated with more than one object, then the additional storage needed is $O(ns)$ if we assume n objects. Since the number of cells s is usually much greater than the number of objects n, the addition of an index to the explicit representation is not as practical as extending the implicit representation with an index of the form object-location as described above. Thus it would appear that the implicit representation is more useful from the point of view of flexibility when taking storage requirements in to account.

The implicit representation can be implemented with access structures other than the array. This is an important consideration when many of the cells are not in any of the objects (i.e., they are empty). The problem is that using the array access structure is wasteful of storage as the array requires an element for each cell regardless of whether or not the cell is associated with any of the objects. In this case, we choose only to keep track of the non-empty cells.

We have two ways to proceed. The first is to use one of a number of multidimensional access structures such as a point quadtree, k-d tree, MX quadtree, etc. as described in [24]. The second is to make use of one of the orderings of space shown in Figure 2 to obtain a mapping from the non-empty contiguous cells to the integers. The result of the mapping serves as the index in one of the familiar tree-like access structures (e.g., binary search tree, range tree, B^+-tree, etc.) to store the address which indicates the physical location where the information about the object associated with the location in space corresponding to the non-empty cell is stored.

3: Blocks

An alternative class of representations of the objects and their environment removes the stipulation that cells making up the object collection be of a unit size and permits their sizes to vary. The resulting cells are termed *blocks* and are usually rectangular with sides parallel to the coordinate axes (this is assumed in our discussion unless explicitly stated otherwise). The volume (e.g., area in two dimensions) of the blocks need not be an integer multiple of that of the unit-size cells, although this is often the case. Observe that when the volumes of the blocks are integer multiples of that of the unit-size cells, then we have two levels of aggregation in the sense that an object consists of an aggregation of blocks which are themselves aggregations of cells. We assume that all the cells in a block belong to the same object or objects. In other words, the situation that some of the cells in the block belong to object o_1 while the others belong to object o_2 (and not to o_1) is not allowed.

The collection of blocks is usually a result of a space decomposition process with a set of rules that guide it. There are many possible decompositions. When the decomposition is recursive, we have the situation that the decomposition occurs in stages and often, although not always, the results of the stages form a containment hierarchy. This means that a block b obtained in stage i is decomposed into a set of blocks b_j that span the same space. Blocks b_j are, in turn, decomposed in stage $i+1$ using the same decomposition rule. Some decomposition rules restrict the possible sizes and shapes of the blocks as well as their placement in space. Some examples include:

- congruent blocks at each stage
- similar blocks at all stages
- all but one side of a block are unit-sized
- all sides of a block are of equal size
- all sides of each block are powers of two
- etc.

Other decomposition rules dispense with the requirement that the blocks be rectangular, while still others do not require that they be orthogonal. In addition, the blocks may be disjoint or be allowed to overlap. Clearly, the choice is large. In the following, we briefly explore some of these decomposition processes.

The simplest decomposition rule is one that permits aggregation of identically-valued cells in only one dimension. It assigns a priority ordering to the various dimensions and then fixes the coordinate values of all but one of the dimensions, say i, and then varies the value of the ith coordinate and aggregates all adjacent cells belonging to the same object into a one-dimensional block. This technique is commonly used in image processing applications where the image is decomposed into rows which are scanned from top to bottom, and each row is scanned from left to right while aggregating all adjacent pixels with the same value into a block. The aggregation into one-dimensional blocks is the basis of *runlength encoding* [22]. Similar techniques are applicable to higher-dimensional data.

The drawback of the decomposition into one-dimensional blocks described above is that all but one side of each block must be of unit width. The most general decomposition removes this restriction along all of the dimensions, thereby permitting aggregation along all dimensions. In other words, the decomposition is arbitrary. The blocks need not be uniform or similar. The only requirement is that the blocks span the space of the environment. We assume that the blocks are disjoint although this need not be the case. We also assume that the blocks are rectangular as well as orthogonal although again this is not absolutely necessary as there exist decompositions using other shapes as well (e.g., triangles, etc.).

It is easy to adapt the explicit representation to deal with blocks resulting from an arbitrary decomposition (which also includes the one that yields one-dimensional blocks). In particular, instead of associating a set with each object o that contains the location in space of each cell that comprises o, we need to associate with each object o the locations in space and sizes of each block that comprises o. This can be done by specifying the coordinate values of the upper-left corner of each block and the sizes of its sides. This format is appropriate, and is the one we use, for the explicit representation of all of the block decompositions described in this section.

Using the explicit representation of blocks, both queries 1 and 2 are answered in essentially the same way as they were for unit-sized cells. The only difference is that for query 2 instead of checking if a particular location l in space is a member of one of the sets of cells associated with the various objects, we must check if l is covered by one of the blocks in the sets of blocks of the various objects. This is a fairly simple process as we know the location in space and size of each of the blocks.

An implementation of an arbitrary decomposition (which also includes the one that results in one-dimensional blocks) using an implicit representation is quite easy as long as the decomposition yields disjoint blocks, which is the case for all of the decompositions discussed in this section. Disjointness is important because it means that only one block can be associated with a location in space. Thus we can build an index based on an easily identifiable location in each block such as its upper-left corner. Therefore, we can, and do, make use of the same techniques that were presented in the discussion of the implicit representation for unit-sized cells in Section 2.

As in the case of unit-size cells, regardless of which tree access structure is used on the identifiable location, we determine the object o associated with a cell at location l by finding the block b that covers l. If b is an empty block, then we exit. Otherwise, we return o. Notice that the search for the block that covers l may be quite complex in the sense that the access structures may not necessarily achieve as much pruning of the search space as in the case of unit-sized cells. In particular, this is the case whenever the space ordering that is applied does not have the property that all of the cells in each block appear in consecutive order. In other words, given the cells in the block e with minimum and maximum values in the ordering, say u and v, there exists at least one cell in block f distinct from e which is mapped to a value w where $u < w < v$. Thus a search for the block b that covers l may require that we visit several subtrees of a particular node in the tree-like access structures.

Perhaps the most widely known decompositions into blocks are those referred to by the general terms *quadtree* and *octree* [23, 24]. They are usually used to describe a class of representations for two and three-dimensional data (and higher as well), respectively, that are the result of a recursive decomposition of the environment (i.e., space) containing the objects into blocks (not necessarily rectangular) until the data in each block satisfies some condition (e.g., with respect to its size, the nature of the objects that comprise it, the number of objects in it, etc.). The positions and/or sizes of the blocks may be restricted or arbitrary. It is interesting to note that quadtrees and octrees may be used with both interior-based and boundary-based representations. Moreover, both explicit and implicit aggregations of the blocks are possible.

There are many variants of quadtrees and octrees, and they are used in numerous application areas including high energy physics, VLSI, finite element analysis, and many others. Below, we focus on *region quadtrees* [15] and *region octrees* [14, 18]. They are specific examples of interior-based representations for two and three-dimensional region data (variants for data of higher dimension also exist), respectively, that permit further aggregation of identically-valued cells.

Region quadtrees and region octrees are instances of a restricted-decomposition rule where the environment containing the objects is recursively decomposed into four or eight, respectively, rectangular congruent blocks until each block is either completely occupied by an object or is empty (such a decomposition process is termed *regular*). For example, Figure 3a is the block decomposition for the region quadtree corresponding to Figure 1. We have labeled the blocks corresponding to object `O` as `Oi` and the blocks that are not in any of the objects as `Wi` using the suffix `i` to distinguish between them in both cases. Notice that in this case, all the blocks are square, have sides whose size is a power of 2, and are located at specific positions. In particular, assuming an origin at the upper-left corner of the image corresponding to the environment containing the objects, then the coordinate values of the upper-left corner of each block (e.g., (a,b) in two dimensions) of size $2^i \times 2^i$ satisfy the property that $a \bmod 2^i = 0$ and $b \bmod 2^i = 0$.

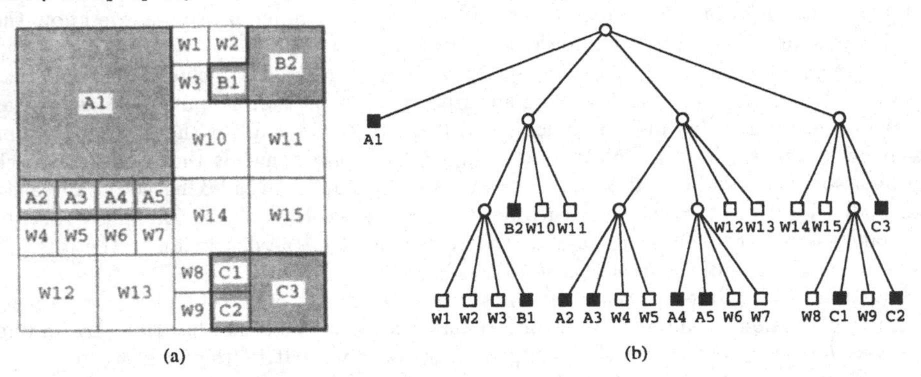

Figure 3: (a) Block decomposition and (b) its tree representation for the collection of objects and cells in Figure 1.

A region quadtree can be implemented using an explicit representation by associating a set with each object o that contains its constituent blocks. Each block is are specified by numbers corresponding to the coordinate values of its upper-left corner and the size of one of its sides. These numbers are stored in the set in the form $(a,b):d$ where (a,b) and d correspond to the coordinate values of the upper-left corner and depth, respectively, of the block. For example, the explicit representation of the collection of blocks n Figure 1 is given by the sets `A={(0,0):2,(0,4):0,(1,4):0,(2,4):0,(3,4):0}`, `B={(5,1):0,(6,0):1}`, and `C={(5,6):0,(5,7):0,(6,6):1}`, which correspond to blocks {A1,A2,A3,A4,A5}, {B1,B2}, and {C1,C2,C3}, respectively.

A region quadtree implementation that makes use of an implicit representation is quite different. First, we allocate an address a in storage for each block b which stores an identifier that indicates the identity of the object (or objects) of which b is a member. Second, it is necessary to impose an access structure on the collection of blocks in the same way as the array was imposed on the collection of unit-sized cells. Such an access structure enables us to determine easily the value associated with any point in the space covered by a cell without resorting to exhaustive search. Note that depending on the nature of the access structure, it's not always necessary to store the location and size of each block with a.

There are many possible access structures. Interestingly, using an array as an access structure is not particularly useful as it defeats the rationale for the aggregation of cells into blocks unless, of course, all the blocks are of a uniform size in which case we have the analog of a two-level grid.

The traditional, and most natural, access structure for a region quadtree corresponding to a d-dimensional image is a tree with a fanout of 2^d (e.g., Figure 3b corresponding to the collection of two-dimensional objects in Figure 1 whose quadtree block decomposition is given in Figure 3a). Each leaf node in the tree corresponds to a different block b and contains the address a in storage where an identifier is stored that indicates the identity of the object (or objects) of which b is a member. As in the case of the array, where we could store the object identifier o in the array element itself instead of allocating a separate address a for o, we could achieve the same savings by storing o in the leaf node of the tree. Each nonleaf node f corresponds to a block whose volume is the union of the blocks corresponding to the 2^d sons of f. In this case, the tree is a containment hierarchy and closely parallels the decomposition in the sense that they are both recursive processes and the blocks corresponding to nodes at different depths of the tree are similar in shape.

Answering query 2 using the tree structure is different from using an array where it is usually achieved by a table lookup having an $O(1)$ cost (unless the array is implemented as a tree which is a possibility [6]). In contrast, query 2 is usually answered in a tree by locating the block that contains the location in space corresponding to the desired cell. This is achieved by a process that starts at the root of the tree and traverses the links to the sons whose corresponding blocks contain the desired location. This process has an $O(m)$ cost where the environment has a maximum of m levels of subdivision (e.g., an environment all of whose sides are of length 2^m).

Using a tree with fanout 2^d as an access structure for a regular decomposition means that there is no need to record the size and location of the blocks. This information can be inferred from knowledge of the size of the underlying space as the 2^d blocks that result from each subdivision step are congruent. For example, in two dimensions, each level of the tree corresponds to a quartering process that yields four congruent blocks (rectangular here, although a triangular decomposition process could also be defined which yields four equilateral triangles; however, in such a case, we are no longer dealing with rectangular cells). Thus as long as we start from the root, we know the location and size of every block.

There are a number of alternative access structures to the tree with fanout 2^d. They are all based on finding a mapping from the domain of the blocks to a subset of the integers (i.e., to one dimension) and then applying one of the familiar tree-like access structures (e.g., a binary search tree, range tree, B^+-tree, etc.). There are many possible mappings (e.g., [23]). The simplest is to use the same technique that we applied to the collection of blocks of arbitrary size. In particular, we can apply one of the orderings of space shown in Figure 2 to obtain a mapping from the coordinate values of the upper-left corner u of each block to the integers.

Since the size of each block b in the region quadtree can be specified with a single number indicating the depth in the tree at which b is found, we can simplify the representation by incorporating the size into the mapping. One mapping simply concatenates the result of interleaving the binary representations of the coordinate values of the upper-left corner (e.g., (a, b) in two dimensions) and i of each block of size 2^i so that i is at the right. The resulting number is termed a *locational code* and is a variant of the Morton order (Figure 2c). Assuming such a mapping and sorting the locational codes in increasing order yields an ordering equivalent to that which would be obtained by traversing the leaf nodes (i.e., blocks) of the tree representation (e.g., Figure 3b) in the order **NW**, **NE**, **SW**, **SE**.

As the dimensionality of the space (i.e., d) increases, each level of decomposition in the region quadtree results in many new blocks as the fanout value 2^d is high. In particular, it is too large for a practical implementation of the tree access structure. In this case, an access structure termed a *bintree* [16, 27, 30] with a fanout value of 2 is used. The bintree is defined in a manner analogous to the region quadtree except that at each subdivision stage, the space is decomposed into two equal-sized parts. In two dimensions, at odd stages we partition along the y axis and at even stages we partition along the x axis.

The region quadtree, as well as the bintree, is a regular decomposition. This means that the blocks are congruent — that is, at each level of decomposition, all of the resulting blocks are of the same shape and size. We can also use decompositions where the sizes of the blocks are not restricted in the sense that the only restriction is that they be rectangular and be a result of a recursive decomposition process. In this case, the representations that we described must be modified so that the sizes of the individual blocks can be obtained. An example of such a structure is an adaptation of the point quadtree [8] to regions. Although the point quadtree was designed to represent points in a higher dimensional space, the blocks resulting from its use to decompose space do correspond to regions. The difference from the region quadtree is that in the point quadtree, the positions of the partitions are arbitrary, whereas they are a result of a partitioning process into 2^d congruent blocks (e.g., quartering in two dimensions) in the case of the region quadtree.

As the dimensionality d of the space increases, each level of decomposition in the point quadtree results in many new blocks since the fanout value 2^d is high. In particular, it is too large for a practical implementation of the tree access structure. Therefore, we use a k-d tree [5] which is an access structure having a fanout of 2 that is an adaptation of the point quadtree to regions. As in the point quadtree, although the k-d tree was designed to represent points in a higher dimensional space, the blocks resulting from its use to decompose space do correspond to regions.

The k-d tree can be further generalized so that the partitions take place on the various axes at an arbitrary order, and, in fact, the partitions need not be made on every coordinate axis. In this case, at each nonleaf node of the k-d tree, we must also record the identity of the axis that is being split. We use the term *generalized k-d tree* to describe this structure. The generalized k-d tree is really an adaptation to regions of the *adaptive k-d tree* [11] and the *LSD tree* [1] which were originally developed for points. It can also be regarded as a special case of the *BSP tree* (denoting *Binary Space Partitioning*) [12]. In particular, in the generalized k-d tree, the partitioning hyperplanes are restricted to be parallel to the axes, whereas in the BSP tree they have an arbitrary orientation. The BSP tree is used in computer graphics to facilitate viewing.

One of the shortcomings of the generalized k-d tree is the fact that we can only decompose the space into two parts along a particular dimension at each step. If we wish to partition a space into p parts along a dimension i, then we must perform $p-1$ successive partitions on dimension i. Once these $p-1$ partitions are complete, we partition along another dimension. The *puzzletree* [7] is a further generalization of the k-d tree that decomposes the space into two or more parts along a particular dimension at each step so that no two successive partitions use the same dimension. In other words, the puzzletree compresses all successive partitions on the same dimension in the generalized k-d tree.

The puzzletree is motivated by a desire to overcome the rigidity in the shape, size, and position of the blocks that result from the bintree (and to an equivalent extent, the region quadtree) partitioning process (because of its regular decomposition). In particular, in many cases, the decomposition rules ignore the homogeneity present in certain regions on account of the need to place the partition lines in particular positions as well as a possible limit on the number of permissible partitions along each dimension at each decomposition step. Often, it is desirable for the block decomposition to follow the perceptual characteristics of the objects as well as reflect their dominant structural features.

For example, consider a front view of a scene containing a table and two chairs. Figures 4a and 4b are the block decompositions resulting from the use of a bintree and a puzzletree, respectively, for this scene, while Figure 4c is the tree access structure corresponding to the puzzletree in Figure 4b. Notice the natural decomposition in the puzzletree of the chair into the legs, seat, and back, and of the table into the top and legs. On the other hand, the blocks in the bintree (and to a greater extent in the region quadtree, although not shown here) do not have this perceptual coherence.

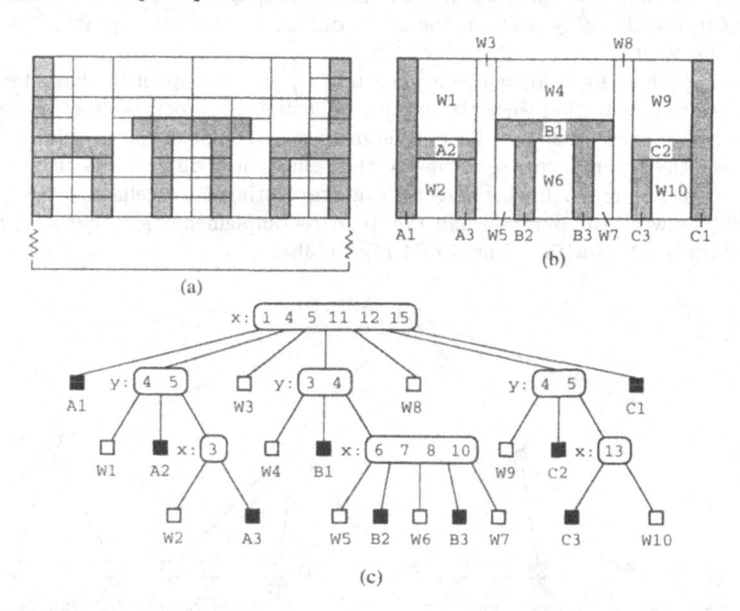

Figure 4: Block decomposition for the (a) bintree and (b) puzzletree corresponding to the front view of a scene containing a table and two chairs; (c) is the tree access structure for the puzzletree in (b).

4: Hierarchical Representations

Assuming the presence of an access structure, the implicit representations described in Sections 2 and 3 are good for finding the objects associated with a particular location or cell (i.e., query 2), while requiring that all cells be examined when determining the locations associated with a particular object (i.e., query 1). In contrast, the explicit representation that we described is good for query 1, while requiring that all objects be examined when trying to respond to query 2. In this section, we focus on representations that enable both queries to be answered without possibly having to examine every cell.

This is achieved by imposing containment hierarchies on the representations. The hierarchies differ depending on whether the hierarchy is of space (i.e., the cells in the space in which the objects are found), or whether the hierarchy is of objects. In the former case, we aggregate space into successively larger-sized chunks (i.e., blocks), while in the latter, we aggregate objects into successively larger groups (in terms of the number of objects that they contain). The former is applicable to implicit (i.e., image-based) representations, while the latter is applicable to explicit (i.e., object-based) representations. Thus, we see again that the distinction is the same as that used in computer graphics to distinguish between algorithms as being image-space or object-space [9], respectively.

The basic idea is that in image-based representations we propagate objects up the hierarchy with the occupied space being implicit to the representation. Thus we retain the property that associated with each cell is an identifier indicating the object of which it is a member. In fact, it is this information that is propagated up the hierarchy so that each element in the hierarchy contains the union of the objects that appear in the elements immediately below it.

The resulting hierarchy is known as a *pyramid* [31] and is frequently characterized as a *multiresolution* representation since the original collection of objects is described at several levels of detail by using cells that have different sizes, though they are similar in shape. Figure 5 shows the pyramid corresponding to the collection of objects and cells in Figure 1 and the labels in Figure 3a. In this case, we are aggregating 2×2 cells and blocks. Notice the similarity between the pyramid and the quadtree implementation that uses an access structure which is a tree with a fanout of 4 Figure 3b).

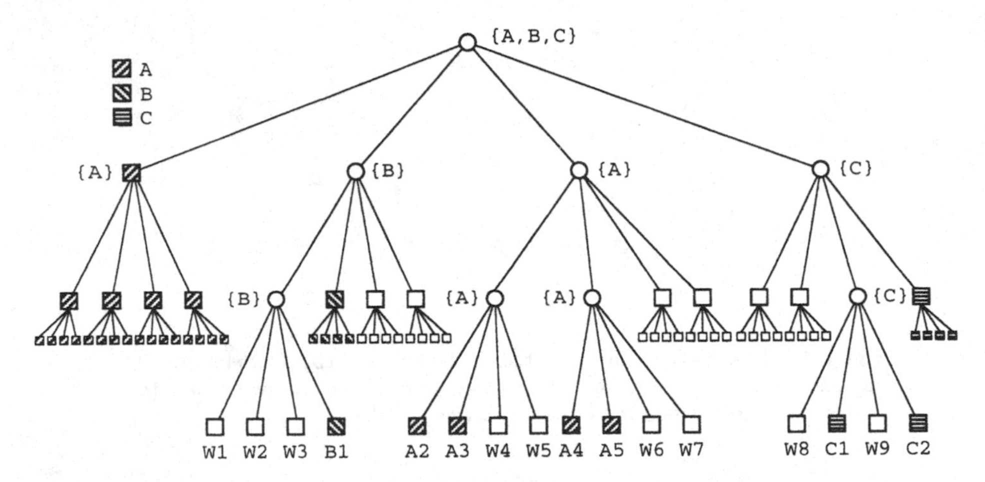

Figure 5: Pyramid for the collection of objects and cells in Figure 1.

Nevertheless, it is important to distinguish the pyramid from the quadtree which, as we recall, is an example of an aggregation into square blocks where the basis of the aggregation is that the cells have identical values (i.e., are associated with the same object or objects if object overlap is permitted). Hence the quadtree is an instance of what is termed a *variable resolution* representation, which, of course, is not limited to rectangular blocks that are square. In particular, it can be used with a limited number of other non-rectangular shapes (most notably, triangles in two dimensions [4, 24]).

The pyramid can be viewed as a complete quadtree (i.e., where no aggregation takes place at the deepest level, or, equivalently, all leaf nodes with zero sons are at maximum depth in the tree). Nevertheless, there are some very important differences. The first is that the quadtree is a variable-resolution representation while the pyramid is a multiresolution representation. The second, and most important, difference is that in the case of the quadtree, the nonleaf nodes just serve as an access structure. They do not include any information about the objects present in the nodes and cells below them. This is why the quadtree, like the array, is not useful for answering query 1. Of course, we could also devise a variant of the quadtree (termed a *truncated-tree pyramid* [25]) which uses the nonleaf nodes to store information about the objects present in the cells and nodes below them. Note that both the pyramid and the truncated-tree pyramid are instances of an implicit representation with a tree access structure.

On the other hand, in object-based representations we propagate the space occupied by the objects up the hierarchy with the identity of the objects being implicit to the representation. Thus we retain the property that associated with each object is a set of locations in space corresponding to the cells that make up the object. Actually, since this information may be rather voluminous, it is often the case that an approximation of the space occupied by the object is propagated up the hierarchy rather than the collection of individual cells that are spanned by the object. The approximation is usually the minimum bounding box for the object that is customarily stored with the explicit representation. Therefore, associated with each element in the hierarchy is a bounding box corresponding to the union of the bounding boxes associated with the elements immediately below it.

The R-tree [13] is an example of an object hierarchy which finds use especially in database applications. The number of objects or bounding boxes that are aggregated in each node is permitted to range between $m \leq \lceil M/2 \rceil$ and M. The root node in an R-tree has at least two entries unless it is a leaf node in which case it has just one entry corresponding to the bounding box of an object. The R-tree is usually built as the objects are encountered rather than waiting until all objects have been input.

Figure 6a is an example R-tree for a collection of rectangle objects with $m = 2$ and $M = 3$. Figure 6b shows the spatial extent of the bounding boxes of the nodes in Figure 6a, with heavy lines denoting the bounding boxes corresponding to the leaf nodes, and broken lines denoting the bounding boxes corresponding to the subtrees rooted at the nonleaf nodes. Note that the R-tree is not unique. Its structure depends heavily on the order in which the individual objects were inserted into (and possibly deleted from) the tree.

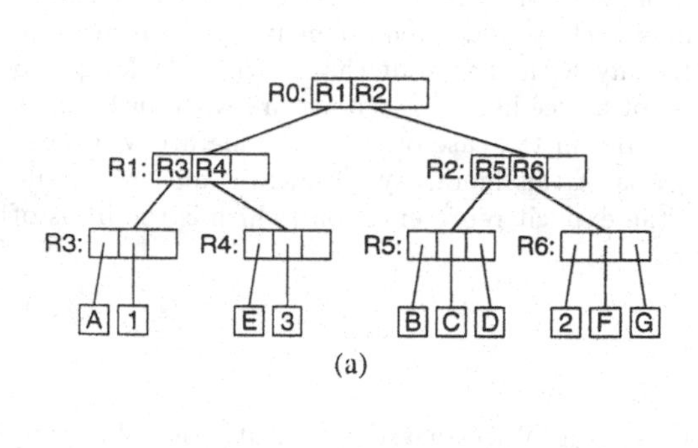

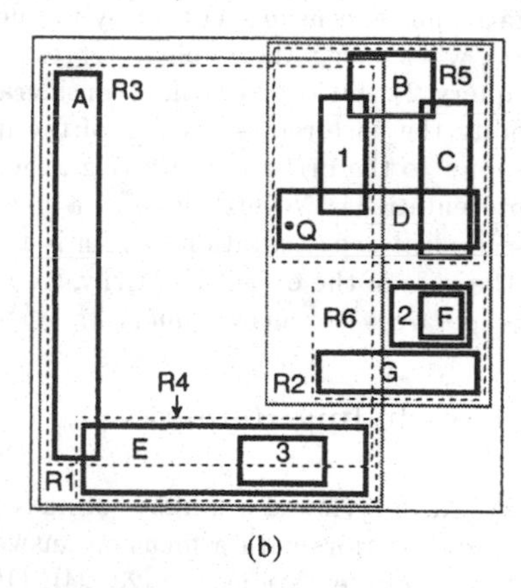

Figure 6: (a) R-tree for a collection of rectangle objects with m=2 and M=3, and (b) the spatial extents of the bounding boxes. Notice that the leaf nodes in the index also store bounding boxes although this is only shown for the nonleaf nodes.

The drawback of the R-tree (and any representation based on an object hierarchy) is that we may have to examine all of the bounding boxes at all levels when attempting to determine the identity of the object o that contains location a (i.e., query 2). This was caused by the fact that the bounding boxes corresponding to different nodes may overlap (i.e., they are not disjoint). The fact that each object is only associated with one node while being contained in possibly many bounding boxes (e.g., in Figure 6, rectangle 1 is contained in its entirety in R1, R2, R3, and R5) means that query 2 may often require several nonleaf nodes to be visited before determining the object that contains a. This can be overcome by decomposing the bounding boxes so that disjointness holds (e.g., the k-d-B-tree [20] and the R^+-tree [29]) which is a form of a space hierarchy in the spirit of the pyramid albeit not a regular decomposition. The drawback of this solution is that an object may be associated with more than one bounding box, which may result in the object being reported as satisfying a particular query more than once. For example, suppose that we want to retrieve all the objects that overlap a particular region (i.e., a window query) rather than a point as is done in query 2.

Note that the presence of the hierarchy does not mean that the alternative query (i.e., query 1 in the case of a space hierarchy and query 2 in the case of an object hierarchy) can be answered immediately. Instead, obtaining the answer usually requires that the hierarchy be descended. The effect is that the order of the execution time needed to obtain the answer is reduced from being linear to being logarithmic. Of course, this is not always the case. For example, the fact that we are using bounding boxes for the space spanned by the objects rather than the exact space occupied by them means that we do not always have a complete answer when reaching the bottom of the hierarchy. In particular, at this point, we may have to resort to a more expensive point-in-polygon test [9].

Furthermore, it is worth repeating that the only reason for imposing the hierarchy is to facilitate responding to the alternative query (i.e., query 1 in the case of a space hierarchy on the implicit representation and query 2 in the case of an object hierarchy on the explicit representation). Thus, usually, the base representation of the hierarchy is still used to answer the original query, because often, when using the hierarchy, the inherently logarithmic overhead incurred by the need to descend the hierarchy may be too expensive (e.g., when using the implicit representation with the array access structure to respond to query 2). Of course, other considerations such as space requirements may cause us to modify the base representation of the hierarchy with the result that it will take longer to respond to the original query (e.g., the use of a tree-like access structure with an implicit representation). Nevertheless, as a general rule, in the case of the space hierarchy, we use the implicit representation (which is the basis of this hierarchy) to answer query 2, while in the case of the object hierarchy, we use the explicit representation (which is the basis of this hierarchy) to answer query 1.

5: Concluding Remarks

We have reviewed a number of image-based and object-based representations of objects by their interiors with a focus on answering queries 1 and 2. For more details about some of these representations, see [23, 24]. Of course, there are also representations based on the boundaries of the objects, such as vectors and chain codes (e.g., [10]), which find much use. Although we do not go into great detail here about these representations, it is important to point out that one of the principal drawbacks of boundary-based representations is the difficulty in determining the value associated with an arbitrary point of the space given by the cell (i.e., query 2) without testing each boundary element using operations such as point-in-polygon tests (e.g., [9]) or finding the nearest boundary element. The problem is that these representations generally just indicate which boundary element is adjacent to which other boundary element rather than their relationship to the space that they occupy. This situation can be remedied by imposing an access structure such as an appropriate variant of a quadtree or octree which provides a way to index the boundary elements. For example, a variant of the PM quadtree (e.g., [28, 19]) can be used for a polygon representation in two dimensions and a PM octree (e.g., [3]) can be used similarly for three dimensions.

References

[1] H. W. Six A. Henrich and P. Widmayer. The LSD tree: spatial access to multidimensional point and non-point data. In P. M. G. Apers and G. Wiederhold, editors, *Proceedings of the Fifteenth International Conference on Very Large Data Bases*, pages 45–53, Amsterdam, August 1989.

[2] W. G. Aref and H. Samet. Efficient processing of window queries in the pyramid data structure. In *Proceedings of the 9th ACM SIGACT-SIGMOD-SIGART Symposium on Principles of Database Systems (PODS)*, pages 265–272, Nashville Tennessee, April 1990. (also *Proceedings of the Fifth Brazilian Symposium on Databases*, Rio de Janeiro, Brazil, April 1990, 15–26).

[3] D. Ayala, P. Brunet, R. Juan, and I. Navazo. Object representation by means of nonminimal division quadtrees and octrees. *ACM Transactions on Graphics*, 4(1):41–59, January 1985.

[4] S. B. M. Bell, B. M. Diaz, F. Holroyd, and M. J. Jackson. Spatially referenced methods of processing raster and vector data. *Image and Vision Computing*, 1(4):211–220, November 1983.

[5] J. L. Bentley. Multidimensional binary search trees used for associative searching. *Communications of the ACM*, 18(9):509–517, September 1975.

[6] R. A. DeMillo, S. C. Eisenstat, and R. J. Lipton. Preserving average proximity in arrays. *Communications of the ACM*, 21(3):228–231, March 1978.

[7] A. Dengel. Self-adapting structuring and representation of space. Technical Report RR-91-22, Deutsches Forschungszentrum für Künstliche Intelligenz, Kaiserslautern, Germany, September 1991.

[8] R. A. Finkel and J. L. Bentley. Quad trees: a data structure for retrieval on composite keys. *Acta Informatica*, 4(1):1–9, 1974.

[9] J. D. Foley, A. van Dam, S. K. Feiner, and J. F. Hughes. *Computer Graphics: Principles and Practice*. Addison-Wesley, Reading, MA, second edition, 1990.

[10] H. Freeman. Computer processing of line-drawing images. *ACM Computing Surveys*, 6(1):57–97, March 1974.

[11] J. H. Friedman, J. L. Bentley, and R. A. Finkel. An algorithm for finding best matches in logarithmic expected time. *ACM Transactions on Mathematical Software*, 3(3):209–226, September 1977.

[12] H. Fuchs, Z. M. Kedem, and B. F. Naylor. On visible surface generation by a priori tree structures. *Computer Graphics*, 14(3):124–133, July 1980. (Also *Proceedings of the SIGGRAPH'80 Conference*, Seattle, WA, July 1980).

[13] A. Guttman. R-trees: a dynamic index structure for spatial searching. In *Proceedings of the SIGMOD Conference*, pages 47–57, Boston, MA, June 1984.

[14] G. M. Hunter. *Efficient computation and data structures for graphics*. PhD thesis, Princeton University, Princeton, NJ, 1978.

[15] A. Klinger. Patterns and search statistics. In J. S. Rustagi, editor, *Optimizing Methods in Statistics*, pages 303–337. Academic Press, New York, 1971.

[16] K. Knowlton. Progressive transmission of grey-scale and binary pictures by simple efficient, and lossless encoding schemes. *Proceedings of the IEEE*, 68(7):885–896, July 1980.

[17] D. E. Knuth. *The Art of Computer Programming vol. 3, Sorting and Searching.* Addison-Wesley, Reading, MA, 1973.

[18] D. Meagher. Geometric modeling using octree encoding. *Computer Graphics and Image Processing*, 19(2):129–147, June 1982.

[19] R. C. Nelson and H. Samet. A consistent hierarchical representation for vector data. *Computer Graphics*, 20(4):197–206, August 1986. (also *Proceedings of the SIGGRAPH'86 Conference*, Dallas, August 1986).

[20] J. T. Robinson. The *k–d–b*–tree: a search structure for large multidimensional dynamic indexes. In *Proceedings of the SIGMOD Conference*, pages 10–18, Ann Arbor, MI, April 1981.

[21] A. Rosenfeld and J. L. Pfaltz. Sequential operations in digital image processing. *Journal of the ACM*, 13(4):471–494, October 1966.

[22] D. Rutovitz. Data structures for operations on digital images. In G. C. Cheng et al., editor, *Pictorial Pattern Recognition*, pages 105–133. Thompson Book Co., Washington, DC, 1968.

[23] H. Samet. *Applications of Spatial Data Structures: Computer Graphics, Image Processing, and GIS.* Addison-Wesley, Reading, MA, 1990.

[24] H. Samet. *The Design and Analysis of Spatial Data Structures.* Addison-Wesley, Reading, MA, 1990.

[25] H. Samet. Spatial data structures. In W. Kim, editor, *Modern Database Systems, The Object Model, Interoperability and Beyond*, pages 361–385. ACM Press and Addison-Wesley, New York, 1995.

[26] H. Samet and W. G. Aref. Spatial data models and query processing. In W. Kim, editor, *Modern Database Systems, The Object Model, Interoperability and Beyond*, pages 338–360. ACM Press and Addison-Wesley, New York, 1995.

[27] H. Samet and M. Tamminen. Efficient component labeling of images of arbitrary dimension represented by linear bintrees. *IEEE Transactions on Pattern Analysis and Machine Intelligence*, 10(4):579–586, July 1988.

[28] H. Samet and R. E. Webber. Storing a collection of polygons using quadtrees. *ACM Transactions on Graphics*, 4(3):182–222, July 1985. (Also *Proceedings of Computer Vision and Pattern Recognition 83*, Washington, DC, June 1983, 127–132; and University of Maryland Computer Science TR–1372).

[29] M. Stonebraker, T. Sellis, and E. Hanson. An analysis of rule indexing implementations in data base systems. In *Proceedings of the First International Conference on Expert Database Systems*, pages 353–364, Charleston, SC, April 1986.

[30] M. Tamminen. Comment on quad– and octtrees. *Communications of the ACM*, 27(3):248–249, March 1984.

[31] S. Tanimoto and T. Pavlidis. A hierarchical data structure for picture processing. *Computer Graphics and Image Processing*, 4(2):104–119, June 1975.

Section 6

Image Understanding in Human and Machine Intelligence

Perceptual Intelligence

Alex Pentland

1 On The Nature of Intelligence

Perhaps the most exciting research project of the last forty-five years has been to make computers intelligent, to work with us, and to be our helpers. Despite much effort, however, computers today are at best idiot savants: good in specialized, often logical domains, but completely lacking in common sense. What is the cause of this failure?

Until recently most researchers have assumed that "intelligence" is roughly synonymous with language-like or logical reasoning, with separate, relatively unintelligent perception and motor modules for input and output. However during the last decade ethological studies of animals [28], studies of biological learning [11], and results from the "artificial life" movement [17] have begun to paint a very different picture of intelligence. They see intelligent behavior as largely composed of perceptual-motor associations learned from personal experience and observation of others.

In this view, the language-like reasoning abilities normally thought of as "intelligence" are only the foam riding on a deep sea of learned perceptual-motor responses. For instance, higher primates have brains and genes that are extremely similar to ours, yet only limited (or no) linguistic or abstract reasoning ability [12]. Despite this lack they live in complex societies, feed and protect themselves, use a wide variety of verbal and hand signs, employ tools, and can adapt to widely varied environments [28, 11, 12]. Clearly they have an intelligence that is directly comparable to human intelligence in all but a very few areas.

At the heart of this new view is what I call *perceptual intelligence*. It is being able to characterize a situation by answering questions such as who, what, when, where, and why, just as writers are taught to do. In the language of cognitive science, perceptual intelligence is the ability to solve the frame problem: it is being able to classify the current situation, so that you know what variables are important, and thus can act appropriately.

The dictionary defines intelligence as the ability to learn and act reasonably. If you have the perceptual intelligence to know who, what, when, where, and why, then even simple learning methods will allow you to determine what aspects of the situation are significant, and to choose a good course of action [17].

The principal problem preventing computer intelligence, therefore, may simply have been that our computers have been deaf and blind, only interacting with the world via text strings. If you imagine raising a child in a closed box with only a telegraph connection to the outside world, you can quickly realize how difficult it has been to make computers become intelligent. Computers have existed in an environment that is almost completely disconnected from the real world, so how can they act intelligently [8]?

2 Toward Perceptual Intelligence

If computers are to act as our helpful assistants they must have roughly the same sort of perceptual intelligence that we do, that is, at every instant their answer to "who, what,

0-8186-7644-2 $5.00 © 1996 IEEE

when, where, and why" must be similar to ours. Otherwise, their behavior will appear to us to be unpredictable and perhaps even irrational. My research goal, therefore, is to give computers (and thereby robots, rooms, cars, desks, tools, and clothes) a human-like sort of perceptual intelligence [24].

Because people are social animals, our answers to "who, what, when, where, and why" are most importantly concerned with other people (as opposed to geometry or inanimate objects which, while important, are certainly less salient). Consequently, I have focused my work on the perception of human behavior. I have concentrated on giving computers the ability to recognize people, to understand our expressions and gestures, and hear the tone and emphasis of the human voice. If I can give computers the ability to perceive and categorize human interactions in a human-like manner, I expect that the computer will be able to work with people in a natural and common-sense manner.

In conducting this project I have discovered that the swiftest and most reliable progress can be made by creatively assembling 2-D image processing modules of the sort developed by Azriel Rosenfeld and other pioneers [7], rather than by developing exotic new mathematical formulations, detailed understanding of photometry, or using sophisticated 3-D representations. Perhaps the success of 2-D methods is not so surprising; after all, what we know of brain function usually looks more like parallel image processing than abstract or logical calculations. That appearance, of course, could be deceiving.

Azriel said in the preface to his 1969 book "Picture Processing by Computer" [26] that his goal was to build up a "body of general-purpose picture processing techniques." The work I survey here demonstrates that this body of picture processing techniques is adequate to build real-time systems that interact with people and exhibit a fair degree of perceptual intelligence. I hope that it is satisfying to Azriel to see this marriage between his long-standing philosophical interest in the nature of human intelligence and his work in computer vision.

The following sections of this paper I will describe how 2-D picture processing modules can be assembled to obtain building blocks of perceptual intelligence, and how they can be used to help build virtual creatures that interact with people in a compelling and lifelike manner. Although the building blocks described here are limited in comparison to human abilities, and only cover a scattering of human perceptual abilities, I believe that they are the first steps down a path to a human-style perceptual intelligence.

2.1 The Modeling and Estimation Framework

The general theoretical approach I have taken is to perform a maximum a *posteriori* (MAP) interpretation on very low-level, 2-D representations of regions of the image data. The appearance of a target class Ω, e.g., the probability distribution function $P(\mathbf{x}|\Omega)$ of its image-level features $\mathbf{x}$, can be characterized by use of a low-dimensional parametric appearance model. Once such a probability distribution function (PDF) has been learned, it is straightforward to use it in a MAP estimator in order to detect and recognize target classes. Behavior recognition is accomplished in a similar manner; these parametric appearance models are tracked over time, and their time evolution $P(\mathbf{x}(t)|\Omega)$ characterized probabilistically to obtain a *spatiotemporal behavior model.* Incoming spatiotemporal data can then be compared to the spatiotemporal PDF of each of the various behavior models using elastic matching methods such as dynamic time warping [10] or hidden Markov modeling [27].

The use of parametric appearance models to characterize the PDF of an object's appearance in the image is related to the idea of view-based representation, as advocated by Ullman [31] and Poggio [25]. As originally developed, the idea of view-based recognition was to accurately describe the spatial structure of the target object by interpolating between various views. However in order to describe natural objects such as faces or hands, I have found it necessary to extend the notion of "view" to include characterizing the range of geometric and feature variation, as well as the likelihood associated with such variation.

This approach is typified by my face recognition research [29, 19], which uses linear combinations of eigenvectors to describe a *space* of target appearances, and then characterize the PDF of the targets appearance within that space. This method has been shown to be very powerful for detection and recognition of human faces, hands, and facial expressions [19]. Other researchers have used extensions of this basic method to recognize industrial objects and household items [20]. Another variation on this approach is to use linear combinations of examples rather than eigenvectors; this type of appearance modeling has demonstrated a power similar to that of the eigenvector-based methods [10, 16], although it necessarily has a lower efficiency.

2.2 The Computer Architecture

Sensing and interpretation of human movement is accomplished by a modular computer architecture, which can be configured in a variety of ways, as more or less interpretation capability is required. The basic element is an SGI Indy computer; these computers are connected by video, audio, ISDN, and ethernet networks. This allows each computer to independently access whatever portion of the audiovisual input it needs, to share control information among the other computers, and to connect with other, distant installations over ISDN phone lines. Normally each computer is dedicated to perform only one type of interpretation. For instance, if video tracking, audio input, and gesture interpretation capabilities were desired, then three computers would be used, one for each task.

3 Building Blocks of Perceptual Intelligence

The following sections will briefly describe each of a set of programs that can find, track, and interpret human behavior [24]. The last section will illustrate how these programs have been assembled into a real-time "virtual dog" that interacts with people in an interesting and lifelike manner. In order of description, the modules are:

- Pfinder, a real-time program that tracks the user, and recognizes a basic set of hand gestures and body postures,
- The face processor, an interactive-time program that recognizes human faces,
- The expression processor, an interactive-time program that uses motion-energy templates to classify people's facial expressions,
- Vision-directed audio, a real-time system that uses the head position information provided by Pfinder to steer a phase-array microphone,
- Behavior recognition, real-time systems that examine users' movements in order to determine what they are doing.

For additional detail the reader should examine the numerous papers, technical reports, interactive demos, and computer code available at our web site [1].

3.1 Pfinder

Pfinder ("person finder") is a real-time system for tracking and interpretation of people. It runs at 10Hz on a standard SGI Indy computer, and has performed reliably on thousands

Figure 1. Analysis of a user in the ALIVE environment. The frame on the left is the video input (n.b. color image shown here in black and white for printing purposes), the center frame shows the support map $s(x, y)$ which segments the user into blobs, and the frame on the right showing a person model reconstructed from blob statistics alone (with contour shape ignored).

of people in many different physical locations [2, 32, 24]. The system uses a multi-class statistical model of color and shape to segment a person from a background scene, and then to find and track people's head and hands in a wide range of viewing conditions. Pfinder produces a real-time representation of a user useful for applications such as wireless interfaces, video databases, and low-bandwidth coding, without cumbersome wires or attached sensors.

Pfinder employs a maximum *a posteriori* (MAP) approach by using simple 2-D models to perform detection and tracking of the human body. It incorporates *a priori* knowledge about people primarily to bootstrap itself and to recover from errors.

In order to present a concise description of Pfinder, I will only describe its representations and operation in the "steady state" case, where it is tracking a person moving around an office environment, neglecting the problems of initially learning a person's description.

3.1.1 Modeling The Person

Pfinder models the human as a connected set of 2-D *blobs*, an old concept, but one to which in recent years little serious attention has been paid in favor of stronger local features like points, lines, and contours. The blob representation that we use was originally developed by me in 1976 [21] for application to multi-spectral satellite (MSS) imagery, and was in part inspired by Azriel's Picture Processing book [26]. The PDF of each blob is characterized by the *joint* distribution of spatial (x, y) and color features. Color is expressed in the YUV space, which is a simple approximation to human color sensitivity.

We define $\mathbf{m}_k$ to be the mean (x, y, Y, U, V) of blob k, and $\mathbf{K}_k$ to be the covariance of that blob's distribution. The mean of a blob expresses the concept "color a at location b," while the covariance expresses how the color and brightness of the blob changes across its surface. For instance, the blob statistics can express that one side is brighter than the other (perhaps due illumination from the side), or that the color changes from top to bottom (perhaps due to light reflected from the floor).

Each blob also has associated with it a *support map* that indicates exactly which image pixels are members of a particular blob. Since the individual support maps indicate which image pixels are members of that particular blob, the aggregate support map $s(x, y)$ over all the blobs represents the segmentation of the image into spatial/color classes.

The statistics of each blob are recursively updated to combine information contained in the most recent measurements with knowledge contained in the current class statistics and the priors. Because the detailed dynamics of each blob are unknown, we use approximate models derived from experience with a wide range of users. For instance, blobs that are near the center of mass have substantial inertia, whereas blobs toward the extremities can move much faster.

3.1.2 Modeling The Scene

It is assumed that the majority of the time Pfinder will be processing a scene that consists of a relatively static situation such as an office, and a single moving person. Consequently, it is appropriate to use different types of model for the scene and for the person.

The surrounding scene is modeled as a texture surface; each point on the texture surface is associated with a mean color value and a distribution about that mean. The color distribution of each pixel is modeled with the Gaussian described by a full covariance matrix. Thus, for instance, a fluttering white curtain in front of a black wall will have a color covariance that is very elongated in the luminance direction, but narrow in the chrominance directions.

In each frame visible pixels have their statistics recursively updated using a simple adaptive filter. This allows us to compensate for changes in lighting and even for object movement. For instance, if a person moves a book it causes the texture map to change in both the locations where the book was, and where it now is. By tracking the person we can know that these areas, although changed, are still part of the texture model and thus update their statistics to the new value. The updating process is done recursively, and even large changes in illumination can be substantially compensated within two or three seconds.

3.1.3 The Analysis Loop

Given a person model and a scene model, Pfinder then acquires a new image, interprets it, and updates the scene and person models. To accomplish this there are several steps:

1. First, the appearance of the users is predicted in the new image using the current state of our model. This is accomplished using a set of Kalman filters with simple Newtonian dynamics that operate on each blob's spatial statistics.
2. Next, for each image pixel the likelihood that it is a member of each of the blob models and the scene model is measured. Self-shadowing and cast shadows are a particular difficulty in measuring this likelihood; this is addressed by normalizing the hue and saturation by the overall brightness.
3. Resolve these pixel-by-pixel likelihoods into a support map, indicating for each pixel whether it is part of one of the blobs or of the background scene. Spatial priors and connectivity constraints are used to accomplish this resolution.
4. Update the statistical models for each blob and for the background scene; also update the dynamic models of the blobs.

For some applications Pfinder's 2-D information must be processed to recover 3-D geometry. For a single calibrated camera this can be accomplished by backprojecting the 2-D image information to produce 3-D position estimates using the assumption that the user is standing on a planar floor. When two or more cameras are available, the hand, head, etc., blobs can be matched to obtain 3-D estimates via triangulation.

Figure 1 illustrates Pfinder's functioning. At the left is the original video frame (shown in black and white rather than the original color). In the middle is the resulting segmentation into head, hands, feet, shirt, and pants. At the right are one-standard-deviation ellipses illustrating the statistical blob descriptions formed for the head, hands, feet, shirt, and

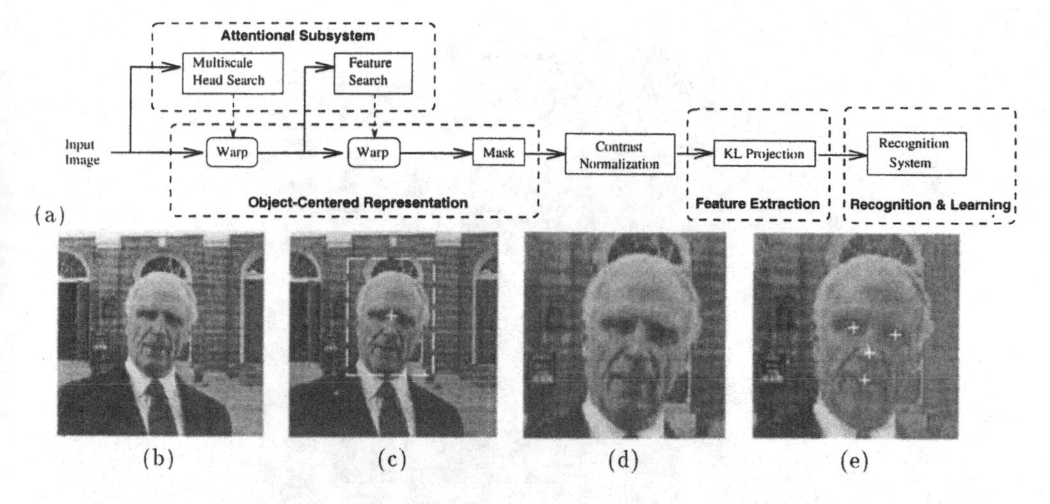

Figure 2. (a) The face processing system, (b) original image, (c) position and scale estimate, (d) normalized head image, (e) position of facial features.

pants. Note that despite having the hands either in front of the face or the body a correct description is still obtained. For additional detail see references [2, 32, 24].

3.2 Face Recognition

Once the rough location of the person's head is known, one can attempt to recognize their face. As with Pfinder, a maximum *a posteriori* (MAP) approach is applied to this problem. This has been accomplished by developing a method for determining the probability distribution function for face images within a low-dimensional eigenspace. Knowledge of this distribution then allows the face and face features to be precisely located, and compared along meaningful dimensions. The following gives a brief description of this system; for additional detail see [19].

3.2.1 Face and Feature Detection

The standard detection paradigm in image processing is that of normalized correlation or template matching. However this approach is only optimal in the simplistic case of a *deterministic* signal embedded in white Gaussian noise. When we begin to consider a target *class* detection problem — *e.g,* finding a generic human face or a human hand in a scene — we must incorporate the underlying probability distribution of the object of interest. Subspace or eigenspace methods, such as the KLT and PCA, are particularly well-suited to such a task since they provide a compact and *parametric* description of the object's appearance and also automatically identify the essential components of the underlying statistical variability.

In particular, the eigenspace formulation leads to a powerful alternative to standard detection techniques such as template matching or normalized correlation. The reconstruction error (or residual) of the KLT expansion is an effective indicator of a match. The residual error is easily computed using the projection coefficients and the original signal energy. This detection strategy is equivalent to matching with a linear combination of *eigentemplates*

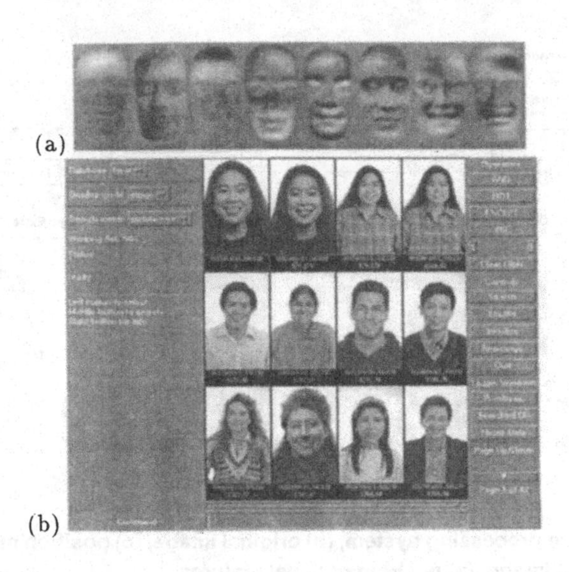

Figure 3. (a) The first 8 eigenfaces, (b) Searching for similar faces in a database, using the Photobook image database tool [23].

and allows for a greater range of distortions in the input signal (including lighting, and moderate rotation and scale). Some of the low-order eigentemplates for a human face are shown in Figure 3(a). In a statistical signal detection framework, the use of eigentemplates has been shown to be orders of magnitude better than standard matched filtering [19].

Using this approach the target detection problem can be reformulated from the point of view of a MAP estimation problem. In particular, given the visual field, estimate the position (and scale) of the subimage which is most representative of a specific target class Ω. Computationally this is achieved by sliding an m-by-n observation window throughout the image and at each location computing the *likelihood* that the given observation $\mathbf{x}$ is an instance of the target class Ω — *i.e,* $P(\mathbf{x}|\Omega)$. After this probability map is computed, the location corresponding to the highest likelihood can be selected as the MAP estimate of the target location.

3.2.2 The Face Processor

This MAP-based face finder has been employed as the basic building block of an automatic face recognition system. The function of the face finder is to very precisely locate the face, since Pfinder produces only low-resolution estimates of head location. The block diagram of the face finder system is shown in Figure 2 which consists of a two-stage object detection and alignment stage, a contrast normalization stage, a feature extraction stage, followed by recognition (and, optionally, facial coding.) Figure 2(b)-(e) illustrates the operation of the detection and alignment stage on a natural image containing a human face.

The first step in this process is illustrated in Figure 2(c) where the MAP estimate of the position and scale of the face are indicated by the cross-hairs and bounding box. Once these regions have been identified, the estimated scale and position are used to normalize for translation and scale, yielding a standard "head-in-the-box" format image (Figure 2(d)). A second feature detection stage operates at this fixed scale to estimate the position of 4

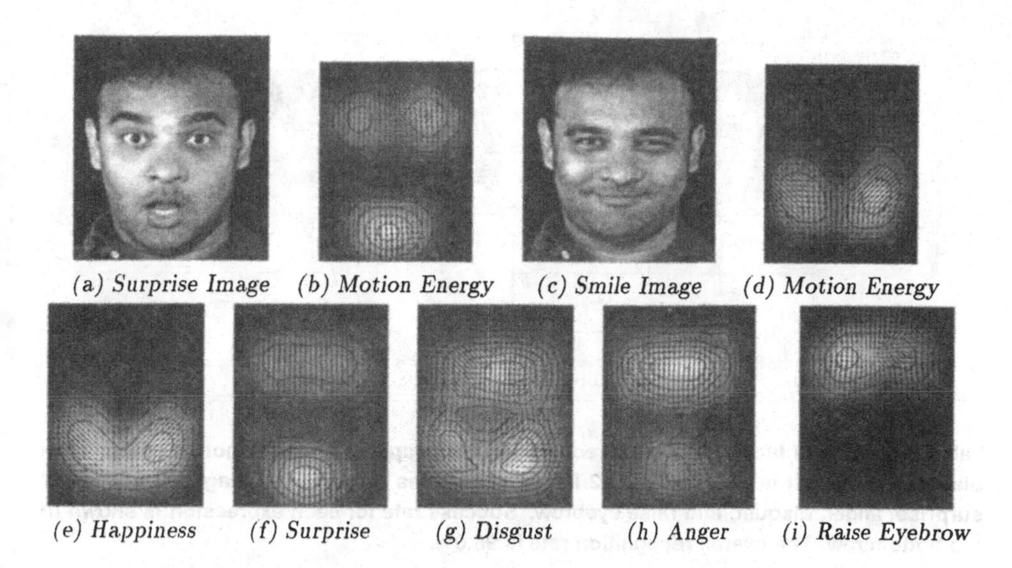

Figure 4. Determining expressions from video sequences. (a) and (c) show expressions of smile and surprise, and (b) and (d) show the corresponding spatio-temporal motion energy pattern. (e) - (i) show spatio-temporal motion-energy templates for various expressions.

facial features: the left and right eyes, the tip of the nose and the center of the mouth (Figure 2(e)). Once the facial features have been detected, the face image is warped to align the geometry and shape of the face with that of a canonical model. Then the facial region is extracted (by applying a fixed mask) and subsequently normalized for contrast. This geometrically aligned and normalized image is then projected onto the set of eigenfaces shown in Figure 3(a).

The projection coefficients obtained by comparison of the normalized face and the eigenfaces form a feature vector which accurately describes the appearance of the face. This feature vector can therefore be used for facial recognition, as well as for facial image coding. Figure 3(b) shows a typical result when using the eigenface feature vector for face recognition. The image in the upper left is the one to be recognized and the remainder are the most similar faces in the database (ranked by facial similarity, left to right, top to bottom). The top three matches in this case are images of the same person taken a month apart and at different scales. The recognition accuracy of this system (defined as the percent correct rank-one matches) is 99% [19].

3.3 Expression Recognition

Once the face and its features have been accurately located, one can begin to analyze its motion to determine the facial expression. This can be done by using spatio-temporal motion-energy templates of the whole face. [13, 14]. For each facial expression, the corresponding template expresses the *peak amount* and *direction* of motion that one would expect to see at each point on the face. Figures 4(a) - (d) show Irfan Essa making two expressions (surprise and smile) and the corresponding motion-energy template descriptions.

These simple, biologically-plausible motion energy "templates" can be used for expression recognition by comparing the motion-energy observed for a particular face to the "average"

Expressions	Smile	Surprise	Anger	Disgust	Raise Brow
Template					
Smile	**12**	0	0	0	0
Surprise	0	**10**	0	0	0
Anger	0	0	**9**	0	0
Disgust	0	0	1	**10**	0
Raise Brow	0	0	0	0	**8**
Success	100%	100%	90%	100%	100%

Table 1. Results of facial expression recognition using spatio-temporal motion energy templates. This result is on based on 12 image sequences of smile, 10 image sequences of surprise, anger, disgust, and raise eyebrow. Success rate for each expression is shown in the bottom row. The overall recognition rate is 98.0%.

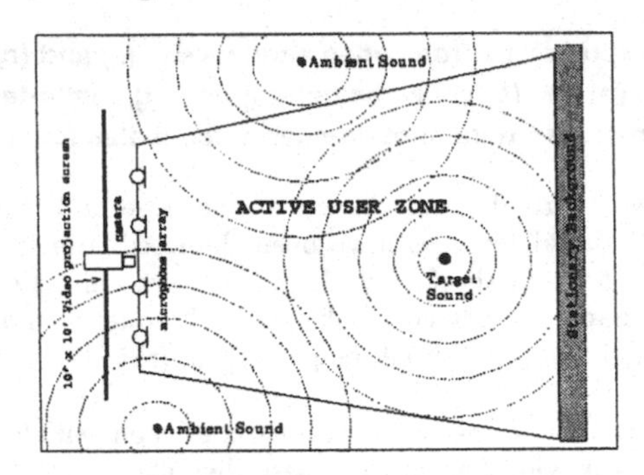

Figure 5. Geometry of the vision-directed phased-array microphone system

template for each expression. Figures 4(e) - (i) show the motion-energy templates for several expressions. To classify an expression one compares the observed facial motion energy with each of these motion energy templates, and then picks the expression with the most similar pattern of motion.

This method of expression recognition has been applied to a database of 52 image sequences of 8 subjects making various expressions. In each image sequence the motion energy was measured, compared to each of the templates, and the expression classified, generating the confusion matrix shown in Table 1. This table shows just one incorrect classification, giving an overall recognition rate of 98.0%. For additional details see [13, 14].

3.4 Vision-Driven Audio

Audio interpretation of people is as important as visual interpretation. Although much work as been done on speech understanding, virtually all of this work assumes a closely-placed microphone for input and a fixed listening position. Speech recognition applications, for instance, typically require near-field ($< 1.5m$) microphone placement for acceptable

(a)

(b)

Figure 6. (a) Real-time reading of American Sign Language (with Thad Starner doing the signing), and (b) real-time classification of driver's actions in a driving simulator.

performance. Beyond this distance the signal-to-noise ratio of the incoming speech affects the performance significantly; most commercial speech-recognition packages typically break down over a 4 to 6 DB range.

The constraint of near-field microphone placement makes audio difficult to integrate into normal human life, so it is necessary to find a solution that allows the user must be able to move around with no noticeable degradation in performance. The solution to this problem is to use the head-tracking ability of Pfinder to steer the audio input so that it focuses on the user's head.

There are several potential ways to steer audio input. One is to use a highly directional microphone that can be panned using a motorized control unit, to track the user's location. However this requires a significant amount of mounting and control hardware, is limited by the speed and accuracy of the drive motors, and can only track one user at a time.

It is therefore preferable to have a directional response that can be steered electronically. This can be done with the well-known technique of beam forming with an array of microphone elements, as illustrated in Figure 5(a). Though several microphones need to be used for this method, they need not be very directional and they can be permanently mounted in the environment. In addition, the signals from the microphones in the array can be combined in as many ways as the available computational power is capable of, allowing for the tracking of multiple moving sound sources from a single microphone array. For additional detail see references [2, 9].

3.5 Recognizing Human Behaviors

Work on recognizing body position, face, expression, and sound are only the first steps toward perceptual intelligence. To make computers really useful, these basic perceptual functions need to be integrated with higher-level models of human behavior, so that we can begin to understand what the person is *doing*.

A general approach to interpreting human behavior is to directly extend the MAP techniques used for recognition of face, expression, and body pose. That is, to track the 2-D parametric appearance models over time, and then probabilistically characterize the time evolution of their parameters. This approach produces the PDF of the behavior as both a function of time and 2-D appearance. Behaviors can then be recognized by comparing them to these learned models using MAP methods. To obtain high accuracy behavior recognition, one must use an elastic spatio-temporal matching technique such as dynamic time warping [10] or hidden Markov modeling [27].

To recognize particular gestures or behaviors, for instance, the person is modeled as a Markov device, with internal mental states that have their own particular distribution of appearance and inter-state transition probabilities. Because the internal states of a human are not directly observable, they must be determined through an indirect estimation process, using the person's movement and vocalizations as measurements. One efficient and robust method of accomplishing this is to use the Viterbi recognition methods developed for use with Hidden Markov Models (HMM).

This general approach is similar to that taken by the speech recognition community. The difference is that here internal state is not thought of as being just words or sentences; the internal states can also be actions or intentions. Moreover, the input is not just audio filter banks but also facial appearance, body movement, and vocal characteristics such as pitch as inputs to infer the user's internal state. Two good examples of that employ this approach to behavior recognition are reading American Sign Language (ASL) [27], and interpreting automobile driver's behavior [22].

The ASL reader is a real-time system that performs high-quality classification of a forty-word subset of ASL using only the hand measurements provided by Pfinder (e.g., hand position, orientation, and width/height ratio). Thad Starner is shown using this system in Figure 6(a). The accurate classification performance of this system is particularly impressive because in ASL the hand movements are rapid and continuous, and exhibit large coarticulation effect.

The second system interprets people's actions while driving a car [22]. In this system the driver's hand and leg motions were observed while driving in the Nissan Cambridge Basic Research Lab's driving simulator (see Figure 6(b)). These observations are used to classify the driver's action as quickly as possible. The goal is to develop safer cars by having the automobile "observe" the driver, continuously estimate the driver's internal state (what action they are taking), and respond appropriately.

I found that the system was surprisingly accurate at identifying which driving maneuver the driver was beginning to execute (e.g., passing, turning, stopping, or changing lane). A recognition accuracy of $95.24\% \pm 3.1\%$ was obtained at 0.5 seconds after of the beginning each maneuver...long before the major, functional parts of the maneuver were executed. This experiment shows that driver's execute *preparatory* movements that are reliable indicators of which maneuver they are beginning to execute.

4 Putting It All Together

Given these building blocks of perceptual intelligence, it is possible to try to "put it all together" and actually build an autonomous creature that displays perceptual intelligence. The ALIVE experiment, which stands for "Artificial Life Interactive Video Environment," is just such an attempt. This experiment, first started by graduate students Trevor Darrell and Bruce Blumberg working with professor Pattie Maes and myself [18], allows wireless full-body interaction between a human participant and a graphical world inhabited by autonomous agents.

It uses a single video camera to obtain a color image of a person, which is then placed (with correct depth and occlusion) within a 3-D graphical world. The resulting image is projected onto a large screen that faces the user and acts as a type of "magic mirror" (Figure 7): the user sees herself surrounded by objects and agents. No goggles, gloves, or wires are needed for interaction with the virtual world. The creatures in this virtual world are autonomous, behaving entities that use the camera and microphones to sense the actions of the participant, interpret them, and react to them in real time.

The creatures in ALIVE are modeled as autonomous semi-intelligent agents using principles drawn from the ethology literature [28]. They have a set of internal needs and motivations, a set of sensors to perceive their environment, a repertoire of activities that they can perform, and a physically-based motor system that allows them to move in and act on the virtual environment. A behavior system decides in real-time which activity to engage in to meet the internal needs of the agent and to take advantage of opportunities presented by the current state of the environment. The agent's kinematic state is updated according to the motor activities associated with the chosen behaviors, and the agent rendered on the graphics display. The user's location and hand and body gestures affect the behavior of the agents, and the user receives visual as well as auditory feedback about the agents' internal state and reactions. The perceptual intelligence of these agents is truly active and purposive [4, 3]

The ALIVE system is built using Blumberg's and Maes's behavior modeling tools for developing semi-intelligent autonomous agents [6, 17]; its goal is to produce agents that select an optimal activity on every time step given their internal needs and motivations, past history, and the perceived environment with its attendant opportunities, challenges and changes. The activities are chosen such that the agents neither dither among multiple activities, nor persist too long in a single activity. The behavior system is capable of interrupting a given activity if a more pressing need or an unforeseen opportunity arises.

When using the behavior toolkit to build an agent, the designer specifies:

- *The motivations or internal needs of the agent.* Internal needs are modeled as variables which may vary over time. For example, the "dog" agent in ALIVE has the internal need to receive attention from the user. Whenever the user pats the dog, this variable will temporarily decrease in value and as a result the dog will be less motivated to seek human attention.
- *The activities and actions of the agent.* An agent's set of activities is organized as a loose hierarchy with the top of the hierarchy representing more general activities and the leaves representing more specific activities. For example, the dog agent has several top-level activities such as Playing, Feeding, Drinking, etc. A top-level activity such as Feeding has several children: Chewing, Preparing-to-Eat, and Searching-for-Food. Searching-for-Food in turn has three children: Wander, Avoid-Obstacles, Move-to-Food, and so on. The lowest-level activities control the motor system, for example, making the dog move a little to the left or right, or making it bark in a certain way.
- *The virtual sensors of the agent.* The sensors of the agent are uniform between the real world and the virtual world. For instance, a distance sensing action aimed at the user will use the camera to estimate distance, whereas the same action aimed at another virtual creature will fire a ray in the direction of other creature and measure the distance to intersection with the other creature's bounding box.

Given the above information, the behavior system selects the activities most relevant to the agent at a particular moment in time given the state of the agent, the situation it finds itself in and its recent behavior history. One of the design goals of the behavior system, and the modeling system in general, was that it be compatible with a physically-based simulated motor system such as those proposed by Raibert [30] or Zeltzer [33]. Consequently, the ALIVE agents use a combination of physically-based and kinematic modeling to implement the motor system; Featherstone's spatial algebra [15] and an adaptive-step-size RKF numerical integrator are used to model the movement of the agent in response to forces

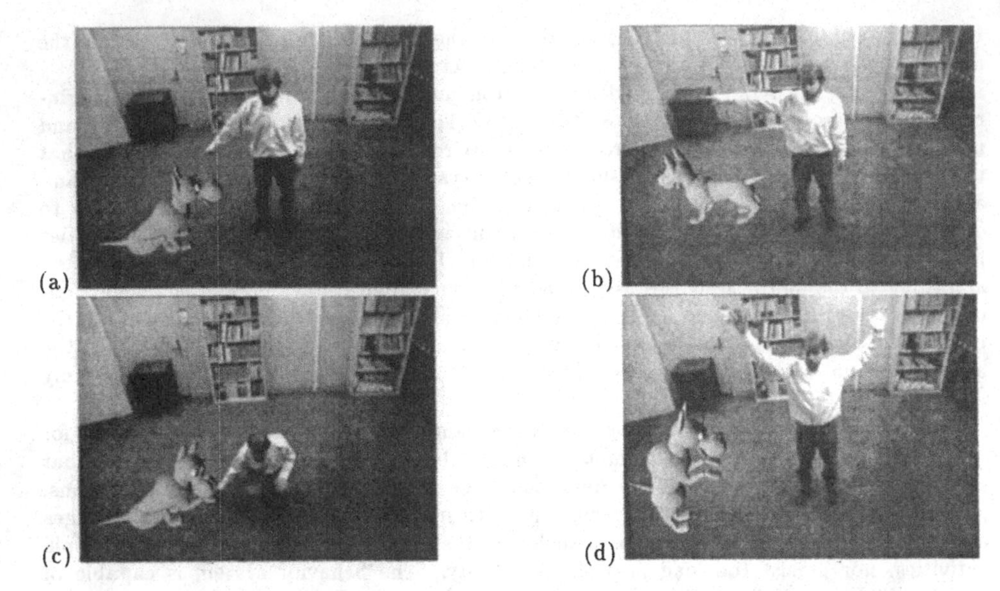

Figure 7. (a) Image of user is composited in 3-D with computer graphics; Here the dog responds to pointing gesture by sitting. (b) Another example of a recognized gesture; the dog walks in direction indicated by user. (c) The dog shakes hands with user; note that this behavior depends on knowing the stance of the user. (d) The dog standing on hind legs to mimic user's gesture.

applied by its driving motor. Details of the behavior and modeling systems are reported upon in [6] and [5].

The ALIVE system represents the user as a "special" sort of 3-D agent. The position and state of the user-agent are based on the information computed by the vision system (e.g., position, gesture) and the auditory system (e.g., speech recognition, pitch recognition [9]). This allows the artificial agents to sense the user using the same virtual sensors that they use to detect virtual objects and other virtual agents. The person agent is rendered in the final image using the live video image of the actual user mapped to the correct place in the 3-D virtual world.

In the most recent version of ALIVE the user interacts with a virtual dog, as shown in Figure 7. The Dog has a repertoire of behaviors that include playing, feeding, drinking, receiving attention and sleeping. The dog also uses auditory input, consisting of simple verbal commands that are recognized using a commercial speech recognition system, and produces auditory output, consisting of a wide variety of prerecorded samples which are played at appropriate times. The dog has both interactive behaviors and autonomous action; while its primary goal is to play with the user, internal motivations (e.g., thirst) will occasionally override.

This system has been installed in several public exhibitions, which has allowed us to gather experience with thousands of users in the system. The overwhelming majority report that they enjoy interacting with the system and consider the actions and reactions of objects and agents believable. I consider people's reactions to these systems to be a sort of informal Turing test; for instance, people attribute all sorts of intentions, goals, and reasoning ability to the dog, that is, they treat it as if it were an intelligent creature.

5 Conclusion

I have argued that intelligent behavior depends more on perceptual intelligence than language-like reasoning abilities. If an agent has the ability to correctly classify the situation via perception, then normally very little else is required to determine which elements are important, and which action to select.

To explore this idea, I have focused on answering the who, what, where, when, and why questions with respect to human behavior. By developing perceptual tools that answer these questions, and coupling them to a simple set of motor reactions, we have been able to build autonomous creatures that interact with people in an interesting and even lifelike manner. Although this experiment has not achieved even dog-level intelligence, it compares very favorably with previous robots and autonomous computer agents. These capabilities and demonstrations are, I believe, the first steps toward endowing computers with a human-like perceptual intelligence.

References

[1] Papers and technical reports on all aspects of this technology are available at http://www-white.media.mit.edu/vismod or by anonymous FTP from whitechapel.media.mit.edu

[2] Azarbayejani, A., Wren, C., and Pentland, A. (1996), "Real-Time 3-D Tracking of the Human Body," ImageCom 96, Bordeaux, France 20-22 May. Also, see Technical Report 374.

[3] Aloimonos Y., ed., (1993) Active Perception, Lawrence Erlbaum Associates

[4] Bajcsy, R., Active Perception, Proc. IEEE, Vol. 76, No. 8, pp. 996-1005, 1988

[5] Blumberg B., (1994) Building Believable Animals, To be published in: Proceedings of the 1994 AAAI Spring Symposium on Believable Agents, Palo Alto, March.

[6] Blumberg B., (1994), Action-Selection in Hamsterdam: Lessons from Ethology, Proceedings of the Third International Conference on the Simulation of Adaptive Behavior, MIT Press, Brighton, August 1994.

[7] Ballard, D., and Brown, C., (1982) Computer Vision, Prentice-Hall, Englewood

[8] Barwise, J., and Perry, J., (1986) "Situation Semantics," MIT Press, Cambridge, MA.

[9] Basu, S., Casey, M.A., Gardner, W., Azarbayejani, A., and Pentland, A.., (1996) "Vision-Steered Audio for Interactive Environments," *Proceedings of IMAGE'COM '96*, Bordeaux, France, May 1996. Also, see Technical Report 373.

[10] Darrell T. and Pentland A., (1993) "Space-Time Gestures," In: IEEE Conference on Vision and Pattern Recognition, NY, NY, June 1993.

[11] Davey, G., (1989) "Ecological Learning Theory," Routledge Inc., London

[12] Eccles, J., (1989) "Evolution of the Brain: Creation of the Self," Routledge and Kegan Paul, London, 1989.

[13] Essa, I., Pentland, A., (1994) A Vision System for Observing and Extracting Facial Action Parameters, *IEEE Conference on Computer Vision and Pattern Recognition*, pp. 76-83, Seattle, WA., June 1994.

[14] Essa, I., and Pentland, A. (1995) "Facial Expression Recognition Using a Dynamic Model and Motion Energy," *Int'l Conference on Computer Vision*, Cambridge, MA, June 20-23 1995.

[15] Featherstone R., (1987) Robot Dynamics Algorithms, Kluwer Academic Publishers, Boston.

[16] Jones, M., and Poggio, T., (1995) "Model-Based Matching of Line Drawings by Linear Combinations of Prototypes," *Int'l Conference on Computer Vision*, Cambridge, MA, June 20-23 1995.

[17] Maes P. (1991), Designing Autonomous Agents: Theory and Practice from Biology to Engineering and Back, Bradford Books/MIT Press.

[18] Maes, P., Blumburg, B., Darrell, T., and Pentland, A., (1995) "The ALIVE System: Full-body Interaction with Autonomous Agents." Proceedings of Computer Animation 95, IEEE Press, April 1995.

[19] Moghaddam, B., and Pentland, A., (1995) "Probabalistic Visual Learning for Object Detection," *Int'l Conference on Computer Vision*, Cambridge, MA, June 20-23 1995.

[20] Murase, H., and Nayar, S., (1994) "Visual Learning and recognition of 3-D objects from appearance," *In'tl Journal of Computer Vision*, 14:5-24

[21] Pentland, A., (1976) "Classification By Clustering," *Proceedings of the Symposium On Machine Processing Of Remotely Sensed Data.* June 1976, IEEE Computer Society Press No. 76.

[22] Pentland, A., and Liu, A., (1995) Toward Augmented Control Systems, *IEEE Intelligent Vehicle Symposium 95*, September 25-26, Detroit, MI.

[23] Pentland, A., Picard, R., Sclaroff, S. (1996) "Photobook: Tools for Content-Based Manipulation of Image Databases," *In'tl Journal of Computer Vision*, Vol. 18, No. 3.

[24] Pentland, A. (1996) " Smart Rooms, Smart Clothes," *Scientific American*, pp. 68-76, June 1996.

[25] Poggio, T., and Edelman, S., (1990) A network that learns to recognize three-dimensional objects," *Nature*, 343:263-266

[26] Rosenfeld, A., (1969) "Picture Processing by Computer," Academic Press, N.Y., N.Y.

[27] Starner, T., and Pentland, A., "Visual Recognition of American Sign Language Using Hidden Markov Models," *Proc. Int'l Workshop on Automatic Face- and Gesture-Recognition 1995*, Zurich, Switzerland, June 26-28, 1995.

[28] Lorenz, K., (1973) "Foundations of Ethology," Springer-Verlag, New York.

[29] Turk, M., and Pentland, A., (1991) *Eigenfaces for Recognition*, Journal of Cognitive Neuroscience, Vol. 3, No. 1, pp. 71-86.

[30] Raibert M. and Hodgins J., (1991) Animation of Dynamic Legged Locomotion, Computer Graphics: Proceedings of SIGGRAPH '91, 25(4), ACM Press, July.

[31] Ullman, S., and Basri, R., (1991) "Recognition by linear combinations of models, " *IEEE Trans. Pattern Analysis and Machine Vision*, 13:992-1006

[32] Wren, C., Azarbayejani, A., Darrell, T.,and Pentland, A., (1995) "Pfinder: Real-Time Tracking of the Human Body," *SPIE Conference on Real-Time Image Processing*, Philadelphia, PA., Oct 27, 1995

[33] Zeltzer D., (1991) Task-level graphical simulation: abstraction, representation and control, in: N.I. Badler, B.A. Barsky and D. Zeltser (editors), Making them move: mechanics, control and animation of articulated figures, Morgan Kauffman, pp. 3-33.

Author Index

http://www.computer.org

Press Activities Board

IEEE Computer Society Press Publications

The world-renowned Computer Society Press publishes, promotes, and distributes a wide variety of authoritative computer science and engineering texts. These books are available in two formats: 100 percent original material by authors preeminent in their field who focus on relevant topics and cutting-edge research, and reprint collections consisting of carefully selected groups of previously published papers with accompanying original introductory and explanatory text.

Submission of proposals: For guidelines and information on CS Press books, send e-mail to cs.books@computer.org or write to the Acquisitions Editor, IEEE Computer Society Press, P.O. Box 3014, 10662 Los Vaqueros Circle, Los Alamitos, CA 90720-1314. Telephone +1 714-821-8380. FAX +1 714-761-1784.

IEEE Computer Society Press Proceedings

The Computer Society Press also produces and actively promotes the proceedings of more than 130 acclaimed international conferences each year in multimedia formats that include hard and softcover books, CD-ROMs, videos, and on-line publications.

For information on CS Press proceedings, send e-mail to cs.books@computer.org or write to Proceedings, IEEE Computer Society Press, P.O. Box 3014, 10662 Los Vaqueros Circle, Los Alamitos, CA 90720-1314. Telephone +1 714-821-8380. FAX +1 714-761-1784.

Additional information regarding the Computer Society, conferences and proceedings, CD-ROMs, videos, and books can also be accessed from our web site at www.computer.org.

6/11/96